Ecosystem Change and Public Health

ECOSYSTEM CHANGE AND PUBLIC HEALTH

A Global Perspective

Edited by

Joan L. Aron, Ph.D.

and

Jonathan A. Patz, M.D., M.P.H.

The Johns Hopkins University Press

Baltimore and London

© 2001 The Johns Hopkins University Press
All rights reserved. Published 2001
Printed in the United States of America on acid-free paper
9 8 7 6 5 4 3 2 1

The Johns Hopkins University Press
2715 North Charles Street
Baltimore, Maryland 21218-4363
www.press.jhu.edu

Library of Congress Cataloging-in-Publication Data

Ecosystem change and public health : a global perspective / edited by
Joan L. Aron and Jonathan A. Patz.
        p.   cm.
    Includes bibliographical references (p.   ) and index.
    ISBN 0-8018-6581-6 (acid-free paper) — ISBN 0-8018-6582-4 (pbk. :
acid-free paper)
    1. Environmental health.   I. Aron, Joan L. (Joan Leslie).   II. Patz,
Jonathan.
RA565.E426   2001
614.4—dc21   00-010214

A catalog record for this book is available from the British Library.

Contents

PART III: CASE STUDIES   325

Contributors

JOAN L. ARON, PH.D., President, Science Communication Studies, Columbia, Maryland, and Associate, Department of Epidemiology, Johns Hopkins School of Hygiene and Public Health, Baltimore, Maryland

JOHN M. BALBUS, M.D., M.P.H., Associate Professor, Department of Environmental and Occupational Health, George Washington University School of Public Health and Health Services, Washington, D.C.

ALFRED A. BUCK, M.D., DR.P.H., Adjunct Professor, Department of Molecular Microbiology and Immunology and Department of International Health, and Senior Associate, Department of Epidemiology, Johns Hopkins School of Hygiene and Public Health, Baltimore, Maryland, and Adjunct Professor, Department of Tropical Medicine, Tulane University, New Orleans, Louisiana

RITA R. COLWELL, PH.D., D.SC., Professor, Center of Marine Biotechnology, University of Maryland Biotechnology Institute, Baltimore, Maryland

ULISSES E. C. CONFALONIERI, M.D., D.V.M., D.SC., Professor, National School of Public Health, Oswaldo Cruz Foundation (FIOCRUZ), Rio de Janeiro, Brazil

NICOLAAS J. P. M. DE GROOT, M.SC., Deputy Director, Water Center for the Humid Tropics of Latin America and the Caribbean (CATHALAC), Panama, Republic of Panama; present address: Senior Advisor for Integrated Water Resources, Water & Coast Department, Resource Analysis, Delft, The Netherlands

J. HUGH ELLIS, PH.D., Professor and Chair, Department of Geography and Environmental Engineering, Johns Hopkins University, Baltimore, Maryland

ERIKA G. FEULNER, M.A., President, Technology Education and Communication Institute, Inc., Bethesda, Maryland

GEORGE W. FISHER, PH.D., Professor of Geology, Department of Earth and Planetary Sciences, Johns Hopkins University, Baltimore, Maryland

GREGORY E. GLASS, PH.D., Associate Professor, Department of Molecular Microbiology and Immunology, Johns Hopkins School of Hygiene and Public Health, Baltimore, Maryland

BENJAMIN F. HOBBS, PH.D., Professor, Department of Geography and Environmental Engineering, Johns Hopkins University, Baltimore, Maryland

ANWAR HUQ, PH.D., Research Associate Professor, Center of Public Issues in Biotechnology, University of Maryland Biotechnology Institute, Baltimore, Maryland

STEVEN A. LLOYD, PH.D., Physical Chemist, Atmospheric and Ionospheric Remote Sensing Group, The Johns Hopkins University Applied Physics Laboratory, Laurel, Maryland

JONATHAN A. PATZ, M.D., M.P.H., Director, Program on Health Effects of Global Environmental Change, and Assistant Professor, Department of Environmental Health Sciences, Johns Hopkins School of Hygiene and Public Health, Baltimore, Maryland

DENNIS C. PIRAGES, PH.D., Professor of Government and Politics and Director, Harrison Program on the Future Global Agenda, University of Maryland, College Park, Maryland

LES ROBERTS, PH.D., Sanitary Engineering Consultant, Martinsburg, West Virginia, and Lecturer, Department of Geography and Environmental Engineering, Johns Hopkins University, Baltimore, Maryland

PAUL J. RUNCI, M.A.L.D., Research Scientist, Global Climate Change Group, Pacific Northwest National Laboratory, Washington, D.C.

R. BRADLEY SACK, M.D., SC.D., Professor, Department of International Health, Johns Hopkins School of Hygiene and Public Health, Baltimore, Maryland

JONATHAN M. SAMET, M.D., M.S., Professor and Chair, Department of Epidemiology, and Director, Risk Sciences and Public Policy Institute, Johns Hopkins School of Hygiene and Public Health, Baltimore, Maryland

CLIVE J. SHIFF, PH.D., Associate Professor, Department of Molecular Microbiology and Immunology, Johns Hopkins School of Hygiene and Public Health, Baltimore, Maryland

ROBERT H. SPRINKLE, M.D., PH.D., Assistant Professor, School of Public Affairs, University of Maryland, College Park, Maryland

MARK L. WILSON, SC.D., Associate Professor, Department of Epidemiology and Department of Biology, University of Michigan, Ann Arbor, Michigan

Foreword

Ecosystem Change and Public Health: A Global Perspective, edited by Joan Aron and Jonathan Patz, is one of those books that has long been due. It will quickly find a place on the shelves and tables of students, teachers, and professionals working in a broad range of disciplines.

Because the theme—the interface between ecosystem change and public health—is so extraordinarily complex, relevant literature and information sources are spread throughout a multitude of different disciplines, from biology, chemistry, and physics all the way to the social, economic, and behavioral sciences. As a consequence, students, faculty, and researchers interested in this area have lacked a primary source of inspiration and reference for their work.

This is no longer the case. A carefully selected list of contributors provides the reader with the navigation tools needed for successful exploration of this interface. Embracing the principle that less is more, the book avoids the temptation to become a source of encyclopedic reference for every discipline. It uses instead modern pedagogical approaches to encourage active learning—where discovery is more important than passive absorption of text—and acts as a platform from which to begin investigating the effects of global ecosystem changes on public health. This innovative approach is undoubtedly good news.

This first textbook on research methods in this area also stresses the combination of investigations into health outcomes with integrated assessment for policy development, which generates insights into the uses and limitations of projections into the future. This is accomplished by structuring the book into three parts, the different chapters of which, being cross-referenced, can be read in any order: (1) approaches to the complex topics of the effects of global change on human health, (2) environmental changes, and (3) case studies, which link facts about the effects of global change with the methods needed to understand and remedy them. There is a great need for further research in this area and for innovative approaches in tackling the complex, multivariate dependent phenomena that are observed— as illustrated, for example, by the ongoing debate on the influence of global climate change on the continuing spread of malaria.

As we enter the twenty-first century, the health implications of a destabilized global ecosystem are an increasing challenge to scientists, physicians, governments, and the general public. This innovative textbook will undoubtedly become a major source of inspiration for those working and researching in this area of such need.

Carlos Morel
Director
Special Programme for Research and Training
in Tropical Diseases, UNDP/World Bank/WHO, Geneva

Preface and Acknowledgments

This textbook, published in the year 2001, introduces a new curriculum for students and professionals studying linkages between global ecosystem change and public health. To the extent that these interactions are highly complex and not at all straightforward, this book could not have been written without the hard work and dedication of the contributing expert authors, to whom we are most grateful. This textbook also might not have been written if not for funding support from the Climate and Policy Assessment Division within the Office of Policy, U.S. Environmental Protection Agency (EPA). Dr. Joel Scheraga, Division Director, has demonstrated insight and vision in his support for activities that address the wide gap in our current knowledge of links between ecosystems and public health. Dr. Scheraga now serves as director of the Global Change Research Program within EPA's Office of Research and Development. Project Manager Anne Grambsch, also originally in the Climate and Policy Assessment Division and now in Dr. Scheraga's new program, was an invaluable help, and her analytical approach ensured a truly "value-added" product. As an economic and policy analyst, she augmented our efforts by providing guidance in the policy relevance of our work and kept us on task to set a new environmental health curriculum for the next millennium.

The key theme of this curriculum is that complex environmental health challenges of today require an integrated and interdisciplinary approach. Multiple perspectives from the standpoint of knowledge and expertise, investigatory methods, and a variety of databases are required. Therefore, it was necessary to call upon the advice and review of many physical, ecological/biological, and social scientists in developing this textbook. We are grateful to these individuals, who participated in e-mail conferences early in the book's development. Excluding the chapter authors, these colleagues comprised four ecologists, three infectious disease specialists, three epidemiologists, two veterinarians, two international health physicians, two environmental epidemiologists, and one each from the disciplines of occupational and environmental medicine, population dynamics, environmental engineering, health policy, climatology, biostatistics, and environmental toxicology.

Our advisory panel played a unique role in seeing that this book does, in fact, take off where current environmental health teaching stops. Dr. Andrew Pope, of the Institute of Medicine of the National Academy of Sciences, previously edited *Environmental Medicine: Integrating a Missing Element into Medical Education* and so could contribute firsthand experience in what had previously been developed for environmental health education for physicians. Dr. John Balbus, from the George Washington University School of Public Health and Health Services, brought a perspective from the field of occupational and environmental medicine and sees this book as a required new textbook for this medical specialty, as well as for students in environmental science and public health.

Our senior science advisor, Prof. Alfred Buck, drew upon his extensive experience in international health and infectious disease epidemiology as a professor at the Johns Hopkins School of Hygiene and Public Health and as an epidemiologist at the World Health Organization. To Professor Buck, the need for an interdisciplinary perspective is not new. In the 1960s, he led a comprehensive health screening program in Peru that included an epidemiologist, anthropologist, entomologist, sanitary engineer, and public health personnel. In the introduction to his book about this project, *Health and Disease in Four Peruvian Villages,* he wrote that "the team concept facilitates the integration and comprehensive interpretation of the extensive field data collected by the various specialists." Our new textbook follows in the path established by Professor Buck over 30 years ago, incorporating new methods of geographic analysis and integrated assessment in the spirit of his comprehensive approach to public health.

Our specialist in education and the use of technologies for gathering information, Erika Feulner, crafted this multiauthored book into a true textbook for new styles of learning for the twenty-first century. Her mission was to assist us in structuring the book to encourage self-learning and exploration, and we believe that she has succeeded even beyond our expectations. We also appreciate the expertise of graphic artist Mark Nardini, who helped to unify the visual style of the graphic elements and enhance the visual clarity of the presentation. A great challenge in any volume, especially one with a diverse set of contributors, is to avoid distracting the reader with abrupt changes in style and presentation.

During the long process of organizing and writing chapters, many people who were not contributors or advisors provided valuable reviews, comments, and sources. We particularly thank Leon Gordis, Mickey Glantz, Mitch Hobish, Steve Connor, Michael Stoto, Peter Winch, Carl Taylor, Robert Zimmerman, John Wiener, Janice Longstreth, Pieter Tans, Arnold Gruber, Bob Larson, Cinzia Cerri, Saskia Nijhof, William Whelan, Haider Taha, Alden Henderson, Carol Rubin, and John Feulner. Many others provided materials from a variety of sources—books or articles not yet published, government reports not readily available, graphic elements from reports, photographs—that have enhanced the quality of the book. They are thanked individually in the appropriate chapters. We apologize to anyone we have missed.

We also greatly appreciate the patience and excellent guidance of Wendy Harris, the public health editor of the Johns Hopkins University Press. During the long path toward production of the manuscript, she never lost faith in the project.

Introduction: How to Use This Book

Purpose

The purpose of this textbook is twofold: (1) to raise awareness of changes in human health related to global ecosystem change and (2) to expand the scope of the traditional curriculum in environmental health to include the interactions of major environmental forces and public health on a global scale. In support of these broad purposes, this textbook incorporates modern pedagogical approaches to encourage active learning and reduce dependence on the lecture format, taking advantage of electronic information accessible on the World Wide Web.

The themes of ecosystem, environment, and ecology appear throughout this book in the context of concerns about planet Earth that developed throughout the twentieth century and will continue to emerge in the twenty-first century. Each of these terms has many meanings in the literature, sometimes generating confusion as these terms are often loosely used as synonyms. The title of the book uses *ecosystem* instead of *environment* or *ecology* because *ecosystem* is a comprehensive term that refers to a system of interacting biotic and abiotic elements applicable to the study of the human population and planet Earth. *Ecosystem* conveys a stronger sense of interactions than does *environment,* which often emphasizes the world external to the human population, such as toxic agents in air, food, or water. *Ecosystem* also conveys a stronger sense of the importance of abiotic elements than does *ecology,* which usually emphasizes the biological world. Admittedly, the distinctions between these terms are not rigid, as usage has evolved in different ways in different groups; the important message is that this textbook emphasizes interactions from a variety of perspectives— biological, chemical, physical, and social.

A global perspective on ecosystem change and public health covers an extraordinarily large and complex array of information. *Global ecosystem change* refers to changes in the earth's ecosystem that are global in extent, including changes in local ecosystems caused by pressures of population and consumption on local resources, which are becoming more widespread. Global changes arise from the interaction of natural and anthropogenic dynamics, involving a variety of factors such as climate and atmosphere, water and land, and the growth and movement of the human population. The scope of investigation is rather extensive and can fill volumes and volumes in print and electronic media. A global understanding of how people are injured, become ill, develop a disability, and die—the core questions in the science of public health—is also a vast subject.

This textbook embraces the principle that "less is more" and avoids any attempt to be an encyclopedic reference for every discipline. As a place to begin investigations of how public health is and may be affected by global ecosystem change, the book focuses on the interface between studies of global ecosystem change and studies of public health. Case studies, overviews of relevant material from many fields, and pointers to additional information help the reader to explore sources of information, develop inquiries, and identify techniques that need to be learned. The methodological aim is to enable the student and researcher to arrive at the intricate picture of interactions among global change and public health by discovery rather than by the absorption of text. This textbook integrates the contributions from multiple authors into a handbook that aids interdisciplinary research and study design (see Appendix A).

The Target Audience

The primary target audience is a master's-level student in public health, especially one with a strong interest in environment and health. This book should be helpful for students in public health seeking to integrate studies of infectious and noninfectious diseases. The scope of the book includes but is not restricted to infectious diseases, which are commonly taught separately from environmental health issues that focus on the toxicity of chemical and physical agents. Another target audience is master's-level students and upper-level undergraduate students in a variety of disciplines—environmental science, climatology, ecology, geography, and social science. This book will foster the development of interdisciplinary courses that bring together students with diverse backgrounds. Guidance on integrating multiple disciplinary perspectives avoids excessive technical jargon and technical notation comprehensible only to specialists within a narrow field. This book can be used as a primary or supplementary text for a course, as well as for independent study. The common element must be a desire to learn more about the study of global change and public health.

Multiple Ways to Use the Chapters

In this book Part I develops approaches for research, Part II describes environmental changes, and Part III provides case studies linking ecosystem change and public health. Although all three parts are interdependent and cross-reference each other, one strategy for using the book is to focus on one part at a time.

Part I: Approaches. Chapters 1–5 present a diverse selection of perspectives and research strategies for approaching complex topics on the effects of global change on human health. Examples illustrate successful applications of various methods, thereby assisting the new researcher in selecting appropriate approaches to research problems.

Part II: Environmental Changes. Chapters 6–10 present a selection of vital issues of today's global change with special emphasis on atmospheric changes and the hydrological cycle. Applying the principle of "less is more," this part of the book does not intend to be complete; rather, it provides selected specific topics on changes in the planet's environment and ecosystems.

Part III: Case Studies. Chapters 11–14 are case studies that emphasize the influences of global change on human health. The examples chosen represent different environmentally related health effects in different geographic areas. Within the context of real-life situations, the case studies link factual knowledge of the effects of global change and methods that are needed to understand and remedy the situation.

An alternative view of the chapters sets them in a web of interconnections. Each successive chapter does not require a thorough comprehension of all preceding chapters, and so the chapters do not have to be read in a linear sequence. All of the chapters cross-reference each other, and different parts of the book may serve as starting points for a course or for independent reading. One option is to start with an overview of the human dimension of global change (Chapter 6) in Part II before using Part I to develop specific approaches for research. Another option is to start with a specific disease, such as the case study on malaria (Chapter 12) in Part III, and then examine linkages with the changes described in Part II; Chapter 12 refers to every chapter in Part II. Yet another option is to start with the case study on global climate change and air pollution (Chapter 13) in Part III and then examine how the issue of global climate change appears in other chapters; Chapter 13 refers to Chapters 6, 7, 9, 10, 11, 12, and 14 in an overview of the potential pathways of the effects of global climate change on public health. A course with a primary interest in water issues may focus in depth on water resources management (Chapter 9) and on water-related health problems (Chapter 14); references in these chapters lead to information in Chapters 2, 4, 5, 8, 10, and 11. Readers with a background in research on global change may want to learn about study designs in epidemiology (Chapter 2), whereas trained epidemiologists may want to concentrate on applying techniques for remote sensing to global change (Chapter 3); both of these chapters refer to Chapter 1 for an overview of information on global change. Another group may want to focus on approaches for linking scientific data to the development of public policy affecting environment and public health (Chapters 4 and 5). And so on.

Electronic References and Information Literacy

Electronic references are references to uniform resource locators (URLs) on the World Wide Web. Although each chapter includes the traditional format of references to published books and articles, most chapters also refer to some URLs. The references to URLs are used like any citation but are set in a different typeface. For example, the notation (Environmental Protection Agency 2000) points to a traditional list of references at the end of the chapter, but the notation (*Environmental Protection Agency 2000*) points to a separate list of electronic references at the end of the chapter. These URLs provide links to reference material that would add considerably to the size of the printed book, including some color images that would be relatively expensive to put in print. Since the information at URLs can be updated regularly, references to URLs help the book to maintain currency. Of course, the addresses of URLs can and do change, but each electronic reference includes details about the owner of the website and the title of the page, which should make it possible to search for new links.

Besides the URLs in individual chapters, Appendix B provides annotation of major websites and a list of online directories and libraries that may serve as gateways into new sources of information. Online libraries may contain copies of traditional print publications, as either abstracts or full-text documents; it is also useful to consult the website of the organization that produces a publication of interest. Another feature of Appendix B is a list of topically arranged websites that are smaller than the annotated websites and have a stronger focus on a particular theme or regional and local concerns. These topically arranged websites are samples of the diverse sources of information available on the World Wide Web that may be of use in interdisciplinary research.

With a plethora of information directly accessible via electronic means, investigators need to develop and apply skills in information literacy. Appendix A contains general guidelines for information literacy in the world of the Internet. Chapter 1 suggests how to search for information about issues of global change, with explicit recognition of information from multiple disciplines. For every theme, Chapter 1 refers to chapters in this book and selected URLs as starting points for inquiry.

The Format of Suggested Study Projects

Each of Chapters 2–14 suggests three study projects that invite the reader to reflect upon the material in the chapter and to extend inquiry beyond the chapter and the book. All chapters explicitly cross-reference other chapters, thereby helping the reader to make full use of the book. The study projects may lead to a variety of written, oral, and multimedia presentations that could be completed by individuals or teams; various combinations may be used to enhance interactivity and communication among course participants (see Appendix A). Since suggested study projects may involve rather extensive research, they should be viewed as options rather than as a set of exercises to be completed to demonstrate knowledge of details in the chapter. The objectives listed for the suggested study projects are useful aids to navigation and help an instructor make use of the chapter in a formal course. However, the list of projects is not comprehensive. Instructors and independent readers may be motivated to design other projects tailored to specific needs.

Conclusion

The problems of global ecosystem change and their effects on public health constitute a growing challenge to scientists, physicians, governments, and the general public. This textbook provides an innovative structure that permits a diversity of approaches to a complex and important subject. The hope is that this book will motivate more and better studies of global change and public health.

APPROACHES

Part I develops approaches for interdisciplinary research on global ecosystem change and public health, beginning with an examination of information on global change from the perspective of multiple disciplines (Chapter 1). A sample site on the World Wide Web is part of the information on each of the major forces of natural and anthropogenic dynamics. The next step is to focus on establishing links with public health through epidemiological analysis that builds on basic concepts of study design—ecological, cross-sectional, case-control, and cohort (Chapter 2). The methodological examples also serve to illustrate a variety of health hazards (e.g., filariasis in Egypt, trachoma in Tanzania, air pollution in China, radiation-induced illness from the Chernobyl nuclear accident, and mortality in refugees from Iraq and Rwanda). More sophisticated tools for the analysis of geographic information can be incorporated into epidemiological studies, taking advantage of relatively new forms of global information from Earth-observing satellites (Chapter 3). The discussion addresses sources of information available on the World Wide Web, including criteria for the evaluation of the quality of datasets.

Empirical studies become part of a process of assessing risks to human health and developing policies to protect the environment and human health. The experience of air pollution in the United States demonstrates four essential concepts of risk assessment—hazard identification, dose-response assessment, exposure assessment, and risk characterization (Chapter 4). Traditional risk assessment is expanded into integrated assessment, which is a broader examination of social, economic, and environmental factors that provides insights into decisions (Chapter 5). The aim is to encourage interdisciplinary research that combines mathematical models of the consequences of different policies with a perspective on integrated assessment as a participatory process.

A multifaceted strategy for research emerges from Part I. Its scope ranges from empirical observations of what has already occurred and what is occurring to the process of integrated assessment that requires anticipation and analysis of possible events in the future. The rationale for this organization is to encourage more and better research in empirical studies, in the analysis of decisions about complex ecosystems, and in the development of better connections between these two areas.

See Part II for background on environmental changes on a global scale. See Part III for case studies on global ecosystem change and public health.

CHAPTER 1

Information on Issues of Global Change

Erika G. Feulner, M.A.

The issues of global change are numerous, of great complexity, and, despite their appearance in diverse settings, strongly interrelated. Information resources on issues of global change are matching this complexity by stretching across different disciplines and geographic borders, especially in the wake of electronic transmission and access. These new demographics of information access and retrieval are having a profound effect on contemporary research strategies. New criteria for selecting and evaluating information are emerging (see Appendix A), and the researcher and student bear a greater responsibility for making the choices among information sources for their research projects.

This chapter will assist the reader in constructing road maps for finding and using appropriate information for documenting environmental and ecosystem changes. It is the natural precedent to the following four chapters, which present appropriate methods for studying the influence of these changes on human health. The complexities of issues of global change demand from the researcher a clear vision in determining the nature of his or her study, knowledge of multiple research infrastructures and information sources, and dexterity in navigating the Internet. Three specific pitfalls in approaching complex research areas are to get overly involved in the details of one issue; to lose sight of its connection to other, equally important issues; and to fail to recognize that all such issues contribute to a larger picture—in this case, global change.

Chapters 6–14 of this volume open windows to the total landscape of complex facts and interacting agents of global ecosystem changes. This chapter gives a bird's-eye view of the information essential to the study of these interacting forces. Projecting the sequential order of chapter headings and subheadings into a cluster with interconnecting lines represents issues of global change as an intricate web of interactions. Using the visual elements of website development, one can produce a site map (Fig. 1.1) to illustrate the interactions that make up global change (see also Fig. 13.3).

The Interaction of Natural and Anthropogenic Dynamics

Eons before the advent of *Homo sapiens,* our planet was shaped by the dynamics of natural forces, which affected the quality of air and water, the formation or loss of land, and the decline or creation of natural resources. As the twenty-first century begins, natural forces are still at work, with the added positive or negative influence of the anthropogenic forces of our global civilizations and cultures. The complexities of studies of global change are rooted in the increase and variety of interactions between natural and anthropogenic dynamics underlying ecosystem change.

Interactions of natural and anthropogenic dynamics can be observed in a great variety of specific instances of change across the globe, and usually the causes of change are not fully clear. Such situations often present themselves as puzzles to the informed and uninformed inquirer alike, thereby inviting potentially dangerous reactions—inertia, rash conclusions, uninformed decision making, and political influence based on lack or misinterpretation of knowledge.

The new possibilities of accessing information through the Internet, which are still evolving and contributing to an escalating information overload, have put stringent demands on a structured approach to information retrieval and evaluation (see Appendix A). Determining the causes of change in a specific situation calls on the researcher's ingenuity and ability to construct an organized approach to the utilization of information within a new technological environment, from retrieving, correlating, and analyzing information found in different sources to integrating new information into sound conclusions.

The goal of an organized approach to information retrieval in the context of this book is the creation of a deeper understanding of ecosystem changes on a global scale. The diagram in Figure 1.1 should be approached like a site map on the World Wide Web that uses this chapter's outline to present a schematic overview of the information and information pointers contained in this book. While the connecting lines in the diagram illustrate the interactive dynamics of natural and anthropogenic forces, the boxes contain selected keywords leading to sources of information about specific phenomena or interactions in a situation of global change.

In the following text, a short paragraph will introduce each force contributing to global change and focus on linkages shown in the diagram. These linkages are supported by information found in this book and in other sources. In each case, the description of one uniform resource loca-

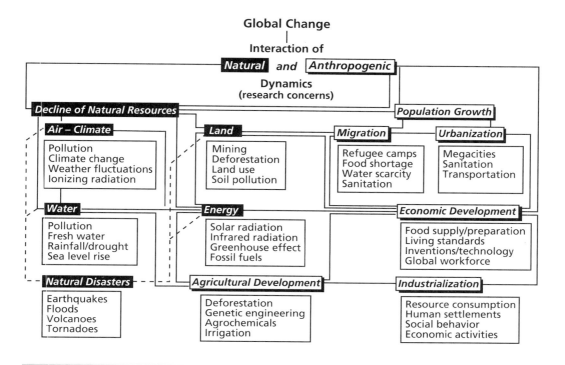

Figure 1.1 Interacting dynamics of global change and selected keywords for information searches. The *solid lines* indicate interactions, and the *broken lines* indicate that natural disasters are subsets of the major classifications of the dynamics of natural resources.

tor (URL) provides an opportunity for the reader to perform a search on the World Wide Web. In this fashion, the design of a road map to information sources will be started for the reader to augment and continue. Extending beyond individual chapter references and the Resource Center of this book (Appendix B), the road map can lead to any available and applicable information.

Research Infrastructures

The global environment is not static. Our planet's interacting and interdependent life systems, including human life, have been forced to adapt or change from their very beginning, millions of years ago. Innate characteristics of resiliency, adjustment, and survival in all life forms cope with change to varying degrees. These characteristics are tested severely when change occurs with increased speed and severity.

Human wellness depends on the intricate balance of environmental forces, which are governed by the cyclical exchange of life-sustaining chemical compounds from the sea to the stratosphere. In planet Earth's at-

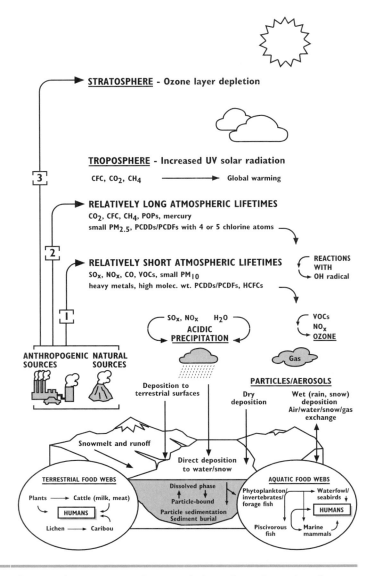

Figure 1.2 Pathways of transport and accumulation of continental pollutants. *Source:* Redrawn with permission from Commission for Environmental Cooperation 1997, Figure 1.

mosphere a delicate interchange and reconstitution of compounds—nitrogen, oxygen, sulfur, argon, and other trace gases—takes place (Warneck 1992). Transboundary pollutants disturb the cycle of chemical exchange and subsequently upset the environmental balance (Fig. 1.2). The disturbance of this interchange is the backdrop of all environmental changes, regardless of whether they are caused by natural or anthropogenic sources.

Therefore, the overarching concern of studies of global change is the interaction of natural and anthropogenic dynamics and their contribution to environmental and global ecosystem changes.

The rate of these changes during the late twentieth and early twenty-first centuries can be represented by an acceleration curve showing the effects of changes, which vary in severity, on ecosystems and human health in different regions of the globe. To assess these effects, to observe the rates of resiliency, adjustment, and survival, to evaluate the increasing body of information, and to find measures for mitigating adverse consequences requires research of greater scope and on a larger scale—interdisciplinary research on the global ecosystem.

Interdisciplinary research depends on access to larger bodies of comprehensive data and the ability to integrate individual research efforts into an expanded scenario. This larger picture of the global environment comprises a multitude of factors derived from scientific studies at local, regional, and global levels. Open-mindedness and acknowledgment of the validity of different professional perspectives are prerequisite for the interdisciplinary researcher. The complexity of environmental research has already invited investigations by researchers from different disciplines of the environmental sciences, but greater participation of the social and information sciences is needed for reaching an in-depth understanding of the complexities involved.

In modern-day research, the information specialist, in particular, must become part of any research project. Different modes of research and information evaluation have to be understood and applied (see Appendix A). What used to be the librarian's venue alone—research assistance based on the collection of, reference to, access to, and evaluation of information— is now largely in the hands of each individual researcher. A researcher needs to absorb the information specialist's skills, especially when accessing and evaluating information from electronic sources, to create a wider perspective in retrieving, cross-referencing, and integrating information as part of the research project (see Appendices A and B).

The purpose of studies of global change is to arrive at a clearer picture and to yield more precise data on the interaction between natural and anthropogenic dynamics that are changing our global ecosystem. The key word is *interaction*, which adds a new dimension to research performance. There is a fundamental difference between the infrastructures of research in the natural and biomedical sciences versus research in the social sciences. Scientific research has become an increasingly specialized investigation of well-defined phenomena designed to produce measurable results. Researchers in the social sciences, based strongly on historical values, are taking the approach of interpreting facts from various points of view by a process of analysis and questioning; arriving at answers by deductive reasoning and reaching consensus define the research results (Barzun and Graff 1985).

The interdisciplinary researcher must be familiar with these two research infrastructures and be able to apply a variety of research approaches during the planning of an initial research strategy (see Chapter 2). The more comprehensive the perspective in approaching a research problem, the easier the selection of methods for shaping the research strategy (see Appendix A, Mental Skills Development).

The Decline of Natural Resources

Different institutions will attach different denotations to the term *natural resources*. It is important to consider this fact when evaluating sources on natural resources. The four categories of natural resources regarded as the basic life-sustaining elements of our planet are air, water, land, and energy. These resources are intertwined and in constant interaction with each other. Drastic changes often occur through heightened activities in one domain or another (e.g., volcanic eruptions, earthquakes, floods, storms), which we accept as natural disasters and as a part of our planet's evolution and growth.

Recently, as Figure 1.1 illustrates, anthropogenic activity has developed into an additional formidable force affecting the primordial life-sustaining cycles of interaction, destruction, renewal, and restoration of our natural resources. Anthropogenic activity has reached a level of interference with natural dynamics that forces researchers to conduct new assessments and include anthropogenic factors in the equation of restoring and safeguarding the interaction of the life-supporting elements on our globe. All chapters in this volume refer to interactions of natural and anthropogenic dynamics.

In this chapter, the four essential natural resources are discussed sequentially. Many references to information sources are found within this book and should be consulted first. Additional references to outside sources augment the perspective of the intricate web of interactions and relationships that shape our global ecosystem. Among the four natural resources mentioned above, greater emphasis is placed on air and water than on land and energy, mainly in correspondence to the four case studies in Part III, which emphasize the environmental factors in air and water that affect ecosystem change.

Air and Climate

A strong focus on atmospheric changes emerged in the scientific community during the late twentieth century. Climate change, stratospheric ozone depletion, and transboundary air pollution have become leading topics in environmental research (see Chapters 7 and 13). In this context, *global cli-*

mate change is quite often used as the generic term for all environmental changes, indicating that all changes in the air caused by pollutants or weather fluctuations are intricately linked to changes in the biosphere and the hydrological and energy cycles (see Chapter 8). The focus on air actually means observing and measuring a multitude of interactions and chemical reactions occurring from the ocean to the stratosphere and beyond.

Nearly all chapters refer to some aspect or effect of climate change, thereby emphasizing this particular issue of global change. Chapter 7 features a description of the intricate patterns of chemical changes in our atmosphere and demonstrates that the four global change phenomena—stratospheric ozone depletion, acid deposition, urban air pollution, and enhanced global warming—are anthropogenically induced. The author argues that, for the "first time in geological history, humanity has changed the earth's environment on a global scale."

Chapter 8 presents an added perspective by establishing links to terrestrial ecosystems and the water, carbon, and energy cycles. The author's outlook on the "global system" as a whole and the focus in Chapter 6 on human activities and their relationship to environmental changes provide a valuable complement to the content of Chapter 7 for understanding the intricacy of interactions.

Chapter 13 concentrates on air pollutants, greenhouse gas emissions from the burning of fossil fuels, and the formation of ground-level ozone—a particular atmospheric disturbance likely to be exacerbated by global climate change. Chapter 13 also stresses the health effects of atmospheric disturbances, noting specific results of global climate change.

Our example URL on this topic (http://climon.wwb.noaa.gov/) belongs to the Network for the Detection of Stratospheric Change (NDSC), a major component of the international effort to conduct research on the upper stratosphere that has been endorsed by national and international scientific agencies. The NDSC lists addresses and telephone numbers of experts who can be contacted for further information.

Water

Water is the signature element of planet Earth. The availability, shortage, pollution, and quality of water are growing concerns of scientists, policymakers, and the public. The hydrological cycle, as information on air and climate reveals, is an integral part of the cyclical exchange of life-sustaining chemical compounds and is the main support of life as we know it (see Chapters 8 and 9). Water is an indispensable resource of the ecosystems of our planet and is closely linked to climate change, as documented in cases of intense rainfall, floods, sea-level rise, and water contamination (see Chapter 9). Many reports on infectious diseases whose appearance or reappearance is traced to ecological change implicate a body of water whose contamination is the primary cause for the onset of the illness or whose presence provides a breeding ground for a disease vector (see Chapters 6, 8, 10, 11, 12, 13, and 14).

As water is the essential environment for sustaining life, it has also become the essential arena for changes in the relationship between infectious agents and their hosts due to altered environmental conditions. Global change phenomena are frequently first noticed in water-related effects.

Chapters 9 and 14 are focused on water. Chapter 9 covers every aspect of the management of water resources—the hydrological cycle, the availability and quality of water, the links between water and climate change, groundwater exploitation, and human intervention in the terrestrial water cycle. With its references Chapter 9 provides an excellent information base for further water-related inquiries. Chapter 14 presents four geographic examples of how the quantity of water affects human health. Water's ecological effects include its scarcity as a result of forced migration, the effects of the diversion of water in the Aral Sea Basin and the Great Basin of California, and the effects of excess water, with a focus on flooding in Brazil.

Our example of a URL on water (http://water.usgs.gov/) belongs to the U.S. Geological Survey, whose mission is to assess the quality and quantity of the natural resources of the United States and to provide information that will assist resource managers and policymakers at federal, state, and local levels. Assessment of water quality is an important part of this mission. The website covers all aspects of water resource management on the national level and is frequently updated. Several programs are monitored, and publications as well as links are available for additional information.

Land

In the interactive chain of natural resources, land is quite frequently the first place where wider audiences observe environmental changes and become more astutely aware of global change. Plants, animals, and humans are visibly in contact with land, food chains and food production are understood in terms of land use, and anthropogenic activities disturbing the interactive cycles of natural resources are started mostly on land. Land features its own complexities in the formation and interdependence of different terrestrial layers—unweathered rock, saturated and unsaturated weathered rock, soil and subsoil cover, and forest cover—components of the ecosphere that play a major role in the dynamics of our ecosystems (see Chapter 8).

Plant growth and food production depend on the quality and health of the soil system. The study of soil systems has grown into its own science, pedology. Soil systems, the *pedosphere,* develop from a dynamic interaction among the atmosphere, biosphere, lithosphere, and hydrosphere.

Many environmental scientists are interested in land use, with special focus on land degradation, desertification, biological diversity, freshwater resources, deforestation, and the spread of infectious disease. *Land use* denotes human activities on the land and will be addressed more specifically under "Economic Development."

Our example of a URL on land (http://www.soils.rr.ualberta.ca/Pedosphere/contents.htm) is an online textbook, "The Pedosphere and Its Dynamics—A Systems Approach to Soil Science," which was created and is maintained by Dr. Noorallah G. Juma, Chris Harland, and Craig Nickel at the University of Alberta, Canada. E-mail addresses are available at the website.

Energy

There are two distinct categories of natural resources for energy, the sun and carbon-based fuels. The primal source of energy is the sun, which interacted with water, air, and land before human evolution and is thereby an intrinsic element of the natural resource. Natural resources for energy are utilized by human intelligence for survival and development.

The sun is fundamental to all life on earth. Viewed from different perspectives, solar energy has been supporting the development of all aspects of human evolution. Chapter 8 explains in simple terms the natural energy cycle—the balance between incoming solar radiation and outgoing infrared radiation. Plentiful references to photovoltaics—the technology that converts light or solar energy into direct current electricity without creating pollution—are usually found under keywords like *solar energy, solar energy systems, passive solar energy,* and *alternative energy.*

The foremost debates on energy, however, concern the natural resources found in the biosphere and lithosphere and converted into energy only by human intervention. Our energy resources have been mostly carbon-based: wood, coal, oil, and natural gas. Their exploitation for human purposes has affected the carbon cycle (see Chapter 8) and caused severe imbalances in the interactive web of natural resources supporting healthy life on Earth (see Chapters 6, 7, 9, and 13).

The URL selected as an example (http://www.pvpower.com/) was developed and has been maintained by Mark Fitzgerald, Science Communications, Inc., with assistance from the photovoltaic industry and government information resources. The most recently updated pages are a history of photovoltaics with a timeline from 1839 to the present.

Population Growth

The impact of human activity on the environment is closely linked to the size and rate of growth of the human population. During the second half of the twentieth century, the entire human population grew to unprecedented levels at unprecedented rates, adding, toward the century's end, 90 million people each year to the world total. The cause of rapid growth was a reduction in the death rate due to improved health conditions worldwide (Cohen 1995, Chap. 4). The per capita rate of growth has slowed somewhat from its peak in 1965–70 as fertility rates have declined in several populous, less-industrialized nations, but the youth of a large proportion of the world population provides momentum for population growth into the twenty-first century (Cohen 1995, Chapters 4 and 17). The growth of the world population is not expected to stabilize until the middle of the twenty-first century (World Resources Institute [WRI] 1994). Population projections are based on the assumption that, in the long term, the fertility rate will come down where it is high and life expectancy will rise where it is low. This assumption, however, is based on assumptions about future human behavior, which cannot always be predicted (WRI 1994).

A larger population puts greater stress on the environment, natural resources, physical infrastructures, and governments and therefore plays a major role in the analysis of threats to ecosystems and human health. This relationship has directed scientists' attention—as stated in Chapter 6—to more specific contemporary changes in the human environment, such as urbanization and migration, which are often linked to patterns of economic development affecting industrialization and agricultural development (see below).

The effects of anthropogenic activities on the environment are extraordinary. As stated in the World Resources Report for 1994–95 (WRI 1994), "of the many human-induced changes that have occurred over the past several centuries, two are especially noteworthy: habitat alteration and pollution." Estimating the rate of anthropogenic change, however, is extremely difficult and opens arenas of much-needed research on specific interactions between the natural forces shaping our environment and anthropogenic forces (WRI 1994).

Urbanization
Human settlement patterns dictate to a large extent the wealth and health of a society (see Chapter 6). Historically, cities have been driving forces in economic and social development and world centers of affluence and political power. Urbanization is associated with overall improvement of the quality of life and has been the nurturing ground for benefits like access to information, diversity, creativity, and innovation (WRI 1996). Demographic shifts estimated by the United Nations project the development of megacities—that, by the year 2015, there will be 26 urban areas of 10 million inhabitants or more, all but 4 of which will be in less-industrialized nations (see Chapters 6 and 9).

The benefits of urban development are counteracted by the stress to the environment and human health caused by urbanization (WRI 1998). Urban environmental problems include lack of access to clean drinking water (see Chapter 9), urban air pollution, and greenhouse gas emissions (see Chapter 13). Chapter 6 is the central information source within this book on anthropogenic changes in our environment; direct and indirect references to urbanization can be found in Chapters 7, 9, and 11–14 (the four case studies).

Our example URL on urbanization (http://royal.okanagan.bc.ca/ mpidwirn/urbanization/urbanization.html) is part of a Canadian site, "Living Landscapes, Thompson-Okanagan: Past, Present, and Future." It presents an excellent summary of the environmental impacts and benefits of urbanization, a treatise on sustainable urban development, and a case study on urbanization and the Okanagan Valley, thereby providing the elements, keywords, and links for extended studies on urbanization.

Migration

Migration—the movement of human populations—increased considerably during the late twentieth century (see Chapter 6). There are several types of migration: rural-to-urban, urban-to-rural, urban-to-urban, and, in some countries, rural-to-rural flows have been recorded (WRI 1996). The causes for migration differ widely, from forced migration due to alteration in the habitat, to the search for economic stability or opportunity, to the avoidance of political, religious, or ethnic discontent. Migration has become a worldwide phenomenon and drawn international attention, which has spurred the establishment of institutions like the International Organization for Migration.

Issues related to migration involve concerns of population growth for individual countries (WRI 1994), immigration and emigration laws, and, above all, the effects of large population movements on health. (See Chapters 2, 6, 10, 11, and 14.) It is stated in Chapter 6 that

> contemporary large-scale population movement is obviously a factor that is changing the balance between human beings and microorganisms. Migrants and refugees frequently live in marginal and unsanitary areas of the world and have poor access to clean water, sufficient food, or health services. Crowded, unsanitary refugee camps are ideal locations for the propagation and spread of diseases such as cholera and typhoid. . . . Thus, future large population movements present serious health risks affecting both migrant populations themselves and recipient countries.

The URL selected as an example for migration issues (http:// migration.ucdavis.edu/mda/mdtxt.htm) is the product of a not-for-profit organization, Migration Dialog, which prepares and distributes timely information on international migration. The organization also sponsors an annual three-day seminar for American and European opinion leaders and maintains a list of speakers willing to discuss migration issues.

Economic Development

While population growth is an anthropogenic phenomenon that raises questions of environmental sustainability, the conservation of resources, and appropriate health services, economic development has a decidedly twofold effect on the environment. On the one hand, it can contribute to high rates of consumption, thereby eroding natural resources and causing environmental ills; on the other hand, it has the potential to find solutions to environmental problems, thereby contributing to restoring the balance of global ecosystems. The direction of economic development depends on policy decisions worldwide and is ultimately a question of human behavior.

High rates of consumption have become a growing concern of scientists, policymakers, and citizens. The need for energy, arable land, and transportation in pursuit of a higher-quality lifestyle has eroded natural resources through modern-day mining practices, deforestation, desertification, construction, and poor waste management. Economic development is the driver of commercial interests, but policy decisions must take into account both economic and environmental interests. Economic growth with consideration of environment protection and sustainability has yet to be worked out.

Chapters 4, 5, 6, 7, 9, and 10 refer specifically to the need for conscientious decision making in balancing economic and environmental interests. WRI reports provide excellent statistics on and summaries of these issues (WRI 1994, 1996, 1998). Two major forces are determining factors in economic development: industrialization and agricultural development.

Industrialization

Industrialization is, in broad terms, a process that entails the transition of human labor skills to operating skills for new tools, new machinery, and inventions or innovations that enhance faster production. This process has a long history and is substantiated by the ideas of world philosophers and experts in many fields. It reached its mark of public recognition in the eighteenth century with the industrial revolution (Drucker 1993). The industrial revolution irrevocably changed society by introducing new relationships into the work force and new definitions of productive work, which rather rapidly changed the physical infrastructures of Western society. The emergence of modern technology has affected society and human thought continuously to the present.

The industrialization process at the beginning of the twenty-first century has a worldwide dimension. The development of industrialization is responsible for vast improvements of many nations' physical infrastructures, has produced the knowledge society, has fostered higher standards for quality of life, and is—in its negative aspects—in ideological conflict with environmental sustainability. Researching industrialization issues in relation to global change requires a historical perspective and a detailed focus on geographically diverse examples. In terms of information sources, the industrialization process is usually embedded in discussions of urbanization and economic development (see references above).

The issues of industrialization, economic development, and a growing awareness of global change appear quite often in the context of complex strategies of community economic development. Hence, the URL selected as an example (http://www.uwex.edu/ces/cced/learn.html) is the University of Wisconsin–Cooperative Extension's Center for Community Development. The center creates, applies, and transfers interdisciplinary knowledge to help people understand community change and identify opportunities. It offers a variety of resources, data resources, publications, and expert contact.

Agricultural Development

The influence of industrialization on agricultural development inspired the heading "agricultural and industrial revolutions" in history texts and a host of publications on the industrialization of agriculture (see the selected URL below as an example). Agricultural development has a specific position in the cross-section of industrial modernization and the preservation of natural resources. The increasing needs of the human population exert pressure on the rate and quantity of food production, but food production also depends to a large extent on healthy ecosystems. Because agricultural practices are a major part of land use or land conversion, our awareness of the consequences of deforestation, desertification, reduced biological diversity, the limits of freshwater resources, and land degradation has grown.

Chapter 6 presents an overview, supported by several references, of "Food Production and Health in the Less-Industrialized World," pointing to the consequences of deforestation and subsequent health effects. Forest conversion is especially associated with higher incidences of certain infectious diseases, including malaria and leishmaniasis (WRI 1998; see also Chapter 12). Other references related to land use issues are found in Chapters 8, 9, 10, 12, and 14.

The URL selected as an example for agricultural development (http://www.agribiz.com/newsind.html) is part of the "News and Articles" within AgriBiz, which provides information for the global agricultural community. The site lists publications, sometimes with abstracts, and extended sources on the industrialization of agriculture.

Conclusion

Global ecosystem changes and environmental issues are drawing increasing attention across disciplinary boundaries. These themes have become concerns of natural scientists, social scientists, educators, medical professionals, policymakers, and citizens. The information sources addressing these issues are mushrooming and exhibit great diversity. Individual sources cover concerns from local to global levels but are often limited by a parochial view and presentation.

The bird's-eye view of information on issues of global change presented in this chapter is a road map to information contained in this book and other locations. This book's Resource Center (Appendix B) lists outside information sources selected through an evaluation process described in Appendix A. The nature of a bird's-eye view is to present ideas and facts in a summary or overview style, but—in the case of an overview of information sources—it will lead to more in-depth information for further exploration.

It is essential at the start of any research to evaluate the authenticity of the information sources used to substantiate facts, hypotheses, and future developments in the interdisciplinary field of global change. The selection of information sources in this chapter and in the Resource Center (Appendix B) is intended to serve as a guide for anyone approaching the task of information collection for a research project.

References

Barzun J, Graff HF. 1985. *The Modern Researcher*. Harcourt Brace Jovanovich Publishers, New York.

Cohen JE. 1995. *How Many People Can the Earth Support?* W. W. Norton & Co., New York.

Commission for Environmental Cooperation. 1997. *Continental Pollutant Pathways*. CEC, Montreal.

Drucker PF. 1993. *Post-Capitalist Society*. Harper Business, New York.

Warneck P. 1992. Atmospheric composition and chemistry. In *Encyclopedia of Physical Science and Technology*, 2d ed. (Meyers RA, ed.). Academic Press, San Diego, Vol. 2, pp. 265–71.

World Resources Institute. 1994. *World Resources, 1994–95*. Oxford University Press, New York.

———. 1996. *World Resources, 1996–97*. Oxford University Press, New York.

———. 1998. *World Resources, 1998–99*. Oxford University Press, New York.

CHAPTER 2

Epidemiological Study Designs

Alfred A. Buck, M.D., Dr.P.H., and Joan L. Aron, Ph.D.

Epidemiology is the basic scientific discipline of public health. Since Frost's first definition of epidemiology 70 years ago, there have been many modifications (Lilienfeld 1978; Last 1995). They reflect the expanding application of epidemiological methods to the study of changing human health problems. A generally accepted version defines epidemiology as the study of the distribution and determinants of health-related states and events in specified populations and the application of this study to the control of health problems (Last 1995; Gordis 1996). An essential objective of epidemiological investigations is the assessment of the magnitude and cause of changes of the health status of people over time and space. This objective may be summarized in terms of the *W* questions: What, Who, When, Where, and Why (see Box 2.1).

The systematic study of global change (see Chapter 1) and health presents several organizational complexities. The populations of interest are large, diverse, and distributed all over the globe. Health effects of global change comprise acute and chronic diseases. The long time scale for chronic diseases to emerge makes them more difficult to study and therefore more likely to be neglected (see Box 2.2). The importance of chronic conditions combined with the long time scale for global change means that studies of global change and health require long periods of observation. Global change also affects infectious diseases along with noninfectious diseases and injuries, areas of specialization that are often separated (see Box 2.3). The collection of sufficient data on multiple health risks to which people of different geographic areas have been exposed concomitantly is obligatory as the basis for in-depth studies. Research in global change and health requires multidisciplinary expertise. In addition to a team of medical specialists, comprising epidemiologists, clinicians, medical laboratory scien-

W Questions for Epidemiology

What?

Changes in the distribution of a health-related state or event related to disease, injury, disability, or death.

Who?

Population affected by the changes. The population is specified in terms of a variety of characteristics, such as age, gender, residence, or occupation.

When?

Timing of the changes. Investigations may examine what has already happened or what might happen in the future.

Where?

Geographic location of the changes. The scale may range from global to local (see Table 2.2).

Why?

Determinants of the changes. Determinants may include any factors in the global ecosystem, ranging from the biomedical, such as genetic characteristics and exposures to infectious agents, to the social, such as economic development.

---

tists, entomologists, and statisticians, the input of other disciplines, including geography, geology, meteorology, social and behavioral sciences, demography, environmental engineering, and veterinary medicine, is essential.

The epidemiological study designs appropriate for global change and health are observational rather than experimental. In experimental studies, the investigator controls one or more factors affecting a group of people by randomly assigning them to different levels of the factor or factors. For example, in a clinical trial of a new vaccine, the investigator randomly assigns subjects to receive either the vaccine or a placebo. The purpose of randomization is to define groups that are known to be similar except for exposure to factors in the study. Observational studies, which are conducted without randomized comparisons, are required for large-scale environmental changes because precise control of exposures is neither feasible nor ethical. However, observational study designs can take advantage of variation caused by natural processes or by human activity to set up contrasts that can be quite powerful in addressing the basic epidemiological questions (see Box 2.1). For example, in an observational study of mortality in different time periods, the investigator compares time periods with many deaths and time periods with few deaths.

Observational studies of global change and health also have limitations simply because the events in the studies have already occurred. The integrated assessment of changes in the global ecosystem and their linkages with public health requires anticipation and analysis of possible events in the future, activities fraught with uncertainty (see Chapter 5). The challenge is to develop stronger linkages between observational studies and the process of integrated assessment in order to improve the overall process. To achieve that aim, it is essential to understand how to conduct and interpret epidemiological studies.

Chronic Diseases

The growing prevalence of disabling "chronic diseases" presents a special challenge to studies of global health because such diseases seem to be particularly sensitive to environmental changes and yet are too often unrecognized and underreported in many developing countries. Chronic disabling conditions may be due to a wide spectrum of infectious causes, such as river blindness and histoplasmosis, and noninfectious causes, such as malnutrition and accidents. These chronic conditions are often the underlying causes of premature death but do not appear in mortality statistics. Some of the chronic infectious diseases require animal hosts and invertebrate vectors for their transmission and persistence (see Chapter 10), so that their geographic distribution and prevalence are highly dependent on the local environment, economy, and cultural characteristics of the people. In each area where endemic disease is prevalent, there is a sensitive equilibrium that is highly susceptible to environmental change. Increases of the level of endemicity of a disease may remain unnoticed for many years because of the long period of clinical inapparency between the time of infection and the delayed development of symptoms, which may require locally unavailable laboratory tests for definitive diagnosis. Unlike the easy recognition of rapid changes in incidence in an outbreak of an acute disease, the insidious spread of chronic infections and diseases after natural or anthropogenic changes of the environment is poorly recognized and remains grossly underreported.

Infectious and Noninfectious Agents of Disease

The complexity of interactions between potential risk factors for health conditions in the intricate triad of environment, human host, and causative agents of disease (biological, chemical, and physical) has played a major role in the development of epidemiology as the basic scientific discipline of public health. However, trends in health risks throughout the twentieth century have encouraged the separation of studies of infectious and noninfectious agents of disease. While infectious diseases remained major causes of mortality and morbidity in the least developed countries, infectious diseases were replaced by noninfectious conditions in economically more advanced countries. The decline of infectious diseases from 1930 to 1980 has invited extrapolations from the success of single disease-control programs to unrealistic forecasts for eliminating others. As the twenty-first century begins, however, the worldwide emergence of new and old infectious diseases, together with the increasing industrialization and urbanization of developing countries, requires a more balanced, holistic approach to environmental epidemiology by considering the health effects of biological, physical, and chemical risk factors together (Last 1995). Therefore, studies on global change and health need to be designed for a broad range of health conditions.

Descriptive Measures of Health Status

Descriptive measures of health status provide a starting point for epidemiological analysis. This section presents basic definitions and sources of data. For studies of global change and health, it is especially important to search for data hidden in obscure locations (see Box 2.4) and to combine multiple sources of data.

The health status of a population is measured in terms of the occurrence of health-related states and events in the population. A health-related state refers to morbidity, which is any departure, subjective or objective, from a state of physiological or psychological well-being (Last 1995). A health-related event can be the onset of morbidity, a vital event such as birth or death, an event related to the cause of an illness such as an exposure to a health hazard, or an event related to the effect of an illness such as a hospitalization.

An incidence rate and a prevalence rate are the two fundamental measures of health status of a population. An incidence rate is the rate at which new events occur in a population, defined as the ratio of the number of new events in a defined period to the population at risk of experiencing the event (Last 1995). In applying this measure to morbidity, the event is typically a new case of a disease or injury. For example, the National Center for Health Statistics (NCHS) reports U.S. influenza statistics for 1994 as 35 new cases of influenza per 100 persons per year, which represents 90.4 million new cases (Adams and Marano 1995). A prevalence rate is the ratio of the total number of individuals who have an attribute or disease at a particular time to the population at risk of having the attribute or disease at that point in time (Last 1995). For example, NCHS reports U.S. asthma statistics for 1994 as 56.1 per 1,000 persons, which represents 14.6 million cases (Adams and Marano 1995). A prevalence rate combines cases that are quite recent in origin and cases that have persisted for years.

Another important measure of the health status of a population is the mortality rate, which is the ratio of the number of deaths during a specified period to the number of persons at risk of dying during the period (Last 1995). A mortality rate is, in effect, an incidence rate for which deaths are counted instead of new cases of a disease. When the entire population is aggregated, the result is called a crude death rate (CDR) or crude mortality rate (CMR). For example, NCHS reports the U.S. CDR in 1995 as 880.0 deaths per 100,000 population, which represents 2,312,132 deaths (Anderson et al. 1997). Mortality rates for specific age groups, with the exception of infants, are calculated in the same format. When calculating an infant mortality rate (IMR), the ratio is the number of deaths of children less than one year of age in a given year to the number of live births in the same year (Last 1995). For example, NCHS reports the IMR for the United

Hidden Data Sources and Climate Change

Many hidden data sources from past investigations and local health statistics can be useful for preparing retrospective studies on environmental health. The development of an epidemiological database for the Onchocerciasis Control Program in 11 West African countries began in 1973 with special efforts to trace the many unpublished records hidden in small health centers or kept in the files of individual investigators. Relevant information from a total of 2,407 communities with an estimated population of three million individuals became available. The data were upgraded to include the exact geographic coordinates of each village, their census population, and the number and proportion of residents who were examined for onchocerciasis. The statistics were adjusted to account for changes in the methods of diagnosis (see text). Mapping of the communities according to the prevalence of onchocerciasis made clear that prevalence was greater when a village was closer to areas of river turbulence, which are the breeding sites for the blackfly that transmits the disease. West Africa is subject to climatic variability that affects rainfall and river discharges in the region (Tourre et al. 1999). Because of the significance of environmental factors in the natural history of onchocerciasis, the disease can serve as an indicator of the health impact of climate change in Africa.

States in 1995 and 1996 as 7.586 and 7.215 infant deaths per 1,000 live births, respectively (Ventura et al. 1997). The IMR is often used as a general indicator of health conditions in international comparisons. Mortality statistics are also usually provided for specific causes of death that are coded according to procedures given in the International Classification of Diseases, now in its 10th revision (Last 1995). For example, NCHS reports the U.S. rate of mortality from malignant neoplasms in 1995 as 204.9 per 100,000 population, which represents 538,455 deaths due to cancer (Anderson et al. 1997).

Most countries maintain morbidity and mortality statistics. Some of these cover decades and even a century. These health data are based on the reports of specific diseases, deaths, surveillance programs (see Box 2.5), military and occupational health records, and an array of insurance and private collections. The diagnostic criteria and case definitions used in these statistics have changed over time because of upgrades of case definitions and requirements for reporting the underlying causes of mortality. There are large variations in the quality of health statistics between countries and between sophisticated urban centers and medically understaffed rural areas. These variations reflect crude differences in diagnostic capacity and competence. Because of these limitations, the use of these health data is often disregarded, although they may be the only available source of urgently needed baseline information for the design of a new study. Sometimes it is possible to verify the existing data in a relatively inexpensive validation or calibration study.

BOX 2.5

Disease Surveillance

Surveillance is the continuing scrutiny of all aspects of occurrence and spread of a disease that are pertinent to effective control (Last 1995). Surveillance is fundamental for guiding decision makers in the development of effective programs for disease prevention and control in specific ecological settings. The scrutinized control activities comprise health education, avoidance and reduction of environmental health risks, and medical interventions (e.g., vaccinations).

Surveillance can be either passive or active. In a passive system, the data available on reportable diseases can be used to monitor changes in their frequency, distribution, and associated mortality rates in different areas or groups of people. While passive surveillance is relatively cheap and easy to develop, underreporting is very likely, especially for diseases and health conditions that require sophisticated diagnostic procedures for case detection and reporting. Active surveillance includes case finding, interviews of patients and of their physicians, search of medical records, travel histories, and sentinel health information systems, in which a limited number of health practitioners report on a specific list of conditions. Maintenance of surveillance programs requires cooperation and communication among all concerned parties.

EXAMPLE: Adjusting for Changes in Diagnostic Criteria

The evolution of the highly successful Onchocerciasis Control Program began in 1973 with a search for existing epidemiological data from the Volta River Basin in West Africa (see Box 2.4). The adult parasitic worms that cause onchocerciasis, also known as river blindness, live mainly in and below the skin, where they form visible nodules in which one or more pairs of worms are coiled. Before 1972, the search for the presence of the characteristic skin nodules was the method of choice in prevalence surveys of the disease. Although nodules are a frequent and characteristic clinical sign of onchocerciasis, persons with mild and early infections remain undetected, making the method less sensitive than the skin snip. The skin snip can detect microfilariae, which are the embryos produced by the female worm. The search for data in local unpublished health statistics and scattered publications from the area made it possible to obtain results from 44 of the 2,407 villages, comprising a population of 13,113 people to whom both diagnostic tests were applied independently. All pairs of these tests were plotted to adjust the observed prevalence based on the presence of nodules alone to the expected frequency of onchocerciasis if skin snips were taken (see Fig. 2.1).

Even with competent diagnostic capacity, cases of a disease may be underreported. Rather than ignore such information, it may be useful to combine multiple sources of data.

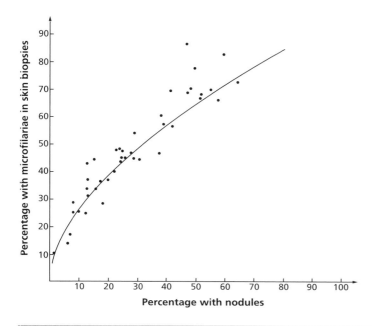

Figure 2.1 Relationship between two measures of the prevalence of onchocerciasis. The percentage with nodules and the percentage with microfilariae in skin biopsies were taken from 44 villages (13,113 inhabitants) in the Volta River Basin of West Africa around 1972. *Source:* Reproduced with permission from United Nations Development Programme/Food and Agriculture Organization/International Bank for Reconstruction and Development/World Health Organization 1973, Figure 4.1.

EXAMPLE: Combining Complementary Sources of Data

The United States has two types of data on the epidemiology of hepatitis A infection (Shapiro et al. 1992). The Centers for Disease Control and Prevention (CDC) is responsible for maintaining a surveillance system (see Box 2.5) for reporting the incidence of hepatitis A disease (i.e., new cases). The reporting of new cases may be accompanied by descriptive information about individual cases, making the system a valuable resource for examining the distribution and sources of infection. However, the surveillance system does not capture all cases, in part because many infections remain clinically inapparent. Another epidemiological portrait of hepatitis A in the U.S. population is obtained using the level of specific antibodies for case definition. Specific antibodies mark a history of infection with the hepatitis A virus regardless of the clinical outcome. The National Health and Nutrition Examination Survey measures the prevalence of hepatitis A antibodies, termed seroprevalence, at one point in time in a sample representative of the U.S. population. Seroprevalence surveys provide a complete picture at any time but cannot identify when infection occurred in the past. The two sources of data taken together complement each other.

The fundamental measures of health status are expressed as per capita rates for a population at risk instead of absolute counts of cases of disease or deaths. The use of per capita rates adjusts for different population sizes when making comparisons of the health status of populations across time and space. Rates rather than absolute counts are important for understanding the determinants of disease. For example, during John Snow's classic 1854 study in London, which demonstrated a high rate of cholera deaths in districts served by the water supply of the Southwark and Vauxhall Company, the aggregate number of cholera deaths was higher in other districts of the city simply because they had many more residents (Gordis 1996).

The measurement of an appropriate population at risk is therefore as important as counts of cases or deaths for avoiding distortions in the calculated rates (see measurement bias under "Systematic Error," below). A rate may be too high if the population denominator is too low or, conversely, the rate may be too low if the population denominator is too high. If the degree of population accuracy varies across time periods or across locations, comparisons of rates may be misleading. Even if population counts are accurate, contention may arise in trying to ensure that the populations are those individuals who are truly at risk of developing the disease of interest. Although some adjustments, such as excluding males from rates of ovarian cancer, are straightforward, the estimation of populations exposed to particular environmental hazards may be difficult. Population censuses are usually aggregated by political subdivisions (country, province, state, city, etc.) that may not correspond to risks presented by environmental hazards. Special subpopulations defined by socioeconomic status, housing conditions, age, or chronic disease may actually be the group at risk.

Simple comparisons of rates for aggregated populations of heterogeneous composition may be very misleading. For example, since mortality is strongly affected by age as well as general health conditions, simple comparisons of mortality rates may not reveal the influence of general health conditions. Despite better health conditions in the United States, it has a slightly higher CDR than Egypt does because the United States has an older population (see Box 2.6). Consequently, mortality rates in populations are routinely adjusted to remove the distortion caused by different age distributions. The direct method of adjustment is to calculate the mortality rates for specific age groups in each population and apply these rates to a single standard age distribution (Gordis 1996). An example of the direct method of adjustment in Table 2.1 shows that the annual age-adjusted mortality rates of 15 per 1,000 and 30 per 1,000 accurately reflect the doubling in

BOX 2.6

Aggregated Population Statistics
May Be Misleading

	United States	Egypt
Crude death rate in 1995	8.80 per 1,000 population[a]	7.61 per 1,000 population[b]
Infant mortality rate in 1996	7.2 per 1,000 live births[c]	57 per 1,000 live births[d]

[a]Anderson et al. 1997.   [c]Ventura et al. 1997.
[b]United Nations Development Programme (UNDP) 1999a.   [d]UNDP 1999b.

A comparison of crude death rates and infant mortality rates for the United States and Egypt illustrates the problems in interpreting differences between aggregated population statistics. The crude death rate aggregates the entire population and is affected by the age distribution of the population as well as general health conditions. The risk of mortality is more properly compared within the same age group. The infant mortality rate, in particular, is widely regarded as a good measure of the overall health status of a population. The infant mortality rate is much lower in the United States. The crude death rate of the United States is slightly larger than that of Egypt because the effect of an older population in the United States outweighs the effect of its better health conditions. The aggregated crude death rate is, in effect, comparing the mortality rates of older Americans and younger Egyptians.

mortality rates across all age groups, whereas the unadjusted rates decline from 21 per 1,000 to 18 per 1,000. It is also good practice to examine the rates in each age group separately because the effects may not be the same across all age groups. Unfortunately, information to adjust for the distribution of age and other factors is not always available.

Analytical Approaches

To address the basic epidemiological questions of investigation (see Box 2.1), we must analyze relationships between health status and other factors in a population. The determinants of changes in health are often called *risk factors*. It is important to search for a variety of sources of information on possible risk factors (see Box 2.4). Most countries have basic data on weather and climate, including temperature, rainfall, wind directions, force, and type (e.g., sandstorm or snowstorm), hours of sunshine, and inversions. Many agricultural areas maintain annual reports about seasonal river flow and related irrigation patterns. Data may exist on the geographic distribution of disease vectors, which are animal species necessary to maintain the transmission of certain infectious diseases (see Chapter 10). Traditional sources of data may be augmented with new databases generated by advances in geographic information systems and satellite-based remote sensing technology (see Chapter 3). A systematic search for information

Table 2.1   Age-Adjusted Mortality Rates in Hypothetical Populations

	Lower Rates		Higher Rates	
	Older Population[a]	Standard Population[b]	Younger Population[a]	Standard Population[b]
Young age group (A)	20,000	50,000	80,000	50,000
Deaths in young (B)	100	250	800	500
Mortality rate in young (C = B/A)	0.005/yr	0.005/yr	0.010/yr	0.010/yr
Old age group (D)	80,000	50,000	20,000	50,000
Deaths in old (E)	2,000	1,250	1,000	2,500
Mortality rate in old (F = E/D)	0.025/yr	0.025/yr	0.050/yr	0.050/yr
Total population (G = A + D)	100,000	100,000	100,000	100,000
Total deaths (H = B + E)	2,100	1,500	1,800	3,000
Mortality rate in total (I = H/G)	0.021/yr[c]	0.015/yr[d]	0.018/yr[c]	0.030/yr[d]

Note: Age adjustment is also referred to as age standardization. The same procedure can be applied to mortality and incidence rates.

[a]The older population has 80% in the old age group and 20% in the young age group. The younger population has 80% in the young age group and 20% in the old age group.

[b]The standard population can have any age distribution in practice. In this simple example, the sizes of the young age group and the old age group are equal.

[c]This a crude mortality rate because the original age distribution is used.

[d]This is an age-adjusted mortality rate because the standard age distribution is used.

on issues of global change requires some familiarity with a variety of disciplines (see Chapter 1).

The application of epidemiological methods to the study of global change and health must combine a wide range of information about exposure to concurrent natural and anthropogenic health hazards, varying in scale from global to individual (see Table 2.2). In general, it will not be possible to estimate the health effects of global change without an understanding of how those effects are influenced by exposure patterns operating at smaller scales. For example, atmospheric problems of greenhouse gases, outdoor air pollution, and indoor air pollution occur together (see Box 2.7).

Four major approaches are described below with examples of applications and methodological difficulties: ecological, cross-sectional, case-control, and cohort. Special issues for studies of infectious diseases are also briefly discussed. The presentation of the approaches is followed by a summary of potential sources of error in epidemiological studies.

Table 2.2   Geographic Scale of Environmental Health Hazards

Scale	Type of Health Hazard
Global	Global warming, climate change, ozone depletion, major natural disasters, geological shifts and subsidence, rise in sea level
Regional	Change of seasonal rainfall and temperature patterns, deforestation, desertification, extreme weather situations, engineered water impoundment, toxic wastes, natural disasters
Community	Polluted water sources, sewage, waste water, solid waste, irrigation systems, breeding sites of invertebrate vectors of disease, burrow pits, rodents, zoonoses of domestic and wild animals, abattoir, air pollution, radiation, natural disasters
Household	Chemical pollution and biological contamination of water and food, poor sanitation, breeding of arthropod vectors indoors and in solid waste, crowding in confined living space, indoor air pollution, farm and domestic animals with zoonoses, rodents, ectoparasites, exposure to radon
Individual	Occupational exposure, travel, habitual substance use, sexual practices

Ecological Studies

In an ecological study, the units of analysis are populations or groups of people, rather than individuals (Last 1995). Ecological studies provide a relatively easy way to compare populations in different geographic areas or in the same area over different time periods because they usually rely on data already available, such as incidence rates or mortality rates. Ecological studies also provide a way to compare populations when the information that can be collected is extremely limited. However, the lack of data on an individual's exposure to environmental health hazards is a source of concern when interpreting the relationship between environment and disease (see ecological fallacy under "Systematic Error," below). The phrase *ecological study* is also used more generally for studies about ecology, the relationship between living organisms and their environment (see Chapter 10). In this broader context, no particular study design is implied.

A related phrase is *population-based study,* for studies on general populations in defined areas, such as a city. This usage includes the epidemiological concept of an ecological study. Other types of study designs may also be population based, depending on how the study population is selected.

EXAMPLE: Heat Wave–Associated Mortality in the United States
In 1980, the United States experienced a summer heat wave when temperatures reached record levels in much of the nation (Kilbourne 1997). In

BOX 2.7

Multiple Levels of Air Pollution Effects in China

China is undergoing rapid industrialization and a subsequent escalating demand for energy, primarily from coal (World Resources Institute [WRI] 1998). Air pollution occurs at multiple geographic levels: indoor, local, regional, and global. Air pollution in China also demonstrates the multiple nature of effects on human health, vegetation, and global climate.

Domestic use of coal for cooking and heating can generate high levels of particulate matter that have been documented to cause high rates of lung cancer among non-smoking women. Ambient air pollution in China's large cities arising from energy, industry, and motor vehicles includes high levels of sulfur dioxide (SO_2) and particulates. The result has been increasing rates of mortality and hospital admissions from respiratory illness (see text).

Regionally, acid rain affects about 40 percent of China's agriculture, according to a survey by the National Environmental Protection Agency (World Bank 1997; Downing et al. 1997). In southeastern China, the annual mean pH of rain is below 4.0, or highly acidic. A study in the southwestern provinces of Sichuan and Guizhou found that approximately two-thirds of the agricultural land in the region is subject to acid precipitation; 16 percent of the crop area suffers significant damage and reduced crop yields as a result. China accounts for 65 percent (22 million tons of SO_2 per year) of total emissions in Asia. Although most of this is deposited within China, 35 percent and 39 percent of the sulfur deposited in North Korea and Vietnam, respectively, originates from Chinese emissions.

Coal burning produces more greenhouse gases, primarily carbon dioxide, than the use of any other fossil fuel. Coal accounts for about 75 percent of China's commercial energy needs, compared with 17 percent in Japan and 27 percent as the world average use of coal (WRI 1998). Demand for coal is projected to increase by 6.5 percent annually (Downing et al. 1997) and is expected to burn 1.5 billion metric tons annually in the year 2000 (WRI 1998). China's share of global carbon emissions has increased from 5.2 percent in 1970 to 11 percent in 1990. Projections are for China's primary energy demand to increase another sevenfold by the year 2050. An increase in emissions of greenhouse gases has raised concerns about global climate change, with possible adverse effects on human health, agriculture, and other sectors (see Chapters 7 and 13).

July of that year, there were some 5,000 deaths in excess of the number expected in the absence of a heat wave. Interestingly, only about 1,700 of those deaths were classified as heat related. An analysis based only on death certificates that list heat as a cause of death systematically underestimates the effect of a heat wave on mortality. A great deal of attention in epidemiological studies is directed at identifying systematic measurement errors (see measurement bias under "Systematic Error," below).

From a public health perspective, it is important to identify subpopulations that are most at risk from heat-related death. Heat waves are associated with an increase in deaths in people with preexisting heart disease, among elderly persons and, to a lesser extent, among very young persons, and in areas of low socioeconomic status. Several studies have also noted that more women than men suffer heat-related mortality. However, such observations have been made in populations where there are many more women than men among the elderly. When comparisons are made within the same age groups, males are actually at somewhat greater risk than females. The apparent effect of gender seen initially was incorrect because another factor, age, was affecting heat-related mortality and relatively more females were in the high-risk age groups. In this situation, age is called a *confounding variable*. A central methodological concern in the interpretation of epidemiological studies is the possible influence of a confounding variable that has not been measured (see under "Confounding"). The comparison of mortality by gender is essentially the same statistical problem as the comparison of mortality by country when the comparison groups have very different age distributions (see Box 2.6).

EXAMPLE: **Mortality in Refugees from Iraq and Rwanda**

A complex emergency describes situations that cause significant excess mortality in large civilian populations because of a combination of war, food shortages, and population displacement (Toole 1997). Mortality rates are specific indicators of the health status of populations affected by emergencies. Since routine reporting mechanisms are not in effect, the number of deaths is estimated from hospital and burial records, community-based surveys, and 24-hour burial-site surveillance. The effect of refugee status on mortality is commonly made by comparing the CMR in the refugee population with the CMR in their country of origin, representing the rate as deaths per 1,000 population per month. For example, the CMR in Iraq of 0.7 per 1,000 per month is compared with the CMR of 12.6 per 1,000 per month for 400,000 Kurdish refugees near the border with Turkey from March through May 1991. Under even harsher conditions (see Chapter 14), Rwandan refugees who fled into the Democratic Republic of the Congo (formerly Zaire) in July 1994 experienced a CMR of 102.0 per 1,000 per month, over 50 times higher than the baseline of 1.8 per 1,000 per month.

There are, of course, serious methodological concerns with data collected under emergency situations. CMRs must be handled cautiously to avoid misleading comparisons (see Box 2.6). It is not possible to determine how well the statistics from the country of origin actually reflect the experience of the refugee population before they fled. It is possible that the displaced population is not representative of the country of origin in terms of age or other factors affecting mortality in the country of origin (see migration bias under "Systematic Error," below). However, the reported CMRs in refugee populations tend, if anything, to be underestimates because deaths are usually undercounted and population size is often exaggerated (see measurement bias under "Systematic Error").

Table 2.3   Households Burning Smoky Coal and Lung Cancer
Mortality in Xuan Wei, Yunnan Province, China

Commune	Households Burning Smoky Coal before 1958 (%)	Lung Cancer Mortality per 100,000[a]
Cheng Guan	100.00	151.8
Lai Bin	89.7	109.3
Rong Cheng	81.9	93.0
Long Tan	78.0	32.1
Long Chang	76.1	33.2
Hai Dai	49.7	10.8
Pu Li	35.2	2.8
Ban Qiao	34.0	16.7
Luo Shui	2.7	1.8
Re Shui	0.0	2.3
Xi Ze	0.0	0.7

Source: Reproduced with permission from Mumford et al. 1987, Table 2. © American Association for the Advancement of Science.

[a]Unadjusted lung cancer mortality in 1973–75.

EXAMPLE: Air Pollution and Respiratory Illness
and Mortality in China

In China, households using coal for domestic cooking and heating are especially at risk for indoor air pollution, since coal can generate a high level of particulate matter smaller than 2.5 microns. Concentrations in these households can reach levels more than 100 times the proposed U.S. standard for ambient air quality (World Resources Institute [WRI] 1998). Perhaps the most compelling example of the health effect of indoor air pollution is the association between lung cancer mortality and the percentage of households burning smoky coal in Xuan Wei, in Yunnan Province (Table 2.3). Fuel for cooking and domestic heating was burned in shallow, unventilated pits, resulting in high levels of indoor air pollution. Indoor air pollution is a likely etiologic agent, since the association between smoky coal and lung cancer mortality was especially strong in women, who rarely used tobacco in this setting (Mumford et al. 1987).

Air pollution in several of China's megacities, such as Shanghai and Shenyang, ranks as among the worst in the world for sulfur dioxide (SO_2) and particulates (WRI 1998). In 1995, more than half of the 88 cities monitored for SO_2 were above World Health Organization (WHO) guidelines, and 85 cities far exceeded WHO guidelines for total suspended particulates (TSP). The cities of Taiyuan and Lanzhou had SO_2 levels nearly 10 times the WHO guideline. According to an epidemiological study in Beijing, the risk of mortality was estimated to increase by 11 percent with each doubling of SO_2 concentration and by 4 percent for each doubling of TSP. Mortality from chronic obstructive pulmonary disease increased 38 percent with a doubling of TSP and 29 percent with SO_2 doubling. In Shenyang, estimated total mortality increased by 2 percent with each 100 micrograms per cubic meter increase in SO_2 and by 1 percent for each 100 micrograms per cubic meter increase in TSP. Overall, it is estimated that 178,000 premature deaths could be avoided each year if China met its air pollution standards (World Bank 1997).

In addition to mortality, urban air pollution in China has resulted in 6.8 million emergency room visits and 346,000 extra hospital admissions for respiratory illness annually. Also, 4.5 million person-years are lost due to illness associated with air pollution (World Bank 1997). Although the energy and industrial sectors are now the largest contributors to urban air pollution in China, their role may soon be surpassed by the transportation sector. The number of motor vehicles has tripled since 1984, reaching 9.4 million in 1994. By 2020, the urban vehicle fleet is expected to be 13–22 times greater than today. A recent study in Beijing showed that all air-monitoring stations within the downtown area exceeded the national standards for carbon monoxide. Also, the levels of nitrogen dioxide have nearly doubled over the past decade. During the summer, concentrations of ground-level ozone repeatedly exceed national standards. Compounding this urban pollution problem is the fact that most Chinese cars are fueled by leaded gasoline (WRI 1998).

Ecological studies can demonstrate the effects of air pollution indoors and outdoors (see also Box 2.7). However, it is difficult to sort out the contribution of each to the etiology of disease when both indoor and outdoor air pollution occur at very high levels.

EXAMPLE: Ultraviolet-B Radiation and Incidence of Skin Cancer in the United States

The principal concern raised by stratospheric ozone depletion is that more ultraviolet-B (UV-B) radiation from the sun will reach the ground (see Chapter 7). One of the potential biological consequences of an increase in UV-B radiation is an increase in the incidence of nonmelanoma skin cancer. The relationship between UV-B radiation and the incidence of skin cancer can be analyzed by comparing different latitudes. In moving from the equator to higher latitudes, the midday sun is lower in the sky and UV-B radiation passes through a longer path of stratospheric ozone. Therefore, higher latitudes experience lower levels of UV-B radiation on the ground. A direct relationship between UV-B radiation and the incidence of nonmelanoma skin cancer leads to a prediction of a steady decline in incidence at higher latitudes (see Fig. 7.5). The incidence of nonmelanoma skin cancer observed among the white population in several U.S. locations in 1971–72 and 1977–78 also reveals a steady decline at higher latitudes (see Fig. 7.5). Within the temperate latitudes of the United States, a shift southward of 1.8 degrees (approximately 400 kilometers) appears to cause a 10 percent increase in incidence.

Since nonmelanoma skin cancer takes years to develop, a simple analysis of incidence by latitude is reasonable if the underlying UV-B conditions have not changed much in a given location. The use of age-adjusted incidence rates is also reasonable under these conditions (see Table 2.1). In recent decades, there has been a general increase in the incidence of skin cancer due to behavioral changes in exposure to the sun. Therefore, incidence at a later period is generally higher, but the comparison of relative changes across latitude is similar within each period of observation. A strength of the study is a comparison within a racial group, the white population, since the incidence of skin cancer varies by race. The presentation of results by gender also demonstrates that males are at higher risk than are females. A weakness of the study is a lack of behavioral data that might indicate that residents at lower latitudes spend more time outdoors (see under "Confounding"). In fact, the latitudinal effect appears to be a little steeper than predicted on the basis of UV-B alone, although this effect is not formally analyzed. This study design is also vulnerable to distortion caused by migration of people across latitudes. Where a person is diagnosed with skin cancer may be quite different from the location of exposures to UV-B early in life (see migration bias under "Systematic Error," below).

EXAMPLE: Rodents and Hantavirus Pulmonary Syndrome in the United States

Hantavirus pulmonary syndrome (HPS) recently has been recognized as an emerging disease in the United States. The causative agent, the sin nombre virus, is closely related to other viruses transmitted from infected rodents to humans (see Chapter 10). Ecological evidence of the relationship between rodents and cases of HPS has been provided in the southwestern United States, where large numbers of cases brought the problem to national attention in 1992–93. During this period, populations of deer mice increased as much as 10-fold in some locations in the area. In late 1993, there was a decline in the number of cases of HPS about the same time as a decline in the populations of deer mice. The increase in the populations of deer mice has been related to the effect of unusually heavy rainfall in the area (Engelthaler et al. 1999), which suggests an association with the climatic pattern known as El Niño (see Chapter 8).

As with many ecological studies, the connection between the supposed risk factor and human cases is indirect. The populations of deer mice had been observed as part of a long-term study at a field station in New Mexico, south of where the cases had been observed. However, the linkage has provided the basis for other studies, such as a case-control study (see under "Case-control Studies," below). The linkage is likely to be complex. The association between cases of HPS and heavy rainfall was strong for the El Niño of 1997–98 (Hjelle and Glass 2000) but not for the El Niño of 1991–92 (Glass et al. 2000). This example demonstrates the importance of long-term studies for identifying trends.

Cross-sectional Studies

Cross-sectional studies are useful to determine the prevalence and distribution of diseases and health conditions in a population. Epidemiological data can be obtained by cross-sectional studies of representative population samples, defined risk groups, or sentinel sites and are especially helpful in areas without good surveillance programs or vital records. In a cross-sectional study, associations between health conditions and possible risk factors may be difficult to interpret in terms of cause and effect because all of the data are collected at the same time. Moreover, the prevalence of disease can be increased not only by new cases, but also by longer duration of disease (e.g., because of a new treatment that prolongs survival without curing), in-migration of cases, or out-migration of healthy people (see migration bias under "Systematic Error," below).

However, cross-sectional studies can be more powerful with data on other related time elements. For example, it is possible to synchronize age-specific population data with the approximate calendar time of events that have affected the health status of a community. Measurements of past experiences and illnesses can also be estimated by interviews and medical histories, by relating historical events to the occurrence of illness in the life history of a person, and by establishing age-specific antibody profiles that demonstrate past exposures to specific infections. Cross-sectional studies may be repeated in the same population and, if identifiers are available to link records from the same person appearing in multiple surveys, may form the basis of cohort studies (see "Cohort Studies," below).

In some situations, it is appropriate to describe a health condition quantitatively instead of merely in terms of presence or absence. For example, the concentration of a chemical in the blood might be expressed in parts per unit of weight or volume, especially when there is no established level that permits a classification into normal or healthy versus abnormal or unhealthy.

EXAMPLE: Vitamin C Deficiency and Scurvy in Refugees
from Ethiopia

Although scurvy is rare in stable populations in developing countries, it
has been reported in displaced populations. Cross-sectional surveys of
Ethiopian refugees in Somalia and Sudan were performed in 1986–87
(CDC 1992). The prevalence of scurvy was as high as 45 percent among females and 36 percent among males; prevalence increased with age. The
prevalence of scurvy was associated with the period of residence in camps
and the time exposed to rations lacking in vitamin C.

This study overcomes the inherent limitation of cross-sectional studies by collecting information about past events related to residence in
camps and consumption of rations lacking in vitamin C. Extensive medical knowledge about vitamin C deficiency and scurvy also provides confidence in critical data obtained retrospectively.

EXAMPLE: Fish Consumption and Mercury in Canada

Industrial activities in North America generate airborne mercury that can
circulate far from its source (Commission for Environmental Cooperation [CEC] 1997). Eastern Canada is particularly affected. Residents of the
Upper Saint Lawrence River basin in the Canadian province of Quebec
were tested for blood levels of mercury (Mahaffey and Mergler 1998). This
area contains two major lakes that are popular for fishing. Increased consumption of lake fish was associated with higher concentrations of blood
mercury. The study concluded that screening for blood mercury should
be conducted more often among people who eat more than 100 grams of
fish from contaminated lakes per day. Typical of studies of toxic chemicals, the outcome variable was a quantitative variable (concentration of
mercury in the blood) and not simply the presence or absence of a condition.

EXAMPLE: Water Contact and Parasitic Disease in Chad

Similar environmental hazards may have quite different effects on the
health conditions of individual communities and households, even among
those located in the same area or village (Buck et al. 1970). Figure 2.2 is a
map of a village in Chad where the pattern of the locally endemic diseases
differed significantly between different groups of houses. The cluster of
families living next to the perennially flowing river (group 1) had very high
prevalence rates of onchocerciasis, also known as river blindness, because
they were close to the breeding sites of the local blackfly vector (*Simulium
damnosum*) of the disease. By contrast, all other households (groups 2–4)
fetched their water from Mbongo Creek, which ran parallel to the road.
Among these families, the prevalence rate of intestinal schistosomiasis

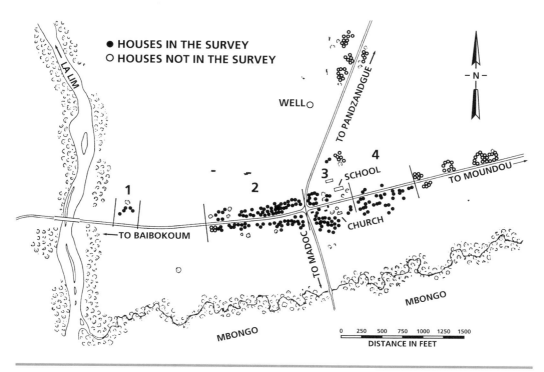

Figure 2.2 Clusters of houses with different sources of water in the village of Ouli Bangala in Chad. *Source:* Reproduced with permission from Buck et al. 1970, Figure 4.11.

(*Schistosoma mansoni*) was three times higher than that found in the cluster near the river. This difference was due to infections acquired by daily, intimate contact with the snail vectors of the parasite while people fetched water from the creek. Another form of schistosomiasis (*Schistosoma hematobium*) was co-endemic in the village but showed no difference between the two groups of families because people were infected with equal frequency outside of the village while fishing in the large Logone River. In addition to these three types of infection, malaria was hyperendemic in the community. This example shows two frequently ignored situations of rural communities in developing countries (i.e., a considerable degree of microepidemiological diversity and concomitance of target diseases that are sensitive to environmental change).

EXAMPLE: Water Usage and Trachoma in Brazil

Trachoma, the leading infectious cause of blindness in the world, had apparently been eradicated in the Brazilian state of São Paulo by the late 1960s. However, in 1982, new cases of inflammatory trachoma, an early stage of the disease, were found in preschool children in Bebedouro, a municipality in the state. In 1985, a cross-sectional study was conducted to determine the prevalence of disease and the risk factors associated with disease (Luna et al. 1992). Households with children 1–10 years old were categorized according to the presence or absence of inflammatory trachoma among those children. Of the total 1,166 households, 84 had cases of inflammatory trachoma. Several household characteristics were associated with the presence of cases. An outdoor source of water, either a pipe outside the doorway or no pipe at all, was an important risk factor. Low water consumption was also a risk factor, even after water source, socio-economic status, and sewage facilities were considered. A large number of children in a household placed it at greater risk. Other risk factors were the lack of daily garbage collection and a head of household without any formal schooling. This kind of study is useful in identifying households for a disease control program.

EXAMPLE: Seroprevalence and Malaria Control Evaluation in Afghanistan

In 1970, after a 10-year uninterrupted program of malaria control in Afghanistan, the country was considered free of the disease and no new cases were reported. During the same year, a cross-sectional survey of multiple diseases was carried out in ecologically contrasting communities of the country (Buck et al. 1972). Among the routinely applied diagnostic methods was a test for the presence of residual malaria antibodies. It was assumed that the results of the study would reflect the success of the control program. It was anticipated that children born during the 10 years of malaria control would have escaped infection and, therefore, would have no antibodies, while significant numbers of adults would show persistent antibody levels reflecting the precampaign level of endemicity. The findings are summarized in Figure 2.3.

In the village of Sayedabad, the results confirmed the concordance between the prevalence of antibodies in persons over 10 years of age and their absence during the corresponding period of malaria control. In contrast, they indicated that a considerable proportion of children in the village of Bula Quchi had acquired infections that produced antibodies to the malaria antigen during the same time. Because of the presumed absence of malaria in the previously large endemic region near the Amu Darya River, the new clinical cases had remained undiagnosed and unreported. Eighteen months after the survey and after a period of unusually heavy rainfalls, an extensive malaria (*Plasmodium vivax*) epidemic occurred in a large band of fertile land along the big river, including the village of Bula Quchi. The example demonstrates the usefulness of cross-sectional surveys to provide valuable information for disease surveillance and follow-up studies.

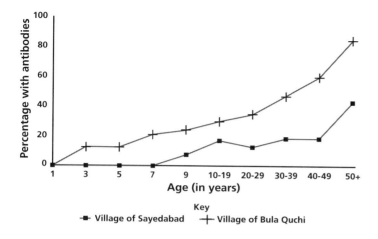

Figure 2.3   Prevalence of antibodies to malaria by age group in the villages of Sayed-abad and Bula Quchi in Afghanistan. Hemagglutinin tests with *Plasmodium knowlesi* reagents were used to detect antibodies to *Plasmodium vivax. Source:* Data from Buck et al. 1972, Chapter 4, and Buck's unpublished records.

EXAMPLE: Incidence Estimates of Late Stages
of Trachoma in Tanzania

Trachoma, caused by repeated infections with *Chlamydia trachomatis,* progresses slowly from inflammatory infections of the eye in early childhood to the late stages of conjunctival scarring, trichiasis (inturned eyelashes), and corneal opacities seen in adults. Estimates of the magnitude of incidence rates of late stages of trachoma are useful for planning needed health services and for providing clues to the progression of disease. Unfortunately, it is difficult to estimate incidence rates directly by following the same people over years and decades. However, cross-sectional data on the prevalence of signs of trachoma, which are more readily obtained, can be used to provide estimates of incidence.

Cross-sectional data on the signs of trachoma were collected in 1988 from 4,898 women 18 years and older in several villages in central Tanzania. The increase in prevalence of signs of trachoma with age was used to infer the probability of developing particular signs of trachoma with age in this population (Muñoz et al. 1997). For example, the probability of developing trachoma-induced corneal opacities over a period of 10 years reached its peak of 2.80 percent in women aged 45–54 years, the oldest group for which estimates were calculated. Such a probability may be interpreted as a 10-year incidence rate (see under "Rates and Risks," below).

The use of age-specific cross-sectional data to construct incidence estimates requires assumptions that should be carefully considered. The most important assumption is that the epidemiological patterns are stable enough that the health status of younger people reflects the experience of the older generations when they were younger, at least for a particular disease. Other assumptions are that a disease state does not disappear naturally, that migration is not related to the disease, and that the particular sign of disease does not cause people to die. In this study, excess mortality associated with blindness related to corneal opacities could mask the true incidence; this effect of mortality was judged to be minor after an analysis of estimates of adult mortality rates and studies linking blindness with increased mortality. Sometimes, as in the study in Tanzania, even the age of the individual must be estimated. The variety of assumptions required to infer incidence estimates from cross-sectional data means that, despite their utility in an area with very limited information, they should not be confused with data obtained from direct observation of new signs and symptoms (see under "Cohort Studies," below).

Case-control Studies

In a case-control study, people with a specific disease (cases) are compared to people without the disease (controls) for differences in their history of exposure to specific environmental factors. Case-control studies are especially suitable for diseases that are relatively rare or that have a long period of latency from exposure to diagnosis of disease. The selection of cases and their control partners can be based on different sources (e.g., hospital records, occupational health statistics, results of cross-sectional studies). In some settings, households with cases and households without cases may be compared in the roles of cases and controls, respectively. A demonstration that history of exposure to a factor is more common among the cases provides some evidence that the factor causes the disease.

However, cases and controls may also differ in characteristics other than the specific target of the study. For example, the cases might be older or poorer on average than the control group. If these other characteristics are related directly or indirectly to the development of the disease in question, they represent confounding variables that distort the effect of the factor of interest (see under "Confounding," below). Matching each case with a control person for the exact age, sex, and other important characteristics likely to cause confounding eliminates the effect of those characteristics in the comparison. Matching can also be applied at the group level so that the proportion of the control subjects with a particular characteristic is the same as the proportion of the cases with that characteristic. The case-control study design is a powerful tool for sorting out multiple determinants of disease, although the selection of the cases and controls is the source of much debate (see selection bias under "Systematic Error").

EXAMPLE: Rodents, Human Behavior, and Hantavirus
Pulmonary Syndrome in the United States

In May 1993, an outbreak of HPS in the southwestern United States was associated with a regional increase in the number of small rodents (see under "Ecological Studies," above, and Chapter 10). A case-control study was conducted to determine the risk factors of individual cases (Zeitz et al. 1995). Seventeen people who had developed HPS were compared with uninfected controls from each patient's own household (98 household controls), uninfected members of neighboring households (70 near controls), and uninfected members of randomly selected households in the community (80 far controls). Children under the age of 10 years were excluded as possible controls. More small rodents were trapped at the case's household than at other households. Comparing within the same household, the case was more likely to have hand plowed or to have cleaned feed storage areas. The comparison with near controls showed that cases were more likely to have planted. The comparison with far controls showed that cases were more likely to have cleaned animal sheds. Compared with any control group, the cases were more likely to have trapped rodents around their household and to have handled dead mice. Therefore, the risk factors associated with HPS are an increased number of small rodents around the house, the conduct of cleaning activities outside of the house, and participation in agricultural activities. This study demonstrates the interaction between ecological change affecting rodent populations and individual activities that place a person at risk of infection.

EXAMPLE: Sun Exposure and Skin Cancer in Australia

Concern about increasing levels of solar UV-B radiation has sparked interest in studies of the effects of sun exposure and skin cancer (see Chapter 7). Basal cell carcinoma (BCC), which is the more common of the two types of nonmelanoma skin cancer, has been associated with sun exposure. But the specific details of exposure in relation to cancer risk are important for risk assessment (see Chapter 4). A case-control study of BCC in western Australia permitted a more detailed investigation of patterns of exposure for individuals (Kricker et al. 1995). Lifetime residence and occupational histories provided a structure for questions about sun exposure in each year of life. The anatomic site of sun exposure and the development of skin cancer were considered. The investigators also collected data on migrant status, age at arrival in Australia, ethnic origin, and pigmentation because these factors have been associated with skin cancer. Risk increased with total lifetime exposure in those who tanned well but not in those who tanned poorly. The target skin cells may be receiving more solar radiation in people who do not tan well than in those who do tan well. This pattern is consistent with other indications that sun exposure beyond a certain level does not further increase the risk of BCC.

However, the effect of sun exposure in this study is not conclusive. The associations were probably affected by measurement error in detailed histories of sun exposure (see under "Random Error," below) and possibly even differential abilities of cases and controls to recall exposure (see recall bias under "Systematic Error"). A history of skin cancer was not considered because of poor records of skin cancer diagnosis and treatment in this population. It is possible that a previous history of sun-related skin disease caused individuals to reduce their sun exposure, leaving individuals with high exposure at apparently low risk.

EXAMPLE: Polychlorinated Biphenyls and Low Birth Weight in Sweden

Polychlorinated biphenyls (PCBs) emitted from industrial processes have dispersed far from their sources and have become particularly concentrated in the tissues of marine animals (CEC 1997). Rylander et al. (1998) examined the effects of exposure to PCBs in Sweden, where a main route of exposure is through the consumption of fatty fish from the Baltic Sea. The design of this study was a built-in or "nested" case-control study using the accumulated data of an ongoing cohort study of low birth weight among infants born to wives of fishermen (see under "Cohort Studies," below). A first analysis of this nested case-control study demonstrated an increased risk of low birth weight among infants born to mothers with high levels of fish consumption.

The next phase was to associate the increased risk with exposure to PCBs. In 1995, samples of blood were collected from 192 women who had given birth during the period 1972–91 and were wives or ex-wives of fishermen from the Swedish east coast, which lies on the Baltic Sea. The 57 cases of low-birth-weight infants were matched with 135 normal-birth-weight infants on gender, calendar year of birth, and parity of the mother. Maternal age and smoking habits were also considered because they affect birth weight. A slightly more than doubled risk of low birth weight was associated with elevated blood levels of a chemical recognized as a biomarker for exposure to PCBs. Two estimations of increased risk considered chemical concentrations above 300 or 400 nanograms per gram of lipid weight, suggesting that chemical levels above a threshold level were required to cause adverse health effects. The effects of limited exposures to a health hazard are an important component of risk assessment (see Chapter 4). This study demonstrates the power of a case-control study to estimate the effects of environmental exposures on adverse health outcomes.

EXAMPLE: Schistosomiasis and Bladder Cancer in Egypt

Cases of bladder cancer in Alexandria, Egypt, from 1994 to 1996 were compared to controls selected from people admitted to the hospital for acute conditions unrelated to cancer or the urinary tract (Bedwani et al. 1998). Individuals were questioned about a variety of factors, including education, occupation, area of residence, smoking, and consumption of coffee. A history of urinary, intestinal, and lymph node schistosomiasis was collected, including age at first diagnosis. A higher proportion of the cases of bladder cancer had a history of urinary schistosomiasis. The association was stronger with a younger age at first diagnosis of schistosomiasis. Smoking and certain industrial occupations were also associated with bladder cancer. Nevertheless, a clinical history of urinary schistosomiasis accounted for some 16 percent of cases of bladder cancer in this Egyptian population. This study demonstrates the relationship between a chronic noninfectious disease (bladder cancer) and a chronic infectious disease (schistosomiasis) whose transmission is strongly influenced by environmental factors (see Chapter 10). This study also demonstrates the importance of analyzing concomitant exposures from infectious diseases, industrial occupations, and personal habits.

Cohort Studies

A cohort study begins with people free of disease (a cohort) who are classified according to their exposure to a possible risk factor for disease. Cases of the disease are then recorded when they are newly diagnosed as the cohort is followed over a period of time. The incidence of disease among people with the exposure is compared to the incidence of disease among people without the exposure. Data on other factors that could affect disease are also acquired over time (see under "Confounding"). This study design may also be applied to monitor the development of symptoms in people who already have a health condition.

A strong advantage of a cohort study is that information about the causation and development of disease is acquired over a period of time instead of by recall, as with the case-control study (see recall bias under "Systematic Error"). A cohort study is also less likely to be subject to selection bias (see selection bias under "Systematic Error"). However, a cohort study is more expensive to conduct. The advantages of cohort studies and case-control studies may be combined in a nested case-control study. Nested studies can be inserted at any time into prospective cohort studies whenever sufficient numbers of new cases have been observed. The control group is selected from the rest of the cohort. The method has also been applied to cases whose diagnosis has required past examinations of biological specimens (blood, stool, urine). When adequately preserved aliquots of such specimens have been kept, previously unavailable new tests can be applied to these *posterity samples* for a retrospective analysis of emerging health conditions, especially those with long preclinical or subclinical conditions. This approach has provided useful data on emerging diseases. Its potential value for long-term studies has been impressively demonstrated by the retrospective identification of different types of human immunodeficiency virus infections in deep-frozen serum samples from different parts of Africa (Quinn et al. 1986; Nzilambi et al. 1988).

Another method to take advantage of historical data is the construction of retrospective cohort studies on exposures and health conditions. Such information can be obtained from health interviews, medical records, and morbidity and mortality statistics. The data can then be synchronized with past environmental changes (e.g., natural catastrophes, weather extremes, agricultural development, migration and resettlement).

The basic design for a cohort study often needs to be expanded to address a variety of environmental and health conditions. Some issues are that the incidence of disease may be strongly affected by multiple factors and that health impairment or environmental risk factors may be measured quantitatively rather than by their presence or absence (see Box 2.8).

EXAMPLE: Health Consequences of the Chernobyl Accident

In 1986, a large radiation accident occurred at the power plant of Chernobyl in the Ukraine. The expected health effects after the explosion of the nuclear reactor covered a broad spectrum of acute and chronic conditions in the populations of large geographic areas in Belarus, the Russian Federation, and Ukraine. Because of the differences in exposure to radiation and in the administration and proficiency of the local health services, a Collaborative International Program on the Health Effects of the Chernobyl Accident (IPHECA) was created under the auspices of WHO (WHO 1995). The two major objectives of IPHECA were (1) the comprehensive assessment of the health effects of the accident and (2) the planning for optimal emergency responses for future accidents of a similar nature. Different types of epidemiological studies were designed to assess the full extent of the health effects in the three countries.

Radiation Sickness. Exposure to radiation had both immediate and chronic health consequences. Among the 444 people who were exposed to large amounts of radiation at the reactor site, 131 (29.5%) were diagnosed as having acute radiation sickness.

Thyroid Cancer. Thyroid cancer required immediate study because the effect on the thyroid gland of long-term exposure to radiation was known from previous experience. Radioactive iodine (iodine-131), with a half-life of eight days, was the most important risk factor during the first

Expanding the Cohort Study Design

The basic design of a cohort study involves classifying individuals according to exposure to a possible risk factor for disease and following those individuals as new cases of the disease are diagnosed. However, many important environmental health problems require more complex elements of study design.

It is especially difficult to study a risk factor like exposure to air pollution in the presence of another major risk factor. A cohort study in the United States followed 8,111 adults in six cities over a period of 14–16 years to determine the effect of urban air pollution on mortality from lung cancer (Dockery et al. 1993). Data obtained on individual smoking habits were critical because mortality was most strongly associated with smoking. The residual effect of urban air pollution was that the rate of dying was 26 percent higher in the most polluted city than in the least polluted city.

The risk of developing hantavirus pulmonary syndrome (HPS) in the southwestern United States has some association with heavy rainfall and the growth of rodent populations (see text). In advance of the El Niño event of 1997–98, which was expected to bring heavy rainfall to the region (see Chapter 8), Hjelle and Glass (2000) prepared to monitor cases of HPS along with local patterns of precipitation and questionnaires about activities that may have led to infection of the cases. This study found an association between HPS and locally heavy rainfall and determined that most exposures occurred in or around the home.

Measurements of health impairment may require quantitative measures. A study of a cohort of children born in and around the lead-smelting town of Port Pirie, Australia, associated measurements of lead concentrations in blood with a decline in measurements of intelligence (Tong et al. 1996). The cohort design could determine that the effect of exposure to lead in the first seven years of life persisted into later childhood.

---

few weeks after the accident. The absorption of radioiodine in the thyroid gland was enhanced because many residents of the affected areas lacked iodine in their daily diet. Preventive action to reduce the intake of the radioactive isotope took the form of distributing (nonradioactive) iodine tablets to the population at risk. However, many children did not receive the tablets, leaving them at a higher risk of developing thyroid cancer. An epidemiological assessment has related the risk of thyroid cancer to the dose of radiation absorbed by children (see Table 2.4). The studies are continuing.

Leukemia. Leukemia, like thyroid cancer, required immediate study because the effect of long-term exposure to radiation was known from previous experience. Cohort studies were designed to detect and treat cases of leukemia in a target population of 270,000 people in the affected areas of the three countries. Based on advanced diagnostic methods for finding cases, the preliminary results have not yet shown a significant increase in the incidence of leukemia that could be linked to the effects of radiation. Presently, case-control studies are under way to investigate associations be-

Table 2.4   Risk of Thyroid Cancer in Children Born between 1971 and 1986 in Parts of the Former Soviet Union

Area	Children (thousands)	Average Thyroid Dose (Gray)[a]	Thyroid Cancer Cases, 1991–95	Observed/ Expected Cases[b]	Excess Absolute Risk[c]
Ukraine					
Zhytomyr oblast[d]	340	0.13	28	4	0.9
Kiev oblast	399	0.18	47	6	1.1
Chernigov oblast	273	0.09	33	6	2.2
Kiev city	581	0.05	67	6	3.8
30-km zone (evacuees)	20	0.92	12	30	1.3
Belarus					
Gomel/Mogilev oblasts[d]	76	0.73	89	56	3.2
Minsk city	357	0.08	41	6	2.3
Gomel city	113	0.40	72	30	3.1
Russia					
Bryansk[e]	169	0.12	31	9	2.7

Source: Reproduced with permission from Jacob et al. 1998, Table 1. © Macmillan Magazines Ltd.

[a]Gray is a unit of absorbed dose, one of many units of measurement in studies of radiation (Wald 1995).

[b]Expected cases are calculated on the basis of the incidence in southern Ukraine of 4.2 cases per 10^6 person-years.

[c]Excess absolute risk is per 10^4 person-years per Gray.

[d]An oblast is an administrative region.

[e]Subarea of the region only.

tween individual exposure doses and leukemia. The major radionuclide (radioactive element) released by the Chernobyl accident was cesium-137. It can enter the body through the consumption of contaminated fresh vegetables, milk, and drinking water or by direct external exposure to contaminated soil and clothing.

Brain Damage in Utero. The risk of brain damage to fetuses of mothers exposed to radiation during the critical developmental stage of pregnancy was studied in 4,210 infants born within a year after the accident. Of these, 2,189 of the newborns were delivered by mothers exposed to radiation. They were compared with 2,021 infants whose mothers resided in nonradiation areas. The two groups were matched for age, socioeconomic status, type of residence, and educational level. The study protocol included assessment of mental retardation, measures of neurological and endocrinal aberrations, and brain mapping by computerized analysis of electroencephalograms. Preliminary results of this cohort study suggest an increased incidence of mental retardation and behavioral disorders in the exposed group. In addition to verification of the preliminary results, the

objectives of the ongoing studies include determining the degree of interference by confounding caused by psychological factors related to the accident (see under "Confounding," below) and performing a detailed dosi-
metric follow-up of exposure and disease (see Fig. 4.4).

Special Issues for Studies of the Transmission of Infectious Disease

The fundamental distinction of infectious diseases is that the spread of an
infectious agent in a population affects the risk of infection to others. For
contagious processes, the susceptibility or immunity of individuals in a
population is a major determinant of risk. Environmental factors then influence the contagious process and interact with the dynamics of transmission. If no infectious agent is present, environmental factors cannot
cause infectious disease; if many infectious agents are present, environmental factors may affect the ability of each to be transmitted. Increasingly,
researchers have recognized that it is important to construct studies that explicitly examine the chain of transmission of an infectious agent. On a small
scale, the investigation of a disease outbreak attempts to reconstruct the history of exposure and transmission. On a larger scale, the analysis of the dynamics of infection in a population may include the levels of infection and
immunity of the population as part of the risk factors. The special issues related to the transmission of infectious diseases are discussed in Chapter 10.

Risks and Rates

Some epidemiological terms for risks and rates can cause confusion. Incidence rates and mortality rates measure the frequency of events in a population over a specified period of time. Risk is the probability that a person will experience an event in a specified period of time. When the risk is
small, the incidence rate may be almost identical to the risk of the health
condition in the specified period of time. For example, if the incidence rate
for a disease is 1/100 per year, then the probability of someone's developing that disease in a year is about 1 percent. However, a rate may increase
without bound and a probability cannot exceed 100 percent. The term *risk*
itself is also used loosely as a synonym for *risk factor* and *hazard*. A *risk factor* is generally an attribute or exposure associated with increased risk of a
disease, but *hazard* is used more specifically in environmental impact assessment. The term *prevalence rate* demonstrates an alternative meaning of
rate as ratio or proportion; in this context, *ratio* is often substituted for *rate*
to reserve the latter for expressing changes over time.

The quantification of risk is conceptually straightforward in a cohort
study (see Fig. 2.4). The risk of developing disease over a period of time is
calculated for people who are exposed to a possible risk factor and people

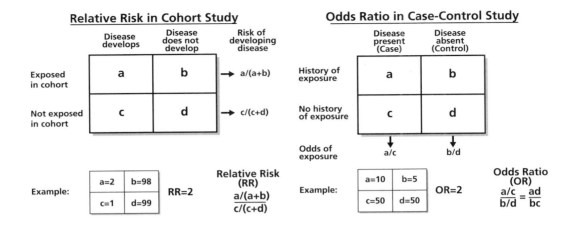

Figure 2.4   Calculations for relative risk and odds ratio.

who are not exposed to that factor. The most common way of expressing the comparison is in terms of the *relative risk,* which is the ratio of the two risks (of people exposed and of people not exposed). If the potential risk factor is associated with disease, then its relative risk is larger than 1. For example, if 2 percent of exposed people develop disease and 1 percent of unexposed people develop disease, the relative risk is 2.

In a case-control study, it is not possible to estimate the relative risk directly because cases rather than exposures are the starting point for the study. However, another method produces a result that is similar to the relative risk (see Fig. 2.4). If an exposure is associated with disease, then the cases are more likely than the controls to have a history of exposure. For example, 50 percent of the cases might have a history of exposure in comparison with only 10 percent of the controls. In epidemiological practice, it is standard to convert the proportion with exposure into the odds of exposure. The odds are defined as the ratio of the probability that something is so to the probability that something is not so (Last 1995). In a case-control study, the odds of exposure are calculated as the ratio of the proportion (or number of people) with a history of exposure to the proportion (or number of people) without a history of exposure. For example, if 10 out of 60 people have a history of exposure, then the odds of exposure are 10:50, or 1:5. The *odds ratio* is the ratio of the odds of exposure for cases and the odds of exposure for controls. For example, if the odds of cases having a history of exposure are 1:5 and the odds of controls having a history of exposure are 1:10, then the odds ratio is 2. An odds ratio larger than 1 indicates that the potential risk factor is associated with disease, in direct analogy with the relative risk. If a disease is relatively rare, such as cancer, then the odds ratio from a case-control study is actually a good estimate of the relative risk. In fact, odds ratios are sometimes (incorrectly) labeled as relative risks in publications.

In environmental impact assessment, the relative risk of disease from an exposure is combined with the probability of exposure in a population to construct the *population attributable risk* (see Chapter 4). This risk measures the proportion of the disease incidence due to the particular exposure and therefore indicates the cases of disease that could be eliminated if the exposure were eliminated.

Potential Sources of Error in Epidemiological Studies

Epidemiological studies require numerous measurements on people and the environment. Unfortunately, errors can be introduced into measurements, affecting the outcome of a study. A *random error* represents the role of chance and is considered nonsystematic error. Systematic error, or *bias,* affects measurements in a particular direction. A related problem is one of *confounding,* in which a factor in a study may have an association with a disease but not be a cause of that disease. In this situation, the true cause lies elsewhere in an unmeasured variable that happens to be associated with the studied factor in question (see "Confounding," below). Good study design can minimize the possible effects of errors. It is also good practice to rely on more than one study in demonstrating any association.

The general remarks below are intended as a guide for consultation with a statistician during study design and analysis. Textbooks on epidemiological methods also provide a wealth of information (e.g., Gordis 1996; Beaglehole et al. 1993).

RANDOM ERROR. *Random error* of a measurement may be due to individual biological variation or an inherent lack of accuracy in a measuring instrument. Measuring instruments include everything from devices that monitor particulates in the air to questionnaires about animal bites. Even if individual measurements are precise and accurate, a sample of individuals from a population may not reflect the composition of the population. Such *sampling error* may arise even if the individuals were truly selected at random. Any kind of random error can lead to an observed association (or lack of association) that is not true (i.e., a factor that seems to cause a disease does not or a factor that seems to be unrelated to a disease actually causes it). Careful attention to individual measurements and the selection of larger samples can reduce random error. Larger sample sizes reduce the chance of missing real associations.

SYSTEMATIC ERROR. *Measurement bias* is one type of systematic error that arises from inaccurate measurements. If the measurement problems have no particular direction, then measurement error is a type of random error, as just described. If the measurements tend to be consistently too high or too low, then the error is systematic. Measurement may also involve the classification of individuals, as in the diagnosis of a disease. In this context, measurement bias is called *classification bias.* The inability to recall information accurately in an interview can introduce measurement bias. If, in addition, the inability to recall information is different in different parts of the study population, the effect produces *recall bias.* For example, people who have been diagnosed with cancer may be more likely to report exposures than are people selected as controls for comparison.

Selection bias is caused by a systematic difference between the people selected for a study and those who are not. Selection bias is related to *surveillance bias,* in which some groups of people or events may be more likely to come to the attention of investigators. For example, children who attend school may be healthier or wealthier than children who do not. The selection of controls in case-control studies is often a point of debate even if the selection of cases is not. *Migration bias* is another way in which the selection of individuals can distort associations between factors and disease. The characteristics of people who migrate from an area may not be representative of the source population, an important issue in monitoring health conditions among migrants. Migration into and out of an area may be associated with the factors and disease under study.

The *ecological fallacy,* or *ecological bias,* is relevant to ecological study designs that rely on group measurements. Group associations may not be the same as associations based on individual data (if such data were available). For example, studies of air pollution usually rely on monitors to sample outdoor air at very few locations. The measurements of pollutants may be higher in one city than in another city, but an individual selected in the first city does not necessarily have a higher personal exposure than an individual selected in the second city. The ecological bias is also called the *aggregation bias.*

Potential biases should be assessed in reports of other studies and in the design of new studies. Great efforts should be made to reduce or eliminate biases.

CONFOUNDING. Even after random errors and biases are considered, a factor may have an association with a disease and yet not be a cause of that disease. The problem is the existence of a *confounding variable* that affects disease and is unequally distributed among the other factors in the study. In the example on mortality during heat waves, old age was a confounding variable in a study that produced an apparent risk due to being female. Elderly persons are well documented to be more affected by heat waves and, in many populations, more women than men are likely to be elderly persons. Confounding is a potential problem for any study design, although it may be worse for ecological studies because fewer data on confounding variables may be available.

For confounding to be controlled, possible confounding variables must be recognized. Comparisons can be made within groups, or *strata,* defined by levels of the confounding variables. For example, the discussion about heat wave mortality refers to studies of mortality of males and females within the same age group (i.e., stratified by age). In case-control studies, cases and controls may be matched on a possible confounding variable, such as age or smoking, to eliminate its effect from the analysis of other factors. Sophisticated statistical procedures to adjust for the effect of confounding variables may be required when data are too limited for direct approaches of *stratification* and *matching.*

The Sequential Application of Epidemiological Methods

The sequential application of different epidemiological methods is required for the solution of global health problems with strong environmental determinants. The following example of a parasitic disease in Egypt demonstrates the process of adapting and combining epidemiological methods, both classical and technically advanced. When new results from a previous study became available, new questions broadened the scope of the investigations. Their scale combined micro- and macroepidemiological situations for which answers were needed to develop an effective program for disease prevention and control. The methods included clinical observations, cross-sectional investigations of selected population samples, case-control studies, and ecological inquiries (ecological in the broader sense used in Chapter 10). These were followed by technically more sophisticated studies using satellite imaging and the development of a geographic information system (GIS) (for details, see Chapter 3).

The disease of concern was lymphatic filariasis, caused by the thread-like worm *Wuchereria bancroftia.* The disease is characterized by acute episodes of fever with infections and swelling of the peripheral lymphatic vessels and lymph nodes, most frequently involving the legs and, in males, the scrotum and testicles (hydrocele). The chronic stage of the disease causes chronic disability due to severe lymphatic obstructions and elephantiasis. The life cycle of filariasis depends on mosquito vectors that are adapted to a wide range of habitats, including canals and ditches heavily contaminated with human and animal wastes. Introduced by mosquito bites into the human host, the worm develops from its larval stages into adults that dwell inside the lymphatic vessels, where the female produces large numbers of mobile larvae, or *microfilariae,* which circulate in the blood of infected individuals and are ingested by biting mosquitoes. After further development in the vector, the life cycle is completed when the infective microfilariae enter the human host with a new mosquito bite. Filariasis is a cumulative infection; the parasite load, measured as the number of worms, increases with the time of exposure in an endemic area.

The gold standard of the diagnosis (i.e., the best available diagnostic criterion) is the microscopic recovery of the microfilariae in small samples of blood. However, many persons with advanced clinical symptoms have no microfilariae in their blood. Therefore, population-based studies of the disease have to include both parasitological and clinical examinations for case detection. The situation is further complicated by the nocturnal periodicity of circulating microfilariae in the peripheral blood (*microfilaremia*). For this reason, population-based studies must be carried out at night, between 10:00 P.M. and 2:00 A.M., when the circulation of microfilariae reaches its peak.

Filariasis has been endemic in the Nile delta of Egypt since time immemorial. Between 1950 and 1965, a large-scale program of filariasis control in the endemic areas included the creation of safe water systems, waste disposal, and improved irrigation practices. The temporary effects of widespread application of insecticides against agricultural pests for higher yields of crops for the rapidly increasing human population also helped to reduce the vector population of *Culex pipiens*. Based on surveillance data from 500,000 persons, the WHO Expert Committee on Filariasis concluded that filariasis was a disappearing disease of no public health significance (WHO 1984). However, at the same time, spot surveys, which are cross-sectional studies conducted rapidly on small populations, and clinical reports indicated that the steady downward trend of filariasis had been reversed. For this reason, the Ministry of Health created a new surveillance program in the southern Nile delta, including population samples from 314 villages with a total of 350,000 persons. The results revealed a dramatic reemergence of the disease with a highly focal distribution. A map of the results is presented in Figure 2.5.

The findings prompted the epidemiologists to ask the *W* questions: What happened, who was affected, when and where did it occur, and, most of all, why did it happen? (See also Box 2.1.) The previous program of filariasis control had been terminated in 1965. Since then, significant environmental and demographic changes had had a profound effect on the reemergence of filariasis. The Aswan High Dam of the Nile was closed, interrupting the annual flooding of the Nile. The dense network of irrigation channels increased in all of the Nile River valley and its delta region. The river water reaching the coast became heavily polluted by agricultural and industrial runoffs and by human and animal waste. In addition to these anthropogenic changes, geological shifts will continue gradually to alter the physical nature of the Nile delta (Stanley and Warne 1993). The sea level is expected to rise, salinization of agricultural land will increase,

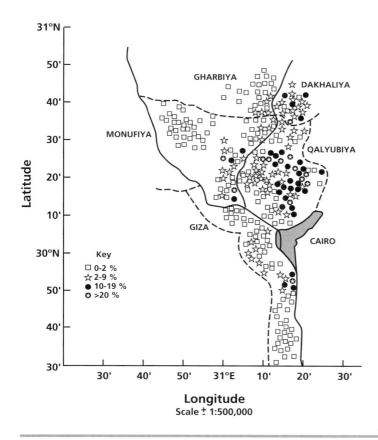

Figure 2.5 Geographic distribution of the prevalence of filariasis in communities in the southern Nile delta. *Source:* Reproduced with permission from Harb et al. 1993, Figure 3.

the ground water will be polluted, and the input of fresh water from the Nile River will be reduced. The estimated growth rate of the population in the affected area is 2.65 percent per annum (see also Chapters 6 and 9). The resulting extension of agricultural irrigation has led to poor drainage of ditches and to the creation of heavily polluted pools and wastewater puddles in which mosquitoes breed. At the same time, the extensive use of pesticides has introduced resistance to pesticides among mosquito vectors.

The observations of the uneven geographic distribution of filariasis led to additional studies comparing villages and households with and without the disease to identify risk factors of exposure. The rationale for conducting a case-control study originated from observations that the cases were unevenly distributed in villages, with clusters in certain family compounds. The study objectives aimed at the identification of those fac-

Table 2.5 Risk Factors of Filariasis

Risk Factor	Percentage with Risk Factor	Odds Ratio for Risk Factor
No toilet in house	52.4	10.6
No safe waste disposal	53.5	9.3
Watch TV outdoors	88.0	5.9
Camels kept inside	78.2	3.9
No piped water supply	11.6	3.7
Milk animals daily	15.0	3.1
Laundry outdoors	69.4	2.6
Buffaloes kept inside	38.2	2.4

Source: Data from Dorgham et al. 1994.

Note: The study population was 468 adolescent girls in Begerum, an Egyptian village in the Nile delta. The overall prevalence of filariasis was 8.8%.

tors (environmental, behavioral, genetic) that were responsible for the different infection rates (Dorgham et al. 1994).

Based on a census, there were 525 young females in the age group of 10–19 years, of whom 468 (89%) participated in the study. Of these, 41 (8.8%) had a history of filariasis. The age distributions of the case and control groups were similar. The frequencies of exposure to selected environmental and behavioral risk factors were then compared between the case and control groups. Exposures showing significant differences between the two groups were then examined in a logistic regression analysis, a statistical procedure that permits the calculation of odds ratios when multiple risk factors are considered simultaneously; the odds ratios are used to determine the ranking in importance of health risks (WHO 1978). The results are shown in Table 2.5. Significantly more households of cases than those of controls used the animal sheds in their compounds in lieu of latrines. The animal shelters of the family compounds provide favorable conditions for the year-round indoor breeding of the vector *Cx. pipiens.* Use of these quarters for defecation and urination increases the exposure to bites by vectors and, thus, to the transmission of disease. The observed difference between cases and controls regarding the frequency and duration of television watching was an unexpected finding. Television watching is mostly done in the evening when family members and their neighbors join in this recreational activity. They may attract large numbers of *Cx. pipiens,* the abundant mosquito vector in the area.

A comparison of households with and without infection also showed that the location of households near stagnant irrigation ditches and their discharge channels or adjacent to standing pools of water constituted a major environmental risk factor for filariasis. However, exceptions to this finding required further study. Detailed entomological examinations in households with multiple cases not situated close to outdoor breeding sites led to the surprising detection of indoor breeding of mosquito vectors throughout the year. The mosquito larvae are breeding in small accumulations of water created and fed by water gently dripping from clay containers on top of the kitchen sinks.

Other investigations were conducted to find causes for the large differences in the infection rates between villages of the same district. Extensive water seepage leading to heavy exposure to mosquito bites was detected as a major environmental hazard. In the lower-lying parts of villages built on clay soil, the bedrooms, kitchens, and toilets of many houses were flooded and their mud-brick walls were soaked. In a large village with 1,250 houses, of which 520 were located in a depression less than 20 meters below the level of the surrounding land, 146 houses were seriously damaged.

Follow-up investigations to assess the causes of water seepage were carried out by environmental engineers. Their study included drilling of an auger hole to explore the soil condition, groundwater table, and aquifer system. They found differences in the topography between the affected villages and the surrounding irrigated land. Geologically, there is a thick layer of clay with poor permeability, leading to accumulations of stagnant pools of water and seepage. The water flow of canals and channels for discharge and drainage is intercepted in the depression. In recent years, many houses have been connected to a piped water system without adequate provisions for sewage and wastewater disposal.

A by-product of the epidemiological studies was the awareness of an increase of other vector-borne diseases transmitted by *Cx. pipiens*, the mosquito vector for filariasis. There is also evidence that children living in the villages affected by water seepage have more frequent episodes of diarrhea and acute respiratory infection than do those living nearby.

These epidemiological findings spurred the planning of a new program of filariasis control. The various components of the program combine environmental improvements for source reduction of mosquito breeding, health education to improve sanitary habits and minimize exposure to mosquitoes, and mass treatment. An important task is the development of a surveillance system to monitor program results. Children from different villages who attend the same school make up ideal sentinel samples (see also Box 2.5 on sentinel health information systems) to monitor the incidence of new infections by annual examinations. Since school

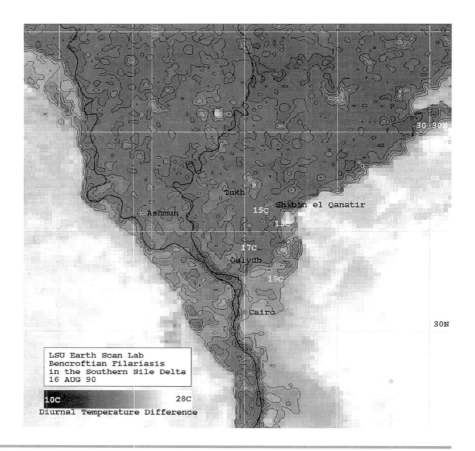

Figure 2.6 Diurnal temperature differences in the southern Nile delta on August 16, 1990. Observations are from the National Oceanic and Atmospheric Administration's Advanced Very High Resolution Radiometer. *Source:* Thompson et al. 1996, figure on p. 234. Courtesy of Donald F. Thompson, Tulane University Medical Center, New Orleans, Louisiana.

attendance is obligatory, they provide a cohort for six years of continuous observation. The children serve as index cases to track the location of new infections in their villages. The widening scope of health conditions affected by the changing environmental conditions demanded that the annual examinations should also include other diseases of local public health importance (e.g., schistosomiasis, intestinal nematodes and protozoa, and arthropod-borne virus infections). For ethical reasons and to increase compliance in the surveillance program, prompt treatment has been given whenever a specific condition is found.

Many communities in the Nile valley and delta at a potential risk for the reemergence of filariasis have not yet been covered by the program. In an effort to narrow the scope of the mass examinations of new communities for filariasis, two new methods were introduced to identify areas of high risk: remote sensing by satellites and the development of a GIS (see Chapter 3). Remotely sensed data based on thermal scanning radiometry were analyzed for diurnal temperature differences (dTs) that indicate surface and subsurface moisture in the soil, including standing water (see Fig. 2.6). The coordinates of each village together with their digitized data on the prevalence of filariasis were superimposed onto the dT map. There was a high degree of correlation between the remotely sensed dT values and the locally corresponding data on the prevalence of disease in the communities on the ground (Thompson et al. 1996). Hand-held Global Positioning System satellite navigators are now being used to increase the precision of the original village coordinates, which were extracted from available maps, most with a scale of 1:10,000 (see Chapter 3 for more details). It is expected that further studies will confirm the predictive value of the dT determinations to establish priorities for selecting communities with high environmental risks of filariasis and other vector-borne diseases to be included in a countrywide surveillance program.

Conclusion

This chapter has introduced basic epidemiological study designs for understanding the linkages between global change and health. This presentation is intended to be a starting point for study designs, since they will inevitably have to be adapted to address unanticipated problems. Advances in science and technology, especially in the area of geographic analysis, provide new tools to strengthen epidemiological analysis (see Chapter 3). Epidemiological analysis in turn must be linked to a larger process of assessing future risks to human health and designing interventions (see Chapters 4 and 5). The process of adapting epidemiological methods to meet new challenges has occurred throughout the history of the discipline. It is hoped that researchers from a variety of disciplines will be motivated to adapt these concepts in new ways.

SUGGESTED STUDY PROJECTS

Suggested study projects provide a set of options for individual or team projects that will enhance interactivity and communication among course participants (see Appendix A). The Resource Center (see Appendix B) and references in all of the chapters provide starting points for inquiries. The process of finding and evaluating sources of information should be based on the principles of information literacy applied to the Internet environment (see Appendix A).

PROJECT 1: Observational Study Designs

PART A. DESCRIPTIVE MEASURES OF HEALTH STATUS. This part of the project will strengthen the ability to find statistics on various health topics, familiarize the student with statistics centers, and reinforce the understanding of rates and risks.

Task 1. Find data on a selected health topic by investigating incidence, prevalence, and mortality rates. Use both morbidity and mortality statistics, if available. Describe the sources, which could include unpublished reports, and the usefulness and credibility of the particular statistics found.

Task 2. On the basis of the section on risks and rates, find examples of cohort and case-control studies to demonstrate the concepts of relative risk and odds ratio.

PART B. ANALYTICAL APPROACHES. The objective of this part of the project is to demonstrate an understanding of a variety of analytical approaches as they are applied in different study designs.

Task 1. Find sources citing examples (not used in this chapter) for

a.   ecological studies
b.   cross-sectional studies
c.   case-control studies
d.   cohort studies

Task 2. Finalize results in written form.

PROJECT 2: "What, Who, When, Where, Why" (see Box 2.1)
The objective of this project is to develop, practice, and refine research skills required for multidisciplinary studies. It comprises seven tasks.

Task 1. Identify and describe one to three examples of changes in the distribution of a health-related state or event related to disease, injury, disability, or death. (WHAT)

Task 2. Describe the population(s) of the example(s) in terms of available statistics on age, gender, occupation, and location. (WHO)

Task 3. Over what period of time have the changes in the population(s) been observed? Do observations lead to future projections? (WHEN)

Task 4. Describe the global-local changes in the geographic location(s) of the example(s). (WHERE)

Task 5. Examine reasons for the changes in the population(s). List and describe the possible environmental determinant(s) of the example(s). (WHY)

Task 6. For this task, more than one example must be available. Compare the findings of at least two different examples. Note similarities and dissimilarities.

Task 7. Summarize.

a.   What did you discover?
b.   What new question(s) were arrived at?
c.   What study design(s) would you use for further research?

PROJECT 3: Sequential Application of Research Methods
The objective of this project is to reinforce the importance of multiple studies in addressing environmental problems.

Task. Design and present an example for sequential application of research methods as described and documented in this chapter.

Acknowledgments

We thank Leon Gordis for reviewing this chapter. We also thank Jonathan Patz for providing material on China, Erika Feulner for suggesting topics for student projects, Mickey Glantz and Yves Tourre for discussing climate variability, and Heidi Curriero and Rebecca Freeman for checking sources of information.

References

Adams PF, Marano MA. 1995. Current estimates from the National Health Interview Survey, 1994. *Vital Health Stat* 10 (193): 81–82.

Anderson RN, Kochanek KD, Murphy SL. 1997. Report of final mortality statistics, 1995. *Monthly Vital Stat Rep* 45 (11), Suppl. 2.

Beaglehole R, Bonita R, Kjellstrom T. 1993. *Basic Epidemiology.* World Health Organization, Geneva.

Bedwani R, Renganathan E, El Kwhsky F, Braga C, Abu Seif HH, Abul Azm T, Zaki A, Franceschi S, Boffetta P, La Vecchia C. 1998. Schistosomiasis and the risk of bladder cancer in Alexandria, Egypt. *Br J Cancer* 77 (7): 1186–89.

Buck AA, Anderson RI, Kawata K, Abrahams IW, Ward RA, Sasaki TT. 1972. *Health and Disease in Rural Afghanistan.* York Press, Baltimore.

Buck AA, Anderson RI, Sasaki TT, Kawata KK. 1970. *Health and Disease in Chad.* Johns Hopkins Press, Baltimore.

Centers for Disease Control and Prevention. 1992. Famine-affected, refugee and displaced populations: Recommendations for public health issues. *MMWR* 41 (RR-13): 1–76.

Commission for Environmental Cooperation. 1997. *Continental Pollutant Pathways.* CEC, Montreal.

Dockery DW, Pope CA, Xu X, Spengler JD, Ware JH, Fay ME, Ferris BG, Speizer FE. 1993. An association between air pollution and mortality in six U.S. cities. *N Engl J Med* 329 (24): 1753–59.

Dorgham LS, Buck AA, Faris R. 1994. Social and cultural risk factors associated with evidence of filariasis among adolescent females of an Egyptian village. In *Proceedings of the Seventeenth International Ain Shams Medical Congress, Cairo, Egypt, 1994.* Ain Shams University, Cairo, pp. 87–94.

Downing RJ, Ramankutty R, Shah JJ, eds. 1997. *RAINS-ASIA: An Assessment Model for Acid Deposition in Asia.* World Bank, Washington.

Engelthaler DM, Mosley DG, Cheek JE, Levy CE, Komatsu KK, Ettestad P, Davis T, Tanda

DT, Miller L, Frampton JW, Porter R, Bryan RT. 1999. Climatic and environmental patterns associated with hantavirus pulmonary syndrome, Four Corners region, United States. *Emerg Infect Dis* 5 (1): 87–94.

Glass GE, Cheek JE, Patz JA, Shields TM, Doyle TJ, Thoroughman DA, Hunt DK, Enscore RE, Gage KL, Irland C, Peters CJ, Bryan R. 2000. Using remotely sensed data to identify areas at risk for hantavirus pulmonary syndrome. *Emerg Infect Dis* 6 (3): 238–47.

Gordis L. 1996. *Epidemiology.* W. B. Saunders Co., Philadelphia.

Harb M, Faris R, Gad AM, Hafez ON, Ramzy RM, Buck AA. 1993. The resurgence of lymphatic filariasis in the Nile delta. *Bull World Health Organ* 71: 49–54.

Hjelle B, Glass GE. 2000. Outbreak of hantavirus infection in the Four Corners region of the United States in the wake of the 1997–1998 El Niño-Southern Oscillation. *J Infect Dis* 181:1569–73.

Jacob P, Goulko G, Heidenreich WF, Likhtarev I, Kairo I, Tronko ND, Bogdanova TI, Kenigsberg J, Buglova E, Drozdovitch V, Golovneva A, Demidchik EP, Balonov M, Zvonova I, Beral V. 1998. Thyroid cancer risk to children calculated. *Nature* 392: 31–32.

Kilbourne EM. 1997. Heat waves and hot environments. In *The Public Health Consequences of Disasters* (Noji EK, ed.). Oxford University Press, New York, Chap. 12.

Kricker A, Armstrong BK, English DR, Heenan PJ. 1995. A dose-response curve for sun exposure and basal cell carcinoma. *Int J Cancer* 60: 482–88.

Last JM. 1995. *A Dictionary of Epidemiology,* 3d ed. Oxford University Press, New York.

Lilienfeld DE. 1978. Definition of epidemiology. *Am J Epidemiol* 107: 87–90.

Luna EJ, Medina NH, Oliveira MB, de Barros OM, Vranjac A, Melles HH, West S, Taylor HR. 1992. Epidemiology of trachoma in Bebedouro State of São Paulo, Brazil: Prevalence and risk factors. *Int J Epidemiol* 21 (1): 169–77.

Mahaffey KR, Mergler D. 1998. Blood levels of total and organic mercury in residents of the upper St. Lawrence River basin, Quebec: Association with age, gender, and fish consumption. *Environ Res [Sect A]* 77: 104–14.

Mumford JL, He XZ, Chapman RS, Cao SR, Harris DB, Li XM, Xian YL, Jiang WZ, Xu CW, Chuang JC, Wilson WE, Cooke M. 1987. Lung cancer and indoor air pollution in Xuan Wei, China. *Science* 235 (4785): 217–20.

Muñoz B, Aron J, Turner V, West S. 1997. Incidence estimates of late stages of trachoma among women in a hyperendemic area of central Tanzania. *Trop Med Int Health* 2 (11): 1030–38.

Nzilambi N, de Cock KM, Fortal DN, Francis H, Ryder RW, Malebe I, Getchell J, Laga M, Piot P, McCormick JB. 1988. The prevalence of infection with human immunodeficiency virus over a 10-year period in rural Zaire. *N Engl J Med* 318 (5): 276–79.

Quinn TC, Mann JM, Curran JW, Piot P. 1986. AIDS in Africa: An epidemiologic paradigm. *Science* 234: 956–63.

Rylander L, Stromberg U, Dyremark E, Ostman C, Nilsson-Ehle P, Hagmar L. 1998. Polychlorinated biphenyls in blood plasma among Swedish female fish consumers in relation to low birth weight. *Am J Epidemiol* 147 (5): 493–502.

Shapiro CN, Coleman PJ, McQuillan GM, Alter MJ, Margolis HS. 1992. Epidemiology of hepatitis A: Seroepidemiology and risk groups in the USA. *Vaccine* 10, Suppl. 1: S59–S62.

Stanley DJ, Warne AG. 1993. Nile delta: Recent geological evolution and human impact. *Science* 260: 628–34.

Thompson DF, Malone JB, Harb M, Faris R, Huh OK, Buck AA, Cline BL. 1996. Dispatches: Bancroftian filariasis and diurnal temperature differences in the southern Nile delta. *Emerg Infect Dis* 2: 234–35.

Tong S, Baghurst P, McMichael A, Sawyer M, Mudge J. 1996. Lifetime exposure to environmental lead and children's intelligence at 11–13 years: The Port Pirie cohort study. *Br Med J* 312 (7046): 1569–75.

Toole MJ. 1997. Complex emergencies: Refugee and other populations. In *The Public Health Consequences of Disasters* (Noji EK, ed.). Oxford University Press, New York, Chap. 20.

Tourre YM, Rajagopalan B, Kushnir Y. 1999. Dominant patterns of climate variability in the Atlantic Ocean during the last 136 years. *J Climate* 12: 2285–99.

United Nations Development Programme/Food and Agriculture Organization/International Bank for Reconstruction and Development/World Health Organization. 1973. *Onchocerciasis Control in the Volta River Basin Area: Report of the Mission for Preparatory Assistance.* World Health Organization, Geneva.

Ventura S, Peters K, Martin J, Maurer J. 1997. *Monthly Vital Stat Rep* 46 (1), September 11: 5.

Wald N, ed. 1995. Ionizing radiation. In *Environmental Medicine: Integrating a Missing Element into Medical Education* (Pope AM, Rall DP, eds.). Committee on Curriculum Development in Environmental Medicine. Division of Health Promotion and Disease Prevention. Institute of Medicine. National Academy Press, Washington, Case Study 37.

World Bank. 1997. *Clear Water, Blue Skies: China's Environment in the New Century.* World Bank, Washington.

World Health Organization. 1978. *Risk Approach for Maternal and Child Care.* WHO Publication No. 39. World Health Organization, Geneva.

———. 1984. *Lymphatic Filariasis: Report of the WHO Expert Committee on Filariasis.* WHO Technical Report Series 720. World Health Organization, Geneva.

———. 1995. *Health Consequences of the Chernobyl Accident: Results of the IPHECA Pilot Projects and Related National Programs. Summary Report.* World Health Organization, Geneva.

World Resources Institute. 1998. *A Guide to the Global Environment: Environmental Change and Human Health.* WRI, Washington.

Zeitz PS, Butler JC, Cheek JE, Samuel MC, Childs JE, Shands LA, Turner RE, Voorhees RE, Sarisky J, Rollin PE, Ksiazek TG, Chapman L, Reef SE, Komatsu KK, Dalton C, Krebs JW, Maupin GO, Gage K, Sewell CM, Breiman RF, Peters CJ. 1995. A case-control study of hantavirus pulmonary syndrome during an outbreak in the southwestern United States. *J Infect Dis* 171: 864–70.

Electronic References

United Nations Development Programme. 1999a. Human Development Indicators: Statistics from the 1998 Human Development Report. Population Trends. http://www.undp.org/hdro/population.htm (Date Last Revised 4/26/1999).

———. 1999b. Human Development Indicators: Statistics from the 1998 Human Development Report. Trends in Human Development. http://www.undp.org/hdro/trends.htm (Date Last Revised 4/26/1999).

CHAPTER 3

Geographic Information Systems

Joan L. Aron, Ph.D., and Gregory E. Glass, Ph.D.

Geographic information systems (GISs) are, in the broadest sense, manual or computer-based sets of procedures that permit users to capture, store, manipulate, retrieve, analyze, and display data with spatial characteristics (Aronoff 1989). The advent of computer hardware and software with the power and storage capabilities to manipulate and analyze large databases has driven the development of computerized GISs during the past 25 years. Compared with manual systems, computerized GISs store and retrieve data at greater speeds and at a lower cost per unit of data. Greater computing power has been accompanied by greater ease of use and greater access to computer technology in general. GISs are now being applied in diverse fields that use spatial or georeferenced data, and the number of applications of GISs in the field of public health continues to grow (Briggs and Elliott 1995; De Savigny and Wijeyaratne 1995; Clarke et al. 1996; Vine et al. 1997; *National Aeronautics and Space Administration (NASA)/Ames 1999a; Agency for Toxic Substances and Disease Registry [ATSDR] 2000;* Beck et al. 2000). A GIS incorporates many technologies and procedures for manipulating multiple types of spatial data and integrating them into a single analysis. Thus, a GIS can be critical in evaluating change and developing planning scenarios and decision models involving spatial data.

Spatial patterns of disease distributions often provide important evidence related to underlying disease processes. Mapping has been standard epidemiological practice since the field's inception (Pyle 1979). Traditional maps include pin maps, on which pins identify locations of cases of a disease, and choropleth maps, on which areas are shaded according to rates of disease incidence, mortality, or other measures of the health status of populations (see Chapter 2). John Snow's map of community water pumps and the deaths due to an outbreak of cholera in London in 1854 remains a classic visual presentation of evidence (Tufte 1997, 31). For practical reasons, these early methods were applied to the study of epidemiological patterns associated with one or at most a few environmental factors.

Public health issues related to ecosystem change are complex, however, and depend on data at many geographic scales. For example, in West Africa, the risk of onchocerciasis, a parasitic disease causing blindness, depends on proximity to areas of turbulence in a river, which in turn are affected by regional variation in rainfall and changes in the global climate system (see Chapters 2, 7, and 8). Exposure to radiation also varies geographically, as demonstrated by the risk of thyroid cancer in relation to radioactive fallout that spread from the Chernobyl nuclear accident and the risk of skin cancer in relation to ultraviolet (UV) radiation flux modulated by latitude and stratospheric ozone depletion (see Chapters 2 and 7). The ability of GISs to use numerous strata of data from various sources simultaneously means that a GIS can be an important tool for understanding and responding to complex ecosystem changes that may adversely affect public health.

This chapter provides an overview of the basic components of a GIS, including related technologies for remote sensing, the global positioning system (GPS), and geographic base file systems. The description of the functions of GISs places emphasis on methods for spatial data analysis and criteria for data quality. Critical issues in the implementation of a GIS are system specification and cost, with aspects related to data sources, computer technology, and human resources.

The Basic Components of Geographic Information Systems

A Generalized Thematic Mapping System

The origins of GISs lie in thematic cartography, in which maps are constructed to display data describing characteristics of geographic features (Clarke 1998). A particular aspect of the descriptive data is referred to as an attribute or, more traditionally, a theme, from which comes the name *thematic cartography* or *thematic mapping*. One example of thematic mapping is a display of the rate of incidence of a disease on a choropleth map (see under "Map Generation," below). The core of a modern GIS is a generalized thematic mapping system that integrates computer hardware, software, and databases to capture, store, manipulate, retrieve, analyze, and display data (Table 3.1).

Vector and Raster Data Formats

The standard model of geographic space partitions a mapped landscape into features, which can be points, lines, or areas (Clarke 1998). A point on the map is a feature with location but no extent, such as a stationary de-

Table 3.1   Six Basic Functions of a Geographic Information System

Function	Description
1. Data capture	Acquisition of data in digital form, including data correction and data conversion for database consistency
2. Data storage	Placement of data where other functions can easily access them
3. Data manipulation	Construction of relationships among various components of data using a database management system
4. Data retrieval	Selection of data according to specified attributes or themes
5. Data analysis	Generation of information using techniques of spatial analysis and statistical analysis
6. Data display	Presentation of thematic maps on a computer screen or on paper

vice to monitor pollution. A line on the map is a feature with one dimension, such as a road. An area on the map is a feature with two dimensions, such as a census tract. The scale of the map affects the classification of features. For example, a city may be represented as a point on a world map and as an area on a local map.

GIS tools represent spatial data in vector format or in raster format. Most GIS software programs have raster-to-vector and vector-to-raster conversion algorithms that permit users to convert data to the more suitable form for a particular application. In either format, spatial information is referenced with respect to a standard geographic coordinate system, such as latitude-longitude, and projected to a plane with Cartesian x-y coordinates. If spatial data are collected from multiple coordinate systems or projections, the data must be converted to a common locational reference system before they are combined for spatial analysis (see *Dana 1997, 1999* and "Sources of Data," below).

In vector format, the precise locations of points, lines, and areas are specified in space. Lines are constructed as a series of line segments connecting points. Areas are constructed as polygons (i.e., regions enclosed by line segments). All of the area in the interior of a polygon is assumed to be uniform, to some specified level of accuracy, for the theme (attribute) being represented. For example, data in vector format show sites in Dade County, Florida, where Black Creek Canal virus was present or absent in populations of cotton rats (Fig. 3.1). Data in vector format can represent quantitative levels at sampling sites by using symbols in different sizes, as shown by a study of chemical and biological water pollution in Puerto Rico (Hunter and Arbona 1995). Data in vector format can also show statistics

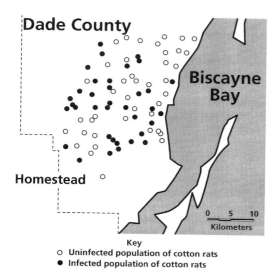

Figure 3.1 Sampling sites of cotton rats in Dade County, Florida, 1994, classified according to the presence (*black circle*) or absence (*white circle*) of rats infected with Black Creek Canal virus. *Source:* Redrawn with permission from Glass et al. 1998, Figure 1.

for a given area, such as a state, county, or census tract (see under "Data Display," below). Mapping in vector format is appropriate for land use data, cadastral data (property records), census data, precise positional data, and transportation networks.

In raster format, spatial data are represented by a regular rectangular array of square cells. Borrowing terminology from digital satellite imagery, a cell is often called a *pixel,* which is a shortened form of picture element. A row number and column number define the location of each cell in space. The value of an attribute is assigned to a cell, so the accuracy of positioning is determined by the size of the cell (see under "Spatial Accuracy and Resolution," below). Data in raster format can show the distribution of forest habitat in Baltimore County, Maryland, which has been analyzed in relationship to the locations of the residences of people who were diagnosed with Lyme disease (see Fig. 3.2 and under "Map Overlay and Comparison," below). Data in raster format are typical in natural resource GISs and maps developed using techniques from remote sensing (see under "Remote Sensing," below) (Burrough and McDonnell 1998; Clarke 1998). In raster format, it is possible to represent fuzzy boundaries between natural resource features, such as the edges of forest. Data in raster format are useful for constructing surfaces that assign a thematic (attribute) value to all geographic points without assuming homogeneity within predetermined district boundaries (see under "Methods for Spatial Data Analysis," below).

Each format has advantages and disadvantages. Data in vector format generate maps similar to hand-drawn maps, which are considered to be more pleasing to the eye. In addition, a map in vector format can usually

Figure 3.2 Spatial distribution of forest habitat, 1989–90, and the locations of residences of people diagnosed with Lyme disease, 1991, in Baltimore County, Maryland. *Source:* Redrawn from Glass 2000b, Figure 9.8. Reprinted with permission from Aspen Publishers, Inc.

be stored more compactly. A system using vector format is usually more efficient at encoding topological relationships, such as lines connecting points, because a system using raster format usually identifies a theme (attribute) at each cell (pixel) in the map regardless of whether the theme changes between adjacent cells. However, management of data in raster format may be more efficient than management of data in vector format if a theme exhibits great spatial variability. Moreover, alternatives to scanning raster data row by row can result in faster rates of data processing (*Goodchild 1997*). Modern GISs employ features of both formats.

A Relational Database Management System

Nonspatial database systems are tools for managing the data in GISs. A relational database management system is the most common user interface, although programmers may use a variety of nonspatial database approaches (*Meyer 1997*). Relational database systems permit complex queries but are easier to use than are programming languages. The basic element is a *data record*, which is also called a tuple in mathematical treatments of the subject. A set of data records defines a *relation* that may be represented as rows of data records in a table. A *data record* is divided into *fields*. One or more fields may be used as a *key* if their values uniquely identify a row. For example, a simple cross-sectional survey of people in a province (see Chapter 2) might generate a data record with three fields: unique identification number (key), place of residence, and history of malaria. Relational database management systems can join two relations (tables) if they share a key field. For example, a survey on malaria and a survey on anemia could be joined through a common key field to analyze how anemia is related to a medical history of malaria. The ability to join two or more relations (tables) by shared key fields provides tremendous flexibility for performing inquiries in response to issues that arise after the initial study design.

Related Technologies

Remote Sensing

Remote sensing refers to the use of electromagnetic radiation sensors to record attributes of the environment (Patterson 1998). The most common types of remote sensing in studies of ecosystem change and public health use instruments on satellites and aircraft. Since the first scientific symposium on satellite-based remote sensing of the environment at the Environmental Research Institute of Michigan in 1962, the remote sensing community has grown to include at least six civilian satellite systems for land remote sensing (Morain 1998). Earth-observing sensors on satellites have deservedly gained attention as a monitor of environmental change on a truly global scale. The origins of techniques for satellite-based observations lie in aerial photography, which continues to make unique contributions to information about the earth.

SENSORS ON SATELLITES. Remote sensing from satellites relies on a technological perspective of using electromagnetic radiation from Earth to observe environmental change in contrast with a scientific perspective of studying the effects of electromagnetic radiation on the global environment (see Chapters 7 and 8). Of course, scientific and technological perspectives are interrelated, since advances in technology depend on gains in scientific knowledge and vice versa.

Features of Earth's land, water, and atmosphere generate variations on the basic concept of energy from the sun warming the earth so that it too radiates energy back to space. According to the theory of black body radiation, the sun and Earth radiate energy at different wavelengths (see Fig. 7.9). Energy from the sun is in the UV, visible, near infrared (NIR), and midinfrared (MIR) portions of the electromagnetic spectrum, while energy radiated from the earth is in the thermal infrared (TIR) portion of the electromagnetic spectrum (Table 3.2).

However, some solar radiation is not absorbed but is instead reflected by objects in the atmosphere, on land, and on sea (Conway 1997; *Short 1999*). Objects vary in their *spectral signatures,* which are the fractions of electromagnetic radiation reflected away from the surface at different wavelengths. The most common techniques for satellite-based remote sensing focus on the identification of objects based on electromagnetic

Table 3.2   Portions of the Electromagnetic Spectrum

Portion of Electromagnetic Spectrum	Wavelength Range (nm)
UV-C (ultraviolet-C)[a]	200–280
UV-B (ultraviolet-B)	280–320
UV-A (ultraviolet-A)	320–400
Visible light	400–700
NIR (near infrared)	700–1,100
MIR (midinfrared)	1,500–2,500
TIR (thermal infrared)[b]	3,000–12,000

Source: Data from Tevini 1993 and Cowen and Jensen 1998.

[a]Shorter wavelength radiation includes x-rays and gamma rays.

[b]Longer wavelength radiation includes microwaves and radio waves.

radiation reflected in the visible, NIR, and MIR portions of the electromagnetic spectrum. Instruments on satellites also monitor temperature through the TIR portion of the spectrum. The primary warming of the earth's surface generated by the absorption of solar radiation is supplemented by a mechanism of secondary warming known as the *greenhouse effect* (see Chapters 7 and 8). Under this mechanism, Earth's atmosphere absorbs infrared (IR) radiation emitted from the surface of the earth and then emits IR radiation of its own, further warming the earth (see Fig. 8.1*a*). The atmospheric concentrations of gases responsible for the greenhouse effect have been increasing since the onset of the industrial revolution, raising concerns about *accelerated global warming* accompanied by scientific uncertainties about this phenomenon (see Chapters 7, 8, and 13). Heat flow from the interior of the earth also contributes to surface heat in some locations, such as active volcanoes and hot springs, but heat from the interior has a negligible effect on the global energy budget (see Fig. 8.1*a*).

Sensors vary according to their spatial and temporal resolutions. For sensors on Earth-observing satellites, spatial resolution measures the smallest area that can be observed and temporal resolution measures the time interval between repeated observations of the same area (see under "Temporal Accuracy and Resolution," below, for other aspects of temporal resolution). Spatial and temporal resolution for some common remote sensing instruments are shown in Table 3.3. Landsat and Système probatoire d'observation de la terre (SPOT) provide greater spatial resolution, whereas the advanced very high resolution radiometer (AVHRR), the

Table 3.3   Spatial and Temporal Resolution for Some Common Remote Sensing Instruments

Remote Sensing Instrument	Spatial Resolution (m)	Temporal Resolution[a] (days)
Landsat 4,5 MSS (multispectral scanner)	79	16
Landsat 4,5 TM (thematic mapper)	30	16
Landsat 7 ETM+ (enhanced thematic mapper plus)	15 (panchromatic)[b]	16
SPOT[c] HRV (système probatoire d'observation de la terre, high-resolution visible)	20 (multispectral) 10 (panchromatic)[b]	26
AVHRR[d] (advanced very high resolution radiometer)	1,100	0.5
MODIS[e] (moderate resolution imaging spectroradiometer)	250 (land) 1,000 (ocean) 1,000 (atmosphere) 1,000 (thermal infrared)	2[f]
Meteosat[g]	2,500 (panchromatic)[b]	0.02

Source: Spatial resolution data, except for Meteosat, from Cowen and Jensen 1998, Figure 8-1; temporal resolution data, except for Meteosat, from *Moody 1996,* Figure 15, and Cowen and Jensen 1998, Figure 8-1. Meteosat data from *NASA/Ames 1998a.*

[a]*Temporal resolution* as used here refers to the time interval between observations.

[b]Panchromatic imaging records all visible light in a single band (Patterson 1998).

[c]SPOT is a French commercial satellite program (Patterson 1998).

[d]AVHRR is on National Oceanic and Atmospheric Administration's polar-orbiting environmental satellites.

[e]MODIS is on the Terra (formerly AM-1) satellite, which is part of the National Aeronautic and Space Administration (NASA)'s Earth Observing System.

[f]The frequency of observations will double when MODIS is operating on both Terra and PM-1 satellites.

[g]European weather satellite that is geostationary, so only part of Earth is observed. The U.S. counterpart is the Geostationary Operational Environmental Satellite.

moderate resolution imaging spectroradiometer (MODIS), and Meteosat provide more frequent observations. In general, there is a tradeoff between the degree of spatial detail and the frequency of observation. Images derived from any sensor require digital processing to enhance features and remove distortions, such as incorrect relative positioning of objects caused by viewing different objects at different angles (Jensen 1996; *Jensen and Schill 1998*).

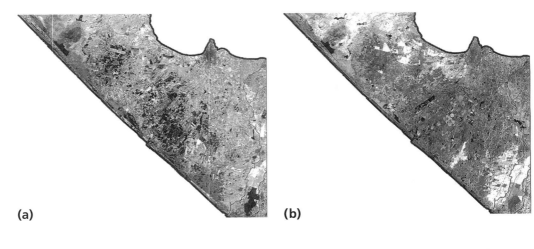

(a)

(b)

(c)

Figure 3.3 *a:* Landsat Thematic Mapper image of the study area in Chiapas, Mexico, acquired on July 8, 1991, during the wet season; the image is a multiband composite of Landsat 5. The spatial resolution of the data is 28.5 meters. The *dark gray tones* are bare soil, water, or urban; the other shades represent various vegetation and crop types. The *diagonal edge* represents the coastline along the Pacific Ocean; the Sierra Madre foothills, located in the *upper right,* have been masked out because the mosquito *Anopheles albimanus,* the primary vector of malaria in the region, is not found there. *b:* Landsat Thematic Mapper image of the study area acquired on March 4, 1992, during the dry season; the image is a multiband composite of Landsat 5. The *bright areas* represent irrigated agriculture, such as tree crops and banana plantations, with the *darker tones* representing bare soil, water, and burned areas. *c:* Land cover classification generated

Reflected radiation and thermal radiation are both useful in applications to public health. For example, patterns of reflected radiation detected by Landsat's thematic mapper were the basis of a classification of land cover in a region of southern Mexico; statistical associations between types of land cover and the abundance of mosquitoes that transmit malaria led to the identification of villages in the region at high risk for malaria (Fig. 3.3) (see Beck et al. 1994, 1997; *NASA/Ames 1999b;* and Chapter 12). Another application of the thematic mapper has demonstrated an association between multiple bands of wavelengths and the risk of hantavirus pulmonary syndrome in New Mexico; however, the study did not find strong

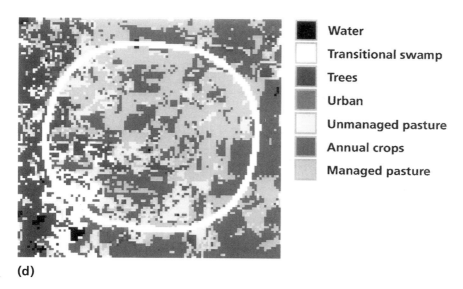

Water
Transitional swamp
Trees
Urban
Unmanaged pasture
Annual crops
Managed pasture

(d)

by combining four bands—bands 3, 4, 5, and 7, which range from visible red to midin-frared in the spectrum—from each of the images used in *a* and *b*. In all, 11 categories of land cover were mapped: mangrove, secondary forest, transitional swamp, riparian vegetation, managed pasture, unmanaged pasture, annual crop, banana plantation, water, burned areas, and cloud/shadow. The *white circles* represent 1-kilometer buffers surrounding the 40 villages sampled for the abundance of the *An. albimanus* mosquito. Within this 1-kilometer radius, the female mosquito must find blood meals, resting sites, and breeding habitat. The proportions of land cover types within the buffers and the mosquito data were used in statistical analyses to relate land cover to mosquito abundance. *d:* An example of a village (*center* of the *circle*) and its surrounding land cover classes. For the purposes of display, the 11 classes were grouped into 7, as shown in the key. Both transitional swamp and unmanaged pasture were found to be predictive of the abundance of *An. albimanus. Source:* All of these photographs are courtesy of Byron Wood of NASA/Ames Research Center, Brad Lobitz of Johnson Controls, and Louisa Beck of California State University/Monterey Bay.

evidence linking the effect to a pathway of increased spring precipitation and increased vegetation growth (see Glass et al. 2000 and Chapters 2 and 10). In other studies of land use and land cover, Landsat and SPOT satellites have provided images of deforested land in the Amazon basin and urban land in the United States (Wood and Skole 1998; Moran and Brondizio 1998; Cowen and Jensen 1998).

Thermal radiation sensors on the AVHRR can reveal diurnal temperature changes; large diurnal temperature changes at locations within the Nile delta indicate the presence of surface and subsurface moisture contained in the soil and plant canopy, which is associated with breeding sites for the mosquitoes that transmit filariasis (see Fig. 2.6). Studies have also combined thermal radiation and reflected radiation measured by the AVHRR (operated by the U.S. National Oceanic and Atmospheric Administration [NOAA]) and thermal radiation measured by Meteosat (operated by the European Space Agency). In a study of the transmission of malaria in Africa, AVHRR's reflectance and thermal data from land were the sources of estimates of vegetation state and land surface temperature, while Meteosat's thermal data from clouds were the sources of estimates of rainfall; all of these environmental variables affect the populations of mosquitoes that transmit malaria (see Thomson et al. 1997 and Chapter 12). The sensors on AVHRR and Meteosat have also provided data for the Famine Early Warning System of the U.S. Agency for International Development on the detection of vegetative and rainfall anomalies likely to be associated with crop stress (Hutchinson 1998). Greater frequency of data collection at lower cost is a major advantage of AVHRR and Meteosat, especially in developing countries.

Other types of remote sensing instruments can contribute to the study of ecosystem change and public health. Some public health problems are directly related to conditions in aquatic environments. Measurements of the color, temperature, and height of the ocean are important applications of remote sensing (see Box 3.1). An estimate of the amount of plankton may be inferred from the reflectance of chlorophyll detected by ocean color scanners (see Fig. 11.4), although such estimates should be viewed with some caution because the underlying computational algorithms may not properly distinguish between chlorophyll and other substances in the water, particularly in coastal regions (*NASA/Goddard Space Flight Center [GSFC] 2000a*). Plankton in the marine and estuarine environment has been linked to the transmission of the vibrio organism that causes cholera (see Chapter 11). The transmission of cholera also seems to be associated with sea surface temperature (SST) as measured by AVHRR (see Fig. 11.5), which in turn is related to the El Niño/Southern Oscillation climatic anomaly (Chapter 8). The reflection of radar signals can be used to monitor surfaces of water, with possible implications for public health (Epstein 1998).

Some public health problems are affected by changes in global processes that are not easily summarized by a measuring instrument. Opportunities exist to link public health data to environmental characteristics that are derived from multiple sources of data and various algorithms. For example, the Center for International Earth Science Information Network (CIESIN) provides a tool for visualizing estimates of the flux of UV radiation at the surface of the earth; these estimates are based on mathematical models and on observations by NASA-sponsored projects that study the loss of stratospheric ozone (see Chapter 7 and *CIESIN 2000*). The

BOX 3.1

Oceanography and Remote Sensing Satellites

Oceanography has distinct requirements for remote sensing. Sensors on satellites are used to measure ocean color, sea surface temperature (SST), sea surface height (SSH), and wave height (*National Aeronautics and Space Administration [NASA]/Goddard Space Flight Center [GSFC] 2000a*).

Ocean color is used to study organic and inorganic particulate matter in the ocean. The coastal zone color scanner was flown aboard the Nimbus-7 satellite and collected ocean color data intermittently from November 1978 to June 1986 with a spatial resolution of 825 meters (*NASA/GSFC 2000b*). The sea-viewing wide field-of-view sensor (SeaWiFS) was launched on August 1, 1997, as a public-private partnership between NASA and Orbital Sciences Corporation (*NASA/GSFC 1999*). SeaWiFS scans approximately 90 percent of the ocean's surface every two days, with a spatial resolution of 1,130 me-

ters. The moderate resolution imaging spectroradiometer (MODIS) will extend ocean color data sets (see Table 3.3).

SST is useful for observing currents and circulation in the oceans and is of particular interest in the study of the El Niño/Southern Oscillation (ENSO) climatic anomaly (see Chapter 8). SST is obtained from the advanced very high resolution radiometer's thermal measurements (see Table 3.3). MODIS will extend SST data sets (see Table 3.3).

Wave height and SSH are obtained by altimetry, which measures the reflection of a radar signal sent from a satellite (*Short 1999*). SSH indicates large-scale patterns of ocean circulation and is important in the study of ENSO (see Chapter 8). Wave height indicates the influence of storms and regions of high variability. Wave height and SSH are monitored by the satellite TOPEX/Poseidon, which is operated jointly by NASA and France's Centre National d'Etudes Spatiales.

Global Precipitation Climatology Project produces data on monthly mean precipitation using multiple sensors on satellites and rain gauges on the ground (*NOAA 2000a*).

Investigators should be aware of the limitations of data collected by sensors on satellites. Several studies of global change have revealed apparent trends later found to be spurious because of various factors, such as changes in the orbital geometry of a satellite, degradation of a sensor over time, and flaws in the algorithms for processing data (*Rasool 1999*). Those who use data from satellites should have some idea of how and how often a sensor is calibrated, as well as how computer algorithms process the raw data. The plan for the validation of the Tropical Rainfall Measuring Mission provides an illustration of the concerns (*NASA/GSFC 2000c*). NASA's Center for Health Applications of Aerospace Related Technologies has an introduction to the evaluation of sensors, with a focus on use by the public health community (*NASA/Ames 1998b*).

The amount and quality of remote sensing data will continue to grow. Landsat 7 (see Table 3.3) was launched on April 15, 1999, and routine acquisition of data began on June 29, 1999. Landsat 7 provides high-spatial-resolution images of one-quarter of the earth's land mass every 16 days, using cloud predictions from the National Weather Service to avoid cloudy areas. MODIS (see Table 3.3) was launched on the Terra satellite on December 18, 1999, and routine acquisition of data began on February 24, 2000. MODIS is the premier instrument in NASA's Earth Observing System (EOS) for conducting research on global change. MODIS views the earth's land, oceans, and atmosphere in the visible and IR portions of the spectrum every two days and extends the datasets of AVHRR and of ocean color (*NASA/GSFC 2000d*). NASA plans a series of launches of new EOS satellites with a variety of instruments over the coming years (*NASA/GSFC 2000e*). NASA's EOS is part of interagency and international networks to develop remote sensing for research on global change (see Table 3.4 and under "Sources of Data," below). Another trend in environmental remote sensing is increasing investment by the private sector, following earlier trends toward the commercialization of communication satellites (*Stoney 1999*). However, the financial costs of the acquisition of remote sensing data could be high, thereby limiting their use in applications to public health.

AERIAL PHOTOGRAPHY. Aerial photography is a more traditional component of remote sensing that remains important. Aerial photography is superior to satellite imagery when very high spatial resolution (less than 1 square meter) is needed to create digital elevation models (DEMs), respond to emergencies, identify built infrastructure (e.g., buildings, property lines, transportation, utilities), assess the demand for energy, or monitor environmental change (Cowen and Jensen 1998; World Bank 1996). As with other forms of remotely sensed data, aerial photographs usually require considerable processing before they are interpreted (*Short 1999; Estes et al. 1999;* Monmonier 1996). In response to the problem that the measurement of distances between objects is distorted if the objects are not directly below the camera, the U.S. Geological Survey (USGS) produces digital orthophoto quadrangles (DOQs), which are digitized aerial photographs that display features in true ground position (*Foote and Lynch 1999*). However, aerial photographs are not typically updated as frequently as are images derived from satellites.

The Global Positioning System

The GPS uses signals from satellites to determine the precise position of a GPS receiver in three dimensions at a precise time (*Dana 2000*). The system is available 24 hours a day anywhere on or above the earth. The space segment of the system consists of 24 GPS satellites controlled by the U.S. Department of Defense. Any civilian can use the standard positioning service without charge or restriction. Until May 2, 2000, the standard posi-

Table 3.4   Principal Participants in Research on Global Change

Type of Participant	Name of Participant
Nation	Canada, France, Germany, Italy, Japan, United Kingdom, United States
Regional international organization	ESA (European Space Agency), EUMETSAT (European Meteorological Satellite organizations), IAI (Inter-American Institute for Global Change Research)
Global international organization	CEOS (Committee on Earth Observation Satellites), ICSU (International Council of Scientific Unions), IGBP (International Geosphere-Biosphere Program), IPCC (Intergovernmental Panel on Climate Change), UNEP (United Nations Environment Programme), WMO (United Nations/World Meteorological Organization), WCRP (World Climate Research Program)

Source: Data from *Hobish 1999; Inter-American Institute for Global Change Research 2000; Stoney 1999.*

Note: Programs in remote sensing are also expected to grow in Russia, India, Australia, Israel, China, and Brazil.

tioning service provided an intentionally degraded signal that was usually accurate to 100 meters horizontally and 156 meters vertically. As of May 2, 2000, the standard positioning service provides a signal that should be accurate to about 22 meters horizontally and 27.7 meters vertically, the parameters of the precise positioning service. The U.S. military restricted the global use of greater levels of precision because of the security threat posed by information used in guidance systems for missiles, but newer technology now permits the degradation of the signal on a regional basis (*Interagency GPS Executive Board 2000*). Civilian receivers are relatively inexpensive, although capabilities for data processing and file storage add to the cost. A GPS receiver uses information from at least four GPS satellites simultaneously to calculate position and time. An unobstructed path between a GPS receiver and GPS satellites is critical, since signals from GPS satellites are blocked by most materials, such as buildings, metal, mountains, or trees.

In public health research, GPS is useful for adding geographic positioning information to datasets, especially in locations lacking accurate digital maps. A study in Egypt started to use hand-held GPS to refine the maps of villages where health surveys had been conducted on filariasis (see Chapter 2). Techniques for differential GPS attempt to improve accuracy by comparing signals between a receiver and a base station at a known position. Differential GPS has been used in a field study of malaria in Kenya to obtain longitudes, latitudes, and altitudes of houses, schools, churches, health care centers, major mosquito breeding sites, borehole wells, shopping areas, major roads, streams, the shore of Lake Victoria, and other geographic features (Hightower et al. 1998). Even with the improvement in the accuracy of the standard positioning service, differential GPS may still benefit some applications.

Geographic Base File Systems

Many applications of GISs rely on geographic base file systems, which permit access to digital maps that are publicly available. The United States has four types that are widely used (*Foote and Lynch 1999*). The Topologically Integrated Geographic Encoding and Referencing System (TIGER), developed by the U.S. Census Bureau, shows census tract boundaries as well as transportation and hydrology networks for the United States, including Puerto Rico, the Virgin Islands, and outlying areas of the Pacific over which the United States has jurisdiction. TIGER also includes feature names, political area codes, and potential address ranges and postal codes for 350 major cities and suburbs across the United States (Aronoff 1989). The USGS supports a Digital Line Graph (DLG) system for the entire country that includes transportation systems, hydrographic features, and political boundaries. The DLG represents by points, lines, and areas the information in traditional USGS paper quadrangle maps. The USGS also has DEMs and, as noted earlier, rectified aerial photographs in the form of DOQs.

The use of geographic base file systems is no guarantee of success. Matching addresses is a particular problem (*Cowen 1997*). There may be typographical errors, multiple spellings of names, and places without street names. And, of course, address files become out of date. Moreover, the postal code address style of matching (ZIP codes in the United States) does not use the actual street address. Instead, the method interpolates the location of an address on the basis of address ranges and the side of the street involved. For example, in Baltimore, Maryland, city blocks are numbered by hundreds (the 600 block, the 700 block, and so on) north and south of Baltimore Street and not all numbers are used (the end of one block may be 618 when the beginning of the next block is 700). The interpolation method assigns locations of addresses from 600 to 618 as though there were addresses from 600 to 699 on the block, thereby placing the addresses in only one-fifth of the block. Other errors may be generated if there are multiple lots with the same address or buildings with numbers out of sequence. Whether such errors in the physical placement of addresses are important depends on the requirements of the specific project. Applications for dispatching emergency health services demand greater accuracy in location than applications for analyzing aggregate data for census tracts. Most georeferencing software allows the user to control how nonmatched records are handled. Thus, geographic base file systems can provide the underpinning for detailed analysis of spatial data. However, the effect of possible errors must be considered within the context of specific applications.

Methods for Spatial Data Analysis

Analysis of the relationship between ecosystem change and public health has three main spatial elements, as shown in Table 3.5 (Briggs 1997). The first element is *map generation,* which is the presentation of data that may or may not require transformation using spatial statistical procedures. The main spatial statistical procedure of interest is interpolation, which assigns values to all locations using data from just a few locations. The second element is *map analysis,* which is an exploration of spatial patterns, such as apparent clustering of cases of disease or "hot spots" of disease incidence. Some patterns in maps may be distorted or misleading because of the methods and the data selected for map generation. The third element is *map overlay and comparison,* which combines and analyzes multiple layers of data using spatial tools and statistical tools. The ordering of the three main elements of analysis does not imply a rigid time sequence for application. The generation of an initial map may lead to an investigation of map overlays, which in turn results in a map analysis, and so forth. The application of methods for spatial data analysis presumes the availability of the other GIS functions of data capture, storage, manipulation, retrieval, and display (see Table 3.1).

Map Generation

DATA DISPLAY. Map generalization is the process of compressing reality onto a two-dimensional picture (Monmonier 1996). Inevitably, the result reflects judgments about the relative importance of mappable features. Generalization issues are important for both geometry and content.

The geometric representations of points, lines, and areas are all affected by the issue of selection (i.e., deciding what details to place on the map) (Monmonier 1996). Problems may occur, especially when the level of detail from a source map is reduced. For example, points may be aggregated, lines may be simplified, and small areas may be dissolved.

The scale of maps is read in terms of fractions. Large-scale maps, with more detail, have scales of 1:24,000 or larger. Small-scale maps, with less detail, have scales of 1:500,000 or smaller. So the loss of detail corresponds to the transfer from a large-scale map to a small-scale map.

Content generalization uses selection and classification of data to filter out irrelevant details (Monmonier 1996). Features deemed irrelevant are not mapped at all. Classification of mapped features makes the map more readily understood by assigning a single type of symbol to similar

Table 3.5   Main Elements of Spatial Data Analysis

Function	Description	Example
Map generation[a]	Display data using maps or interactive computer technology	Sites indicating the presence or absence of infection (Fig. 3.1)
Map generation with spatial statistical transformation	Transform data using spatial interpolation procedures before map generation	Isopleths (contour lines) of the acidity of rain (Fig. 7.7)
Map analysis	Explore spatial patterns of data, such as clustering of cases of disease, for possible distorted or misleading effects	Clustering of disease by residence less important than clustering by workplace (murine typhus discussion in text)
Map overlay and comparison[b]	Compute and analyze spatial relationships, such as distances,[c] using multiple layers of data	Risk of Lyme disease related to proximity to forest (Fig. 3.2 and discussion in text)

[a]Unless otherwise specified, the term is used to indicate presentation without spatial statistical transformation.

[b]The statistical procedures for analysis, such as regression, are sometimes grouped separately from map overlay tools for analysis, such as distance computation.

[c]Distance may be used between individual (point) locations or to construct a buffer zone (area) around features, such as rivers or industrial facilities.

features. For example, a choropleth map of the incidence of malaria in Rondônia, Brazil, classifies incidence for each municipality into a gray scale of three levels (Fig. 3.4a,b). However, a classification of incidence data that assigns an equal number of municipalities to each level (*equal counts* in Fig. 3.4a) has far more high-incidence municipalities than a classification of the same data that divides the entire range of incidence into three equal parts (*equal ranges* in Fig. 3.4b). Malaria seems to be a much greater public health problem in the first map. The presentation of the number of cases rather than the per capita rate of incidence also creates a very different impression of the geographic distribution of malaria whether equal counts or equal ranges are used to construct the gray scale (Fig. 3.4c,d). Therefore, it is important to remember that apparently straightforward decisions about the selection and classification of data can have unintended consequences for interpretation and can even be deliberate to create a particular impression. A thorough analysis should examine more than one way of selecting and classifying data.

Pickle and Herrmann (1995) reviewed cognitive issues of statistical mapping in the context of the practical demands of designing a national

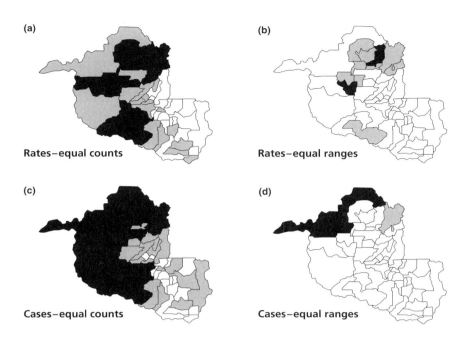

Figure 3.4 Incidence of malaria by municipality in Rondônia, Brazil, 1996. Incidence for each municipality, measured as the number of cases per population, is grouped into three levels marked by the intensity of gray scale. For each level described, the number in parentheses shows the number of municipalities at that level. In *a*, the classification is based on equal counts: 0.000–0.005 (18) are *white;* 0.005–0.110 (17) are *gray;* 0.110–0.944 (17) are *black.* In *b*, the classification is based on equal ranges: 0.000–0.315 (41) are *white;* 0.315–0.630 (9) are *gray;* 0.630–0.944 (2) are *black.* The number of cases in each municipality is also grouped into three intervals using the same two methods. In *c*, the classification is based on equal counts: 0–100 (17) are *white;* 100–1,500 (17) are *gray;* 1,500–25,100 (18) are *black.* In *d*, the classification is based on equal ranges: 0–8,400 (50) are *white;* 8,400–16,800 (1) is *gray;* 16,800–25,100 (1) is *black.* See text for discussion. *Source:* Courtesy of Saskia Nijhof of the International Centre for Integrative Studies in Maastricht, Netherlands, based on data from the Fundação Nacional de Saúde (National Health Foundation), Rondônia, Brazil.

mortality atlas using data from all counties in the United States. No single map design is superior in all respects. They made six important conclusions about the accurate interpretation of maps:

1. The accuracy of reading a rate from a map decreases as the number of visual categories increases because the difficulty in discriminating between levels is compounded by the necessity of alternately gazing back and forth at the map and the legend.

2. Schemes that use multiple colors result in greater accuracy of reading individual rates because colors allow greater discrimination and are more easily remembered.
3. Monochrome (gray scale) schemes are superior for pattern-recognition tasks because it may be easier to identify areas with similar rates.
4. Classed maps, in which data are classified into categories, can be confusing to readers who do not understand how perception depends on the classification scheme (see discussion of Fig. 3.4).
5. Unclassed maps, in which size or shading increases in direct proportion to the value of the underlying statistic, create cognitive burdens for both novices and experts.
6. Classed maps lead to greater ability to rank the rate of an area relative to other areas, but unclassed maps lead to greater accuracy for estimating the absolute value of a rate.

In addition, familiarity with the style of presentation is very important. Readers of maps are more comfortable when the geographic areas correspond to conventional maps, such as states, provinces, or counties. Epidemiologists are very comfortable with classed choropleth maps, such as Figure 3.4, but different disciplines have different preferences, posing possible barriers to the interdisciplinary cooperation required for studies of global ecosystem change and public health.

An important topic is the representation of statistical reliability of quantitative data. A classic problem is that rates for areas with small populations are statistically unreliable and liable to fluctuate greatly from year to year. For geographically large reporting areas, it is possible to place a symbol, such as an asterisk, in an area to denote significance (and hence reliability) according to a specified statistical test. A more general graphic approach allows colors to represent the rates but uses crosshatching or reduced saturation (intensity) to indicate unreliable areas. Alternatively, map smoothing allows statistical precision to be shared across areas on the basis of prior assumptions about the statistical distributions involved (see "Map Analysis," below). However, readers with little or no statistical training may not understand how to interpret the statistical procedures used in map generation. Lack of training is also an issue in the understanding of statistical procedures used in presentation of incidence or mortality rates, such as age adjustment (see Chapter 2).

The development of interactive computer technologies offers new opportunities for flexibility in viewing statistical data on maps, such as dynamically changing the categories of the data or the time periods of the data (Pickle and Herrmann 1995; Monmonier 1996). Animation can be highly effective at showing the geographic spread and disappearance of infectious diseases (Clarke et al. 1996). More generally, images, video, graphics, text, and sound are combined in multimedia presentations, and virtual reality tools permit even greater interactivity (*Taylor 1997*). Technologies for scientific visualization are expanding rapidly (see Chapter 5 and *NASA/ Ames 2000*). As new technologies are developed for the presentation of data, the ability of users to understand data through these new technologies will need to be carefully evaluated. Old and new technologies alike can benefit from sound principles of visual design (Tufte 1983, 1990, 1997).

SPATIAL INTERPOLATION. Some form of spatial transformation is required when thematic (attribute) data are available only at a fixed number of sampling points and information about other points is desired (Briggs and Elliott 1995; Briggs 1997). Spatial transformation permits estimation at points that have not actually been sampled. For example, interpolation, which fills in areas between sampling points, is useful for mapping the concentration of a pollutant over an entire country (see Fig. 7.7). Extrapolation, which is extending the map beyond the sampling points, may also be desired. Another reason for spatial transformation is combining multiple layers of data for analysis; map overlay and comparison are often easier to perform if all layers have a raster format. As with simpler issues of data presentation, good use of spatial transformation depends on judgment and an understanding that a variety of methods should be investigated.

The most widely available interpolation method within GISs is voronoi tessellation, which is the construction of Thiessen polygons (voronoi) around each sampling point (locus) of data. The polygon is defined such that any location within the polygon is closer to its associated locus than to any of the neighboring loci. The entire area within the polygon is assigned the value at its locus. Consequently, the values may change abruptly at the borders of polygons. Loci that are far apart generate larger polygons that have a disproportionate effect; the closer the loci, the better the interpolation.

Another widely available interpolation method within GISs is contouring, which is the construction of contour lines that represent a constant level of the interpolated variable (see Figs. 7.7 and 8.2*b*). The algorithms for contouring in a GIS often use triangulated irregular networks (TINs). Contouring in TINs, which may be referred to as *triangulation,* defines mathematical functions that depend on the nearest three locations with data. The shape of the contour line is sensitive to data points that are unusually high or low in value. The results also depend on the selection of the mathematical function and on the arrangements of the data points. As with voronoi tessellation, the denser the set of locations with data, the better the interpolation.

Kriging is a more statistically sophisticated interpolation technique, available in some GISs, which incorporates more locations with attribute data values in the computational process. Kriging is based on the concept of spatial correlation in which the value of an attribute at a selected point is likely to be closer to values at points nearby than values at points far away. Central to the development of kriging is the *semivariance,* which is calculated with a formula using differences between the value observed at a selected site and the values at sites a fixed distance away from the selected site. The semivariance is relatively small for close points and grows with distance. Once a certain distance is reached (depending on the application), further increases in distance do not change the semivariance. The graph of the value of the semivariance with increasing distance from the starting point is called the *semivariogram.* The semivariogram itself is then described by a mathematical formula (spherical, exponential, or power) that is used to provide values at points where data were not collected. The surface estimated may be displayed using contour lines, which show where the surface has a constant value. In general, kriging has two major advantages over contouring using triangulation: kriging is less sensitive to unusually high or low data points, and it provides its own error estimates. The values generated by kriging, however, depend strongly upon the selection of the mathematical formula describing the semivariogram.

The various methods of interpolation can produce very different results. An example of 79 sampling points of nitrogen dioxide (NO_2) levels within an area of 300 square kilometers in Huddersfield, England, demonstrates the application of simple averaging (ignoring any spatial aspects), voronoi tessellation, triangulation, and kriging (Briggs and Elliott 1995; Briggs 1997). The unit of measurement for the concentration of NO_2 was micrograms per cubic meter. The estimate of the mean concentration of NO_2 ranged from 27.6 to 30.2, while the estimate of the 98th percentile was even more sensitive, ranging from 40.1 to 56.2. These differences reflect implementation as well as theoretical differences because the algorithms for voronoi tessellation in two different commercial systems (SPANS and ARC/INFO) produced different results. In fact, the voronoi tessellation in SPANS produced the highest estimate for the 98th percentile, and the voronoi tessellation in ARC/INFO produced the highest estimate for the mean.

Map Analysis

With map analysis one aims to determine whether there is a spatial pattern in a map of a particular dataset (Briggs 1997). Spatial techniques address point patterns, line patterns, area patterns, and surface or contour patterns (Nobre and Carvalho 1995).

Point patterns display the locations of health events, such as the residence of a person diagnosed with a particular disease in a defined time period. A primary objective of point analysis is the detection of clustering in space and time. A space-time analysis of an outbreak of dengue in Florida,

Puerto Rico, showed that the spatial pattern of cases of dengue was similar to that of the community as a whole and not clustered strongly within individual households (Morrison et al. 1998). The public health implication was that control measures should be applied to the entire community rather than to areas immediately surrounding the houses of cases. The possible clustering of illness near identifiable sources of pollution has generated a great deal of interest, although the identification of "hot spots" on the basis of point data is a difficult and controversial statistical problem (Briggs and Elliott 1995). If distances from specific sources of pollution are used in calculations, then the approach uses information from multiple layers of data (see under "Map Overlay and Comparison," below).

Line patterns show the links between origin and destination points. Line patterns can convey the dynamics of spread of a disease on an otherwise static map (see Fig. 11.1). An analysis of distances between points may be useful in making decisions about the sites of new facilities or the distribution of materials, but these applications are not reviewed here. Chapter 5 discusses decision theory in the context of integrated assessment.

Area patterns provide thematic (attribute) data for specified areas, such as rates of the incidence of a disease in administrative health districts. The identification of hot spots in area data raises issues similar to those in point data. Some statistics detect hot spots using a neighborhood approach, where neighboring areas are usually defined in terms of sharing a common boundary or having points within a specified distance of each other (Anselin 1996). The G statistic of this type has been used to identify foci of the transmission of Eastern equine encephalitis in Michigan (Kitron 1998). This statistic can also measure clustering around known or suspected foci of transmission (see Kitron 1998 and "Map Overlay and Comparison," below). Another approach for examining structure is map smoothing. A study of colorectal cancer in Sheffield, England, smoothed data using an empirical Bayes method, which adjusts for the statistical unreliability of rates in districts with small populations (Haining 1996). The benefit of smoothing is that underlying spatial patterns may be clearer, although too much smoothing may blur important detail about areas at extremely high or extremely low risk. The problem is essentially one of choosing an appropriate level of aggregation in a map (see under "Data Display," above).

Surface and contour data provide thematic (attribute) data for points in a region without assuming that all points in a predetermined area, such as an administrative district, have the same value. (It may be advantageous to overlay surface data with a map of political or administrative boundaries when presenting results.) Surface and contour data are central to the field of geostatistics, which developed initially to estimate the concentration of ores for mining purposes (Oliver 1996). Kriging, which is used for spatial interpolation, is one of the fundamental tools for geostatistics and can be extended to provide more tools for exploratory spatial data analysis. A study of 605 cases of childhood cancer distributed among 840 small districts in the West Midlands of England estimated spatial variation in the risk of childhood cancer (Oliver 1996). The structure of the original data was transformed from area data to point data by using the centroid (geometric center) of each district; the resulting estimates of risk did not assign the same risk to all locations within a district. A study of soil salinity in Israel estimated spatial variation in the probability of exceeding a statutory limit for salinity (Oliver 1996).

Map analysis should also investigate whether alternative sources of data might provide a different picture. No amount of statistical manipulation can make up for the omission of important data. For example, the discovery of the mode of transmission of an infectious agent can be aided by consideration of multiple possibilities during data collection and analysis. A classic study of cases of murine typhus in Montgomery, Alabama, from 1922 to 1925 found that spatial clustering was more important by the place of occupation than by the place of residence (Maxcy 1926). The study's original hypothesis that human lice were the vectors of transmission led to the prediction that close contact within households would generate clustering within households. The observed pattern of occupational clustering, especially among food service workers, led to a new hypothesis that a disease vector associated with food services was responsible. It was later confirmed that the vectors of murine typhus were fleas in rat-infested food facilities (Woodward 1970).

Map Overlay and Comparison

Map overlay and comparison use multiple layers of data in GISs and are central to the study of ecosystem change and public health (Box 3.2). Many studies use the spatial tools of GISs to construct risk factors but rely on nonspatial statistical techniques to measure association. A relatively simple overlay analyzed the time series of clinical cases of malaria for a country, Niger, in relation to the time series of estimates of vegetation cover from satellite data of the Africa Real Time Environmental Monitoring System of the Food and Agriculture Organization of the United Nations (Thomson et al. 1997). In this situation, the analysis seemed to be useful even though the estimate of vegetation was based on observations at a single location in the country.

In most situations, the spatial relationships between the data on risk factors and data on human health must be examined more carefully. For example, a standard GIS tool of constructing a circular buffer zone around a point source of chemical pollution or radiation can provide an indicator for the exposed area, but actual exposures are likely to depend on how

BOX 3.2

Epidemiological Study Designs and Geographic Information Systems

Epidemiological study designs to elucidate the cause of disease (see Chapter 2) relate directly to the functions of map overlay and comparison in geographic information systems (GISs). Different layers of data correspond to environmental risk factors and health outcomes. The simplest analyses can examine aggregate data on environment and health, such as the average elevation of a region and its incidence of disease. GISs can also link data about a person's health to detailed geographic data. For example, a global positioning system can be used to assign a geographic locator to the residences of study participants, making it possible to link personal health data and georeferenced environmental data. Geographic data on individuals can be incorporated into cross-sectional studies, case-control studies, and cohort studies. The example of Lyme disease in the text shows a progression of study designs using GISs.

weather affects dispersion (Briggs and Elliott 1995). Since the radioactive fallout from the Chernobyl nuclear accident had moved northward, Jacob et al. (1998) considered the residents of southern Ukraine to be normal for the comparative analysis of health effects. It may also be difficult to link a person to a point of exposure to a hazard if that person moves from place to place during the course of a day (or other relevant unit of time). As noted under "Map Analysis" (above), the place of occupation turned out to be more important than the place of residence in a study of murine typhus. Long time intervals between exposure to a risk factor and the appearance of disease also make it difficult to link disease to a place of exposure.

GIS overlay and comparison techniques were used for a series of studies of Lyme disease in Maryland (see Chapter 10 for more on Lyme disease). In the northeastern United States, human contact with the tick species responsible for transmission occurs as part of daily activities near the place of residence. A simple geographic analysis showed that Baltimore County was the county reporting the highest number of cases in Maryland every year. More analytical epidemiological studies were needed to elucidate the pattern of risk within the county. Using layers of data for the place of residence and forest habitat (see Fig. 3.2), it was possible to determine that most people diagnosed with Lyme disease lived within 100 meters of the forest edge. However, this statistic is difficult to interpret without knowledge about the typical characteristics of residences in Baltimore County. Glass et al. (1995) then compared cases of Lyme disease in Baltimore County with a "control" group of individuals randomly selected from a sample of residential addresses within the county (see under "Case-control Studies" in Chapter 2). The study linked the locations of the residences to several types of geographic data, including land use characteristics, soil types, elevation, and distance from forests. The analysis of environmental risk factors used (nonspatial) logistic regression techniques, a standard procedure in epidemiology. The risk factors became the basis for designating levels of risk for areas within the county. A prospective study confirmed that the risk of being diagnosed with Lyme disease was at least 16 times higher in the areas designated at high risk than in the areas designated at low risk.

This procedure was subsequently used to identify and enroll people at high risk for Lyme disease in a trial of a new vaccine to protect against Lyme disease (Steere et al. 1998). If higher-risk populations can be identified, fewer subjects need to be enrolled in a trial to demonstrate the efficacy of a vaccine. Although the crude incidence of Lyme disease in Maryland in 1995 was approximately 9/100,000, the areas selected for the vaccine trial experienced an annual incidence rate of 2,000/100,000 (in the comparison group that did not receive the vaccine). That incidence rate was above the target of 500/100,000 required for the vaccine trial and was sufficient to demonstrate vaccine efficacy.

A major area of research in map overlay and comparison is the integration of aspects of spatial correlation used by kriging (see under "Spatial Interpolation" and "Map Analysis," above) with nonspatial epidemiological techniques. Spatial statistical tools that truly integrate these features are more powerful than the sequential application of spatial (distance) computation and nonspatial epidemiological techniques (such as logistic regression). The latter approach was effective in studies of human cases of Lyme disease because cases were far apart, the infectious agent was not transmitted from person to person, and data on environmental risk factors were readily available for residences. When the focus shifts to data on the abundance of tick vectors on white-tailed deer, it is desirable to analyze environmental influences while taking spatial correlation into account (Das et al. 2000). The main effect of ignoring spatial correlation is to create a false sense of precision because correlated variables are treated as though they were independent (Breslow 1984; Griffith 1987). Recent geographic studies of various health conditions discuss trends in research on spatial methods in epidemiology (Clarke et al. 1996; Thomson et al. 1997; Kitron 1998; *ATSDR 2000;* Glass 2000a; McMichael and Kovats 2000; Briggs et al. 2000; Elliott et al. 2000). Advances in exploratory spatial data analysis will also contribute to the application of spatial statistical techniques (e.g., *Haining and Wise 1997*). Other important areas of research for GISs are integration with tools for analyzing decisions and models that permit more complex dynamics. Chapters 4 and 5 develop these topics within the context of integrated assessment.

The Quality of Data

The quality of data in GISs is fundamental to the quality of analysis that can be performed using those data. Although the quality of data has always been an issue in database development, recent trends have increased the threat of poor quality data (*Veregin 1998*). First, greater access to data over the Internet and other electronic pathways has led to greater reliance on sources of data that were not collected by the investigator. Reliance on other sources of data, of course, can be of great benefit; to realize that benefit, the prospective user must carefully investigate the quality of data. Second, the private sector is producing more data without the strict quality control traditionally enforced by national agencies such as the USGS. Private-sector products are not necessarily a problem, but again the prospective user must be vigilant. Third, the demands on the use of data are growing to include more support for decisions. Data that are adequate for one purpose may simply be inadequate for another. In an actual case involving a map of administrative districts in Britain, the creators of the map placed high priority on thematic accuracy and low priority on positional accuracy, with the result that secondary users were surprised to find some point locations out in the North Sea (*Hunter 1998*). The public and private sectors recognize the need to set standards (see Box 3.3).

Spatial Accuracy and Resolution

Spatial accuracy means that the location encoded for data is correct to some agreed-upon level of precision in the coordinate system in use. Positioning errors can be introduced in many ways—poor addresses in a geographic base file system, an unrectified aerial photograph, or malfunctioning GPS.

Spatial resolution is the amount of spatial detail that can be observed. Spatial resolution is the minimal mapping unit for vector data and the dimensions of the pixel (or cell) for raster data. For example, the spatial resolution of different sources of data from satellites can vary by orders of magnitude, such as a SPOT resolution of 10 meters and AVHRR resolution of 1,100 meters (see Table 3.3).

Temporal Accuracy and Resolution

Temporal accuracy means that the time encoded for data is correct to some agreed-upon level of precision (*Veregin 1998*). Observations on the exact sequence of events are often critical for developing models of causation of disease. For infectious diseases, knowledge about incubation periods and accurate reporting of when symptoms appear may permit the identification of a source of infection. *Temporal resolution* is the minimum duration of an event that can be observed. For example, world surveys on nutrition conducted during 1994–96 were reported for the entire period rather than for a single year (see Table 6.3). Agricultural and meteorological services report many variables by dekad, an interval of 10 days; Thomson et al. (1997) constructed a composite index of vegetation cover by dekad.

Setting Standards for Geographic Information Systems

Setting standards for geographic information systems (GISs) helps to maintain the quality of data and, therefore, the quality of data analysis. Standards specify the accuracy, resolution, completeness, and consistency of data (see text). Standards apply not only to datasets, but also to metadata, which provide descriptive information about datasets. Guidelines for the quality of data and metadata for GISs have some overlap with guidelines for evaluating the quality of information available on the World Wide Web (see Appendix A). If metadata do not clearly explain how a dataset was constructed, the dataset should probably not be used for analysis. The integration of databases from different sources also requires standards for data structures and technological specifications.

In the United States, interest in improving the ability to share geographic data across agencies in the federal government led to the development of the Federal Geographic Data Committee (FGDC) to promote GIS standards (*FGDC 2000a*). Issues for public health data are handled by the FGDC's Subcommittee on Cultural and Demographic Data (*FGDC 2000b*). The European Umbrella Organisation for Geographic Information (EUROGI) coordinates a parallel effort to promote GIS standards in Europe (*EUROGI 2000*). The Open GIS Consortium (OGC), led by commercial vendors of GIS products, also aims to facilitate the sharing of data across systems by developing technical solutions for interoperability (*OGC 2000*). Monitoring the activities of these groups (with information easily accessible on the Internet) is a good way to keep informed about emerging standards in GISs.

Temporal sampling rate refers to the rate at which observations are made, such as 24 frames per second in motion pictures. The time interval between observations, which is the inverse of the temporal sampling rate, is called *temporal resolution* in the field of satellite-based remote sensing (see Table 3.3). Different sources of data from satellites vary by over an order of magnitude in the period of time it takes to repeat observations in an area. AVHRR on NOAA's weather satellites returns twice a day, while Landsat 5 has a scheduled interval of 16 days (see Table 3.3). The actual intervals may be longer in practice if viewing equipment is turned off or cloud cover blocks the view.

Temporal resolution may also refer to how often managers need to update information, especially for applications in remote sensing (Cowen and Jensen 1998). Management requirements and the actual rates of change of geographic conditions determine how current data must be. For example, digital elevation information should be updated once or twice a decade, whereas weather predictions should be updated every 30 minutes.

Thematic Accuracy and Resolution

Thematic accuracy is the correctness of the attribute values in the database at an agreed-upon level of precision (*Veregin 1998*). Quantitative variables like atmospheric concentrations of a pollutant are subject to measurement error from the instrumentation (see Chapter 2). If variables are categorical, such as urban versus rural, then the possible errors are in misclassification. Sensors on satellites add a new dimension to the problem of thematic accuracy because the raw data simply represent the recording of electromagnetic radiation in different wavelengths. Ground-truthing is the procedure for determining how well those recordings may be used to infer land cover, surface temperature, or other variables of interest. Thematic accuracy may also be called *interpretive accuracy*.

Thematic resolution refers to the precision of the measurements of the attributes. For quantitative data, thematic resolution will depend on the measuring device; for categorical data, thematic resolution is determined by the number of categories. The field of remote sensing has terminology specific to sensors of electromagnetic radiation on satellites (*Moody 1996*). Spectral resolution refers to the number, spacing, and width of wavelength bands of electromagnetic energy that a sensor can detect. Radiometric resolution refers to the number of different output levels of radiant energy that can be recorded from a sensor. Directional resolution refers to the range of viewing angles from which the ground can be sampled. The viewing angle affects how objects are observed to reflect electromagnetic radiation, so that more viewing angles on the same object provide more detailed information for identification. In general, greater resolution should lead to greater thematic (interpretive) accuracy.

Consistency and Completeness

Consistency means the absence of contradictions within the database in terms of space, time, or theme (*Veregin 1998*). For example, a person can be in only one place at a given time. *Completeness* means that the database is not missing information. Completeness must be assessed relative to a particular database design because geographic mapping and analysis always involve the selection of information to present (see under "Data Display," above).

Issues in Implementation

The first step in the implementation of a GIS is the specification of system requirements. The problem to be solved should determine the technology required instead of letting the technology define the problem. The specification should consider the six basic functions of GISs (see Table 3.1) in defining exactly how the GIS will be used. For example, the data-capture component might require the GPS, and the data-manipulation component might have to generate buffer zones around the edges of lakes. In research on ecosystem change and public health, the tools for spatial data analysis are a prime consideration (see Table 3.5). The relationship between the selection of an epidemiological study design and the tools in map overlay and comparison is especially important (see Box 3.2). If the system is expected to be in operation for a long time, it is necessary to project how system requirements might change over time.

An estimate of the cost of implementing a system must accompany the process of system specification. One should also consider estimates of the costs of maintenance and future expansion, if appropriate. When the cost of a design far exceeds the resources available, the design must be changed or great energy must be devoted to raising funds. The major considerations are data, computer technology, and human resources. It would be helpful to review other teaching material (e.g., Clarke 1998; Glass 2000b; *Rushton and Armstrong 1997; National Center for Geographic Information and Analysis 2000*) as well as discussion groups, conferences, and workshops on the use of GISs in public health (*Croner 2000; ATSDR 2000*). Contact with other organizations, such as the European Umbrella Organisation for Geographic Information (EUROGI) and the World Health Organization (WHO), may prove useful as well (*EUROGI 2000; WHO 2000*).

Sources of Data

The specification of data quality is central to the design of GISs (see "The Quality of Data," above). Requirements for high levels of accuracy and resolution in space, time, and theme increase the cost. Linking databases from multiple projects may entail additional costs in data conversion to establish compatibility in spatial organization and computer technology. Using databases that adhere to standards developed for GISs should make linkage easier (see Box 3.3). However, improvements in resolution and the technology for linkage have been accompanied by growing concerns about privacy (Rindfuss and Stern 1998). The design of databases must take into account legal and ethical restrictions on the use of data.

A major decision is whether to use existing databases or to generate new databases. The generation of new databases represents the single greatest cost in developing GISs (Aronoff 1989). That cost can be avoided through the use of public and commercial databases in digital format, which have proliferated in number as the field has matured. However, there may be hidden costs in using and converting databases (*Foote and Lynch 1999*). Ascertaining the quality of a database may be time consuming. Conversion may be necessary because the database does not have the exact structure desired or depends upon multiple sources sold by different vendors in a variety of formats.

Investigators need to understand basic geographic properties of datasets. Systems for describing precise geographic positions take into account the flattening of the earth at its poles and irregularities in its surface (*Dana 1997*). Different nations and international agencies use different reference systems. The U.S. Geodetic Survey is responsible for developing the National Spatial Reference System used by the United States (*NOAA*

2000b). Coordinate systems also depend on projections of the surface of the earth onto a flat plane, which distort areas, angles, shapes, distances, and directions (Monmonier 1996). Knowledge about map projections is especially important for studies involving large areas of the globe because all projections distort large (e.g., continental) shapes. Many useful projections are either *conformal,* which means preserving local angles and shapes, or *equal-area,* which means preserving relative sizes of areas. These properties are mutually exclusive; conformal maps severely distort areas and equal-area maps severely distort shapes. Dana (*1999*) compares an Albers equal-area projection and a Lambert conformal conic projection for North America.

Different projections can be tailored for different purposes. For example, the USGS uses the Universal Transverse Mercator (UTM) projection to produce a series of 1:24,000-scale topographic maps for most of the United States and its territories but uses a polar stereographic projection to produce a series of 1:250,000-scale topographic maps in support of the U.S. Antarctic Research Program (*USGS 2000*). The UTM projection is a conformal projection based on zones around 60 equally spaced meridians, and stereographic projections, which are used in polar regions, maintain true directions from a central point (*Dana 1999*).

Another consideration is whether to use data routinely collected by agencies or from special research studies. Routine data collection, as in disease surveillance, has the advantage of providing data over long periods of time, but the quality of the results may vary (see Box 2.5). The use of data from other research studies benefits from close cooperation with the investigators. There is also an inherent tension between the requirements for spatial structure in epidemiological analysis and the requirements for spatial structure in the management functions of agencies that perform routine data collection (Haining 1996). Fragmented areas that may be epidemiologically important are not convenient for administrators. For example, the existence of locations at high risk of filariasis in Egypt was obscured when the case reports were aggregated in large administrative districts (see Chapter 2).

Studies of ecosystem change and public health face an additional challenge in the need to examine data at multiple geographic and temporal scales (see Table 2.2). At a minimum, data specifications appropriate for each scale must be defined. Chapter 5 discusses approaches for addressing problems of multiple scales in the context of integrated assessment.

One good starting point for finding data is the federal government of the United States, especially the USGS, NASA, National Climatic Data Center, U.S. Environmental Protection Agency, U.S. Census Bureau, U.S. Department of Health and Human Services, and the Federal Geographic Data Committee (see Table 3.6 and *Foote and Lynch 1999*). Guidelines for

Table 3.6   Selected Data Sources in the U.S. Federal Government

Agency	Data Sources
USGS (U.S. Geological Survey)	Data in formats of Digital Line Graphs, Digital Elevation Models, and Digital Orthophoto Quadrangles; Landsat imagery at Earth Resources Observation Systems Data Center
NASA (National Aeronautics and Space Administration)	NASA/Goddard Space Flight Center (GSFC) has links to satellite data for Earth Observing System and Pathfinder.[a] NASA/Ames has health applications of remote sensing.[b]
NCDC[c] (National Climatic Data Center)	Climate data from ground-based observations and weather satellites
EPA (Environmental Protection Agency)	Locations of regulated facilities and the toxics release inventory
Census Bureau	Topologically Integrated Geographic Encoding and Referencing System files of census tracts (and street addresses in populous areas) with associated social, demographic, and economic information
HHS[d] (Department of Health and Human Services)	Vital statistics, national health surveys, disease surveillance, cancer registries
FGDC[e] (Federal Geographic Data Committee)	Geospatial data clearinghouse, standards for sharing data, and information on geographic data in the federal government and cooperation with state, local, and tribal governments

Source: Data from *USGS 2000; NASA/GSFC 2000d; NASA/Ames 1999c; National Oceanic and Atmospheric Administration (NOAA) 2000c; EPA 1999; Census Bureau 2000; HHS 2000; FGDC 2000a.*

[a]The Pathfinder program provides access to long time series of large remote-sensing datasets applicable to global change research, such as tropical deforestation.

[b]See Center for Health Applications of Aerospace Related Technologies, which is part of the program on Ecosystem Science and Technology.

[c]NCDC is part of NOAA.

[d]Of special note are Agency of Toxic Substances and Disease Registry; Centers for Disease Control, including the National Center for Health Statistics, Epidemiology Program Office and Office of Global Health; and National Institutes of Health, including the National Cancer Institute with cancer registries in the Surveillance, Epidemiology and End Results Program.

[e]FGDC is an interagency committee.

how to locate and evaluate information on global change may be found in Chapter 1. Many types of data, such as population and health, must be obtained within the country of interest. Detailed information about land holdings and land use may require contact at the very local level. In addition, software vendors have formed partnerships with providers of data, so the search for data should be coordinated with the selection of GIS technology.

Computer Hardware and Software

The system configuration must include the number of users and the number of work stations, the operating system, whether or not the system is networked, and the level of technical support available. How the system connects to the Internet is also becoming more important as GISs are structured to use new forms of distributed processing. The software must be able to handle the functions identified in the specification of system requirements in a manner compatible with the chosen sources of data, such as remote sensing, GPS, and geographic base file maps (see under "Related Technologies," above). Ease of use is also a consideration. Additional information may be obtained from GIS sourcebooks or GIS trade publications, such as GeoInfo Systems, which review products annually. Some software packages that have been used in public health are MapInfo, IDRISI, and ARC/INFO.

Human Resources

Planning for GISs must include the number of potential users as well as their technical background and need for further training. Staff cost is a major factor in the high cost of database development. Facilities should also be made available for training. Some international organizations, such as the United Nations Institute for Training and Research, the Inter-American Institute for Global Change Research, and the WHO, provide support for scientists and public health professionals from developing countries, whose participation is especially important for studies of global change and public health.

Conclusion

GISs have already made important contributions to the field of epidemiology. Continuing advances in technology, especially in sensors on Earth-observing satellites and the computing power to handle and visualize large quantities of data, offer major opportunities to link epidemiological study designs (Chapter 2) with the study of ecosystem change at a global level. This chapter has provided guidelines for implementing GISs with a focus on the problems of global change and public health. GISs should be use-

ful for planning new investigations that contribute to the assessment of
future risks to human health (see Chapters 4 and 5).

SUGGESTED STUDY PROJECTS

Suggested study projects provide a set of options for individual or team
projects that will enhance interactivity and communication among course
participants (see Appendix A). The Resource Center (see Appendix B) and
references in all of the chapters provide starting points for inquiries. The
process of finding and evaluating sources of information should be based
on the principles of information literacy applied to the Internet environment (see Appendix A).

PROJECT 1: GIS-Related Technologies: Remote Sensing (Sensors
on Satellites, Aerial Photography), Global Positioning System,
Geographic Base File Systems

The objective of this project is to demonstrate an understanding of the
contribution of GIS-related technologies to applications of GISs to ecosystem change and public health. It comprises five tasks:

Task 1. Find an example in the literature (not used in this chapter) on
applications of GISs to ecosystem change and public health that uses one
or more of the following GIS-related technologies: sensors on satellites,
aerial photography, global positioning system, geographic base file systems.

Task 2. Describe the objective of the application.

Task 3. Describe the data used in the application.

Task 4. Relate the analytical methods used in the application to the
framework of methods for spatial data analysis in the chapter (see Table
3.5).

Task 5. Summarize your results.

PROJECT 2: Data Quality

The objective of this project is to practice evaluating the quality of data and
metadata used in applications of GISs to ecosystem change and public
health. It comprises four tasks:

Task 1. Find an example of a dataset relevant to ecosystem change and
public health.

Task 2. Evaluate the quality of data and metadata using the criteria presented in the chapter.

Task 3. Consider how the quality of the data and metadata could be
improved.

Task 4. Summarize your results.

PROJECT 3: Project Design for GIS Applications to Ecosystem Change and Public Health

The objective of this project is to develop, practice, and refine skills for designing a GIS application for ecosystem change and public health. It comprises one integrated task:

Task. Design and present an example for applying GISs to ecosystem change and public health. Take into account issues of implementation, data quality, and analysis described in the text.

Acknowledgments

We thank Chuck Croner of the National Center for Health Statistics and Bill Henriques of the Agency for Toxic Substances and Disease Registry for suggesting references on geographic information systems. We also thank Byron Wood of the NASA/Ames Research Center and Carla Evans and Blanche Meeson of the NASA/Goddard Space Flight Center for providing information on remote sensing. The black-and-white images in Figure 3.3 were prepared by Bradley M. Lobitz of Johnson Control World Services at the NASA/Ames Research Center based on a research study conducted by Louisa R. Beck of California State University/Monterey Bay, who prepared the accompanying text in the legend. The maps of malaria in Figure 3.4 are redrawn from a figure prepared by Saskia Nijhof of the International Centre for Integrative Studies in Maastricht, Netherlands, using data from the National Health Foundation in Rondônia, Brazil. We also thank Mitch Hobish, Steve Connor, and Phil Arkin for reviewing the manuscript.

References

Anselin L. 1996. The Moran scatterplot as an ESDA tool to assess local instability in spatial association. In *Spatial Analytical Perspectives on GIS* (Fischer M, Scholten HJ, Unwin D, eds.). Taylor & Francis, London, Chap. 8.

Aronoff S. 1989. *Geographic Information Systems: A Management Perspective.* WDL Publications, Ottawa.

Beck LR, Lobitz BM, Wood BL. 2000. Remote sensing and human health: New sensors and new opportunities. *Emerg Infect Dis* 6 (3): 217–27.

Beck LR, Rodriguez MH, Dister SW, Rodriguez AD, Rejmánková E, Ulloa A, Meza RA, Roberts DR, Paris JF, Spanner MA, Washino RK, Hacker C, Legters LJ. 1994. Remote sensing as a landscape epidemiologic tool to identify villages at high risk for malaria transmission. *Am J Trop Med Hyg* 51: 271–80.

Beck LR, Rodriguez MH, Dister SW, Rodriguez AD, Washino RK, Roberts DR, Spanner MA. 1997. Assessment of a remote sensing-based model for predicting malaria transmission risk in villages of Chiapas, Mexico. *Am J Trop Med Hyg* 56: 99–106.

Breslow N. 1984. Extra-Poisson variation in log-linear models. *Appl Stat* 33: 38–44.

Briggs DJ. 1997. Using geographical information systems to link environment and health data. In *Linkage Methods for Environment and Health Analysis: Technical Guidelines* (Corvalan C, Nurminen M, Pastides H, eds.). Report WHO/EHG/97.11 of the Health and Environment Analysis for Decision-Making (HEADLAMP) Project, Office of Global and Integrated Environmental Health. World Health Organization, Geneva, Chap. 5.

Briggs DJ, de Hoogh C, Gulliver J, Wills J, Elliott P, Kingham S, Smallbone K. 2000. A regression-based method for mapping traffic-related air pollution: Application and testing in four contrasting urban environments. *Sci Total Environ* 253 (1–3, May 15): 151–67.

Briggs DJ, Elliott P. 1995. The use of geographical information systems in studies on environment and health. *World Health Stat Q* 48: 85–94.

Burrough PA, McDonnell R. 1998. *Principles of Geographical Information Systems.* Oxford University Press, Oxford.

Clarke KC. 1998. *Getting Started with Geographic Information Systems,* 2d ed. Prentice Hall, Upper Saddle River, N.J.

Clarke KC, McLafferty SL, Tempalski BJ. 1996. On epidemiology and geographic information systems: A review and discussion of future directions. *Emerg Infect Dis* 2 (2): 85–92.

Conway ED. 1997. *An Introduction to Satellite Image Interpretation.* Johns Hopkins University Press, Baltimore.

Cowen DJ, Jensen JR. 1998. Extraction and modeling of urban attributes using remote sensing technology. In *People and Pixels: Linking Remote Sensing and Social Science* (Liverman D, Moran EF, Rindfuss RR, Stern PC, eds.). National Academy Press, Washington, Chap. 8.

Das A, Lele SR, Glass GE, Shields T, Patz JA. 2000. Spatial modeling for discrete data using generalized linear mixed models. In *Accuracy 2000* (Heuvelink GBM, Lemmens MJPM, eds.). Delft University Press, Delft, The Netherlands, pp. 125–33.

De Savigny D, Wijeyaratne P. 1995. *GIS for Health and the Environment.* International Development Research Centre, Ottawa.

Elliott P, Wakefield JC, Best NG, Briggs DJ. 2000. *Spatial Epidemiology: Methods and Applications.* Oxford University Press, New York.

Epstein PR. 1998. Health applications of remote sensing and climate modeling. In *People and Pixels: Linking Remote Sensing and Social Science* (Liverman D, Moran EF, Rindfuss RR, Stern PC, eds.). National Academy Press, Washington, Chap. 10.

Glass GE. 2000a. Spatial aspects of epidemiology: The interface with medical geography. *Epidemiol Rev* 22 (1): 136–39.

———. 2000b. Geographic information systems. In *Infectious Disease Epidemiology: Theory and Practice* (Nelson KE, Williams CM, Graham NMH, eds.). Aspen Publishers, Gaithersburg, Md., Chap. 9.

Glass GE, Cheek JE, Patz JA, Shields TM, Doyle TJ, Thoroughman DA, Hunt DK, Enscore RE, Gage KL, Irland C, Peters CJ, Bryan R. 2000. Using remotely sensed data to identify areas at risk for hantavirus pulmonary syndrome. *Emerg Infect Dis* 6 (3): 238–47.

Glass GE, Livingstone W, Mills JN, Hlady WG, Fine JB, Biggler W, Coke T, Frazier D, Atherley S, Rollin PE, Ksiazek TG, Peters CJ, Childs JE. 1998. Black Creek Canal Virus infection in *Sigmodon hispidus* in southern Florida. *Am J Trop Med Hyg* 59 (5): 699–703.

Glass GE, Schwartz BS, Morgan JM, Johnson DT, McNoy P, Israel E. 1995. Environmental factors for Lyme disease identified with geographic information systems. *Am J Public Health* 85: 944–48.

Griffith DA. 1987. *Spatial Autocorrelation: A Primer.* Association of American Geographers, Washington, D.C.

Haining R. 1996. Designing a health needs GIS with spatial analysis capability. In *Spatial Analytical Perspectives on GIS* (Fischer M, Scholten HJ, Unwin D, eds.). Taylor & Francis, London, Chap. 4.

Hightower AW, Ombok M, Otieno R, Odhiambo R, Oloo AJ, Lal AA, Nahlen BL, Hawley

WA. 1998. A geographic information system applied to a malaria field study in western Kenya. *Am J Trop Med Hyg* 58 (3): 266–72.

Hunter JM, Arbona SI. 1995. Paradise lost: An introduction to the geography of water pollution in Puerto Rico. *Soc Sci Med* 40 (10): 1331–55.

Hutchinson CF. 1998. Social science and remote sensing in famine early warning. In *People and Pixels: Linking Remote Sensing and Social Science* (Liverman D, Moran EF, Rindfuss RR, Stern PC, eds.). National Academy Press, Washington, Chap. 9.

Jacob P, Goulko G, Heidenreich WF, Likhtarev I, Kairo I, Tronko ND, Bogdanova TI, Kenigsberg J, Buglova E, Drozdovitch V, Golovneva A, Demidchik EP, Balonov M, Zvonova I, Beral V. 1998. Thyroid cancer risk to children calculated. *Nature* 392: 31–32.

Jensen JR. 1996. *Introductory Digital Image Processing: A Remote Sensing Perspective.* Prentice Hall, Upper Saddle River, N.J.

Kitron U. 1998. Landscape ecology and epidemiology of vector-borne diseases: Tools for spatial analysis. *J Med Entomol* 35 (4): 435–45.

Maxcy KF. 1926. An epidemiological study of endemic typhus (Brill's disease) in the southeastern United States with special reference to its mode of transmission. *Public Health Rep* 41: 2967–95.

McMichael AJ, Kovats RS. 2000. Strategies for assessing health impacts of global environmental change. In *Implementing Ecological Integrity: Restoring Regional and Global Environmental and Human Health* (Crabbe P, Westra L, eds.). Kluwer Academic Publishers, Dordrecht, pp. 217–31.

Monmonier M. 1996. *How to Lie with Maps,* 2d ed. University of Chicago Press, Chicago.

Morain SA. 1998. A brief history of remote sensing applications, with emphasis on Landsat. In *People and Pixels: Linking Remote Sensing and Social Science* (Liverman D, Moran EF, Rindfuss RR, Stern PC, eds.). National Academy Press, Washington, Chap. 2.

Moran EF, Brondizio E. 1998. Land-use change after deforestation in Amazonia. In *People and Pixels: Linking Remote Sensing and Social Science* (Liverman D, Moran EF, Rindfuss RR, Stern PC, eds.). National Academy Press, Washington, Chap. 5.

Morrison AC, Getis A, Santiago M, Rigan-Perez JG, Reiter P. 1998. Exploratory space-time analysis of reported dengue cases during an outbreak in Florida, Puerto Rico, 1991–1992. *Am J Trop Med Hyg* 58 (3): 287–98.

Nobre FF, Carvalho MS. 1995. Spatial and temporal analysis of epidemiological data. In *GIS for Health and the Environment* (De Savigny D, Wijeyaratne P, eds.). International Development Research Centre, Ottawa, pp. 21–31.

Oliver MA. 1996. Geostatistics, rare disease and the environment. In *Spatial Analytical Perspectives on GIS* (Fischer M, Scholten HJ, Unwin D, eds.). Taylor & Francis, London, Chap. 5.

Patterson M. 1998. Glossary. In *People and Pixels: Linking Remote Sensing and Social Science* (Liverman D, Moran EF, Rindfuss RR, Stern PC, eds.). National Academy Press, Washington, Appendix B.

Pickle LW, Herrmann DJ, eds. 1995. *Cognitive Aspects of Statistical Mapping.* Cognitive Methods Staff Working Paper Series, No. 18. NCHS Office of Research and Methodology, Hyattsville, Md.

Pyle GF. 1979. *Applied Medical Geography.* V. H. Winston & Sons, Washington, D.C.

Rasool SI. 1999. Scientific responsibility in global climate change research. *Science* 283 (5404): 940–41.

Rindfuss RR, Stern PC. 1998. Linking remote sensing and social science: The need and the challenges. In *People and Pixels: Linking Remote Sensing and Social Science* (Liverman D, Moran EF, Rindfuss RR, Stern PC, eds.). National Academy Press, Washington, Chap. 1.

Steere AC, Sikand VK, Meurice F, Parenti DL, Fikrig E, Schoen RT, Nowakowski J, Schmid CH, Laukamp S, Buscarino C, Krause DS. 1998. Vaccination against Lyme disease with recombinant *Borrelia burgdorferi* outer-surface lipoprotein A with adjuvant. *N Engl J Med* 339 (4): 209–15.

Tevini M, ed. 1993. *UV-B Radiation and Ozone Depletion: Effects on Humans, Animals, Plants, Microorganisms, and Materials.* Lewis Pubs., Boca Raton, Fla., p. iii.

Thomson MC, Connor SJ, Milligan P, Flasse SP. 1997. Mapping malaria risk in Africa: What can satellite data contribute? *Parasitol Today* 13 (8): 313–18.

Tufte ER. 1983. *The Visual Display of Quantitative Information.* Graphics Press, Cheshire, Conn.

———. 1990. *Envisioning Information.* Graphics Press, Cheshire, Conn.

———. 1997. *Visual Explanations: Images and Quantities, Evidence and Narrative.* Graphics Press, Cheshire, Conn.

Vine MF, Degnan D, Hanchette C. 1997. Geographic information systems: Their use in environmental epidemiologic research. *Environ Health Perspect* 105: 598–605.

Wood CH, Skole D. 1998. Linking satellite, census, and survey data to study deforestation in the Brazilian Amazon. In *People and Pixels: Linking Remote Sensing and Social Science* (Liverman D, Moran EF, Rindfuss RR, Stern PC, eds.). National Academy Press, Washington, Chap. 4.

Woodward TE. 1970. President's address: Typhus verdict in American history. *Trans Am Clin Climatol Assoc* 82: 1–8.

World Bank. 1996. *Geographic Information Systems for Environmental Assessment and Review.* Environmental Assessment Sourcebook Update No. 3. Environment Department, World Bank, Washington, D.C.

Electronic References

Agency for Toxic Substances and Disease Registry. 2000. Proceedings of the 1998 GIS in Public Health Conference. http://www.atsdr.cdc.gov/GIS/conference98/proceedings/proceedings.html (Date Last Revised 5/18/2000).

Census Bureau. 2000. U.S. Census Bureau. U.S. Department of Commerce. http://www.census.gov (Date Last Revised 3/28/2000).

Center for International Earth Science Information Network. 2000. UV Interactive Service. http://www.ciesin.org (Date Last Revised 1/31/2000).

Cowen DJ. 1997. Unit 016—Discrete Georeferencing. NCGIA Core Curriculum in GIScience. http://www.ncgia.ucsb.edu/giscc/units/u016/u016.html (Date Last Revised 12/18/1997).

Croner CM, ed. 2000. *Public Health GIS News and Information, 1994-Present.* National Center for Health Statistics, Centers for Disease Control and Prevention. http://www.cdc.gov/nchs/about/otheract/gis/gis_publichealthinfo.htm (Date Last Revised 7/13/2000).

Dana PH. 1997. Unit 013—Coordinate Systems Overview. NCGIA Core Curriculum in GIScience. http://www.ncgia.ucsb.edu/giscc/units/u013/u013.html (Date Last Revised 7/4/1997).

———. 1999. Map Projection Overview. The Geographer's Craft Project, Department of Geography, University of Texas at Austin. http://www.utexas.edu/depts/grg/gcraft/notes/mapproj/mapproj.html (Date Last Revised 12/15/1999).

———. 2000. Global Positioning System Overview. The Geographer's Craft Project, Department of Geography, University of Texas at Austin. http://www.utexas.edu/depts/grg/gcraft/notes/gps/gps.html (Date Last Revised 3/30/2000).

Environmental Protection Agency. 1999. United States Environmental Protection Agency Geospatial Data Clearinghouse. http://www.epa.gov/nsdi (Date Last Revised 3/30/1999).

Estes JE, McGwire KC, Kline KD, eds. 1999. Air Photo Interpretation and Photogrammetry. Remote Sensing Core Curriculum, Vol. 1. http://umbc7.umbc.edu/tbenja1/santabar/rscc.html (Date Last Revised 8/10/1999).

European Umbrella Organisation for Geographic Information. 2000. European Umbrella Organisation for Geographic Information. http://www.eurogi.org (Date Last Revised 2/24/2000).

Federal Geographic Data Committee. 2000a. Federal Geographic Data Committee. http://www.fgdc.gov (Date Last Revised 3/28/2000).

———. 2000b. Subcommittee on Cultural and Demographic Data. Federal Geographic Data Committee. http://www.census.gov/geo/www/standards/scdd (Date Last Revised 1/20/2000).

Foote KE, Lynch M. 1999. Data Sources for GIS. The Geographer's Craft Project, Department of Geography, University of Texas at Austin. http://www.utexas.edu/depts/grg/gcraft/notes/sources/sources.html (Date Last Revised 10/14/1999).

Goodchild MF. 1997. Unit 057—Quadtrees and Scan Orders. NCGIA Core Curriculum. http://www.ncgia.ucsb.edu/giscc/units/u057 (Date Last Revised 10/23/1997).

Haining R, Wise S. 1997. Exploratory Spatial Data Analysis. NCGIA Core Curriculum in GIScience. http://www.ncgia.ucsb.edu/giscc/units/u128/u128.html (Date Last Revised 12/5/1997).

Health and Human Services. 2000. Department of Health and Human Services. http://www.hhs.gov/ (Date Last Revised 3/20/2000).

Hobish MK. 1999. Earth System Science, Mission to Planet Earth, and the Earth Observing System. In The Remote Sensing Tutorial, Section 16. http://rst.gsfc.nasa.gov (Date Last Revised 12/1999).

Hunter GJ. 1998. Managing Uncertainty in GIS. In NCGIA Core Curriculum in GIScience. http://www.ncgia.ucsb.edu/giscc/units/u187/u187.html (Date Last Revised 2/3/1998).

Interagency GPS Executive Board. 2000. Interagency GPS Executive Board Home Page. http://www.igeb.gov (Date Last Revised 5/8/2000).

Inter-American Institute for Global Change Research. 2000. IAI. Inter-American Institute for Global Change Research. http://www.iai.int (Date Last Revised 3/16/2000).

Jensen JR, Schill SR, eds. 1998. Introductory Digital Image Processing. Remote Sensing Core Curriculum, Vol. 3. http://www.cla.sc.edu/geog/rslab/rsccnew/index.html (Date Last Revised 5/28/1998).

Meyer TH. 1997. Non-spatial Database Models. NCGIA Core Curriculum in GIScience. http://www.ncgia.ucsb.edu/giscc/units/u045 (Date Last Revised 11/19/1997).

Moody A. 1996. Concepts of Spatial, Spectral, Temporal, and Radiometric Resolution. In Remote Sensing Core Curriculum, Vol. 2 (N Faust, ed.). http://grouchy.geog.ucsb.edu/rscc/vol2/rsccvol2.html (Date Last Revised 7/30/1996).

National Aeronautics and Space Administration/Ames. 1998a. CHAART. Sensor Specifications: Meteosat. http://geo.arc.nasa.gov/sge/health/sensor/sensors/meteosat.html (Date Last Revised 10/27/1998).

———. 1998b. CHAART. Sensor Evaluation. Introduction. http://geo.arc.nasa.gov/sge/health/sensor/sensor.html (Date Last Revised 10/30/1998).

———. 1999a. CHAART. Remote Sensing/GIS and Human Health: A Partial Bibliography. http://geo.arc.nasa.gov/sge/health/rsgisbib.html (Date Last Revised 11/3/1999).

———. 1999b. Malaria in Chiapas, Mexico. http://geo.arc.nasa.gov/sge/health/gmhh/mexico.html (Date Last Revised 1/22/1999).

———. 1999c. Center for Health Applications of Aerospace Related Technologies. http://geo.arc.nasa.gov/sge/health/chaart.html (Date Last Revised 4/19/1999).

———. 2000. Numerical Aerospace Simulation Facility. Scientific Visualization Sites. http://science.nas.nasa.gov/Groups/VisTech/visWeblets.html (Date Last Revised 3/14/2000).

National Aeronautics and Space Administration/Goddard Space Flight Center. 1999. Ocean Color Data & Resources. Data Set Readme for the Sea-viewing Wide Field-of-View Sensor (SeaWiFS). http://daac.gsfc.nasa.gov/CAMPAIGN_DOCS/OCDST/seawifs_readme.html (Date Last Revised 11/22/1999).

———. 2000a. Ocean Color Data & Resources. Recently Asked Questions about SeaWiFS and Ocean Color. http://daac.gsfc.nasa.gov/CAMPAIGN_DOCS/OCDST/seawifs_raq.html (Date Last Revised 1/12/2000).

———. 2000b. Coastal Zone Color Scanner (CZCS) Instrument Guide. http://eosdata.gsfc.nasa.gov/SENSOR_DOCS/CZCS_Sensor.html (Date Last Revised 2/22/2000).

———. 2000c. Tropical Rainfall Measuring Mission. http://trmm.gsfc.nasa.gov/ (Date Last Revised 3/28/2000).

———. 2000d. GSFC Earth Sciences (GES). Distributed Active Archive Center. http://daac.gsfc.nasa.gov/ (Date Last Revised 3/20/2000).

———. 2000e. NASA's Earth Observing System. http://eospso.gsfc.nasa.gov (Date Last Revised 3/22/2000).

National Center for Geographic Information and Analysis. 2000. National Center for Geographic Information & Analysis. http://www.ncgia.ucsb.edu/ (Date Last Revised 3/30/2000).

National Oceanic and Atmospheric Administration. 2000a. Global Precipitation Climatology Project (GPCP) Home Page. http://orbit-net.nesdis.noaa.gov/arad/gpcp/ (Date Last Revised 2000).

———. 2000b. National Geodetic Survey. http://www.ngs.noaa.gov (Date Last Revised 3/28/2000).

———. 2000c. National Climatic Data Center. http://www.ncdc.noaa.gov (Date Last Revised 3/2/2000).

Open GIS Consortium. 2000. Open GIS Consortium, Inc. http://www.opengis.org (Date Last Revised 3/23/2000).

Rushton G, Armstrong MP. 1997. Improving Public Health through Geographical Information Systems: An Instructional Guide to Major Concepts and Their Implementation. University of Iowa. http://www.uiowa.edu/geog/health/ (Date Last Revised 12/1997).

Short NM. 1999. The Remote Sensing Tutorial. http://rst.gsfc.nasa.gov (Date Last Revised 12/1999).

Stoney WE. 1999. Remote Sensing into the 21st Century. In The Remote Sensing Tutorial, Section 20. http://rst.gsfc.nasa.gov (Date Last Revised 12/1999).

Taylor G. 1997. Multimedia and Virtual Reality. In NCGIA Core Curriculum in GI-Science. http://www.ncgia.ucsb.edu/giscc/units/u131/u131.html (Date Last Revised 12/17/1997).

U.S. Geological Survey. 2000. USGS Geospatial Data Clearinghouse National Mapping and Remotely Sensed Data. http://mapping.usgs.gov/nsdi (Date Last Revised 2/22/2000).

Veregin H. 1998. Data Quality Measurement and Assessment. NCGIA Core Curriculum in GIScience. http://www.ncgia.ucsb.edu/giscc/units/u100/u100.html (Date Last Revised 3/23/1998).

World Health Organization. 2000. World Health Organization/Organisation Mondiale de la Santé. http://www.who.int/ (Date Last Revised 3/31/2000).

CHAPTER 4

The Science/Policy Interface

Jonathan M. Samet, M.D., M.S.

In confronting and controlling the consequences of global ecosystem change for public health, we face a problem of nearly unprecedented size and complexity (see Fig. 1.1 and case studies in Chapters 11, 12, 13, and 14). A long time scale and global scope challenge our capabilities for characterizing the problem, and any characterization of the adverse consequences will inevitably be subject to numerous and possibly crippling uncertainties (Last 1993; McMichael 1993). Of necessity, much of the evidence used in policy decisions will be based on a conceptual model of relationships between driving factors and outcomes, as many of the paths from changes in the global ecosystem to public health consequences are on time frames longer than the spans of most conventional observational research. Yet concern for degradation of the public's health may be one of the key motivations for limiting the pace of global ecosystem change, and estimates of effects on public health at the global and more local levels will be made to motivate the implementation of control strategies and to guide their development.

This chapter provides a perspective on the interfaces between scientific evidence and policy development that may serve as a point of departure for assessing global ecosystem change and its public health consequences. The focus is on an evaluation of approaches currently used for translation at these interfaces in the area of environmental health. Emphasis is placed on quantitative risk assessment, a tool for assembling scientific evidence on risk for the purpose of policy development. Some analogies between more conventional environmental health issues and global ecosystem change are suggested, setting the stage for a broader discussion of integrated assessment in Chapter 5.

Science/Policy Interfaces

This chapter focuses on the intersection of scientific evidence with policy development, the science/policy interface. Looking across the myriad of such interfaces, we find a wide range, extending from decisions made by individuals based on their own interpretations of evidence to national and now global problems that need science-based decisions. At the level of nations, there are several specified approaches for amassing and evaluating the relevant evidence for policy implications. In the United States, specific legislative acts even mandate the application of risk assessment for problem characterization (National Research Council [NRC] 1994; Samet and Burke 1998).

Environmental pollution offers a tangible example. Air pollution, for instance, is an almost unavoidable consequence of power generation, transportation, and manufacturing. Depending on the sources, control strategies enacted at the local level may not suffice, and regional or national programs may be needed to protect air quality. The sequence from source of pollution to adverse health effect (Fig. 4.1) provides a framework for identifying the evidence base needed to formulate policy for the control of air pollution (NRC 1991a, 1998a). In this sequence, the implementation of policy would be motivated by the demonstration of adverse effects linked to exposure. Confidence in the evidence of adverse effects would be bolstered by an understanding of the toxicologic basis of the adverse effects, including the delivery of doses of the toxic material to target sites in the body and the mechanism of injury. For the purpose of developing control strategies, the link from sources to population exposure also needs to be described.

In the example of air pollution, the development of policy in the United States proceeds at national, state, and local levels. At the national level, the Clean Air Act specifies the approaches to be followed in translating from science to policy (*Environmental Protection Agency [EPA] 1999a*). For one series of prevalent pollutants (particles, carbon monoxide, sulfur dioxide, nitrogen dioxide, ground-level ozone, and lead), the act mandates a comprehensive review with preparation of a "criteria document," leading to the designation of this group of pollutants as "criteria" pollutants. The EPA, which holds regulatory authority over air pollution, prepares the criteria document and then uses the evidence in a now clearly scripted sequence of steps (Fig. 4.2). The EPA considers policy options within a framework set by the evidence in the criteria document. The act offers guidance to policymakers on the degree of risk to be allowed; for the criteria pollutants, the act specifies that protection will be afforded against adverse effects within "an adequate margin of safety." The final standard for

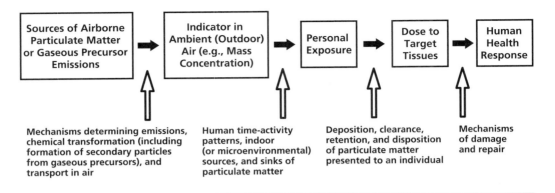

Figure 4.1 General framework for integrating particulate-matter research. *Source:* Reproduced with permission from National Research Council 1998a, Figure 3.1.

a pollutant is to reflect not only the health risk posed by the air pollutant, but also feasibility of control. The language of the Clean Air Act prohibits primary consideration of costs, but the EPA does conduct cost-benefit and cost-effectiveness analyses.

Another group of pollutants, the air toxics, is also covered by the Clean Air Act, but with different guidance. These pollutants are largely carcinogens or irritants, and come mostly, although not necessarily, from point sources, like power plants and factories. Examples include vinyl chloride, dioxin, and benzene. The 1990 amendments to the act offer explicit guidance to the EPA on the use of scientific evidence in approaching regulation of toxic air pollutants, calling for a process that uses risk assessment rather than the weight-of-evidence approach used for the criteria pollutants. A two-step approach of "technology first, then risk," which requires the maximum achievable reduction in toxic emissions followed by assessment of health and environmental risks, has reduced toxic emissions severalfold (*EPA 1999b*).

This particular example of a single science/policy interface, while lacking generality, does point to some characteristics of a science/policy interface for which scientific evidence is at the core of the process that links science to policy (Fig. 4.3). In the example of the regulatory apparatus for air pollution, the findings on adverse health effects motivate the process and the prevention of those effects is the goal of the policy. Translation of the scientific findings to policy depends typically on a synthesizing process that is generally reductionistic. For many complex problems, a predictive model is derived from the evidence, so that the consequences of policy options can be weighed and evaluated against criteria that might include societal values and ethics, costs, and political considerations. In the example

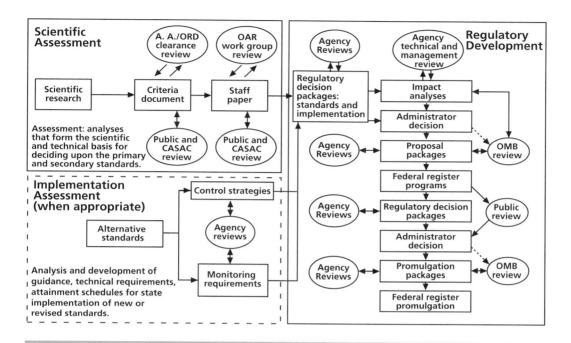

Figure 4.2 Process for review and promulgation of U.S. National Ambient Air Quality Standards. *A.A./ORD,* Assistant Administrator, Office of Research and Development; *OAR,* Office of Air and Radiation; *CASAC,* Clean Air Scientific Advisory Committee; *OMB,* Office of Management and Budget. *Source:* Reproduced with permission from McClellan and Miller 1997, Figure 1.

of the Clean Air Act in the United States, specific guidance is provided on the level of public health protection to be achieved by a national ambient air quality standard (NAAQS). The act also inveighs against consideration of costs to meet the NAAQS.

Decision makers will also be influenced by the strength of the evidence and by uncertainty, which is its complement. Gaps in the scientific evidence inevitably plague decision making, particularly for complex environmental problems. Uncertainty—what is not known—may restrain decision making while new scientific studies are undertaken to meet critical information needs (Morgan and Henrion 1990). Uncertainties may also cloud the choices among policy options, leaving decision makers without a clear preference. Nonetheless, policy decisions can be made in the face of uncertainty; in fact, complete certainty is illusory for contemporary, complex problems involving the environment and health. Rather, uncertainties need to be acknowledged and quantified to the extent possible. Methods are now evolving for this purpose.

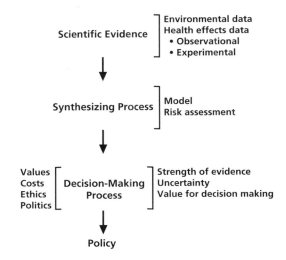

Figure 4.3 Science/policy interface for environmental health.

Models and the Science/Policy Interface

Models are likely to figure centrally in the development of policies to limit possible adverse effects of global ecosystem change on public health. Of course, the word *model* has multiple meanings, and it is widely and loosely applied. In the context of environmental decision making, a model is a representation of the elements of the problem under consideration. A model might be a straightforward depiction of the relationship between a factor and the response, such as the essentially linear (straight-line) relationship between the number of cigarettes smoked and the risk of developing lung cancer, or a formulation of a more complex relationship, like the steps linking sources of regional air pollution to hospitalizations of children due to the exacerbation of asthma.

For problems related to the environment and health, models generally describe the occurrence of a response to one or more environmental factors. Figure 4.1 is one example of a model of this type, relating sources of air pollution to the risk of adverse health effects through a sequence of linkages. Each element of the overall model may itself be considered as a model: the emissions of pollutants from sources, the generation of concentrations of air pollutants by the emissions, the interactions of people with these concentrations that lead to human exposure, the delivery of doses of toxic materials consequent to these exposures to target sites of biological action, and the mechanisms of injury by the agent that leads to the toxic effect.

Any one of the elements in this sequence may itself demand representation by a complex model. Large-scale models, for example, have been developed to describe the generation of acidic air pollution and its dispersal across the northeastern United States and Canada. Models for ozone in the lower atmosphere (troposphere) relate concentrations to the use of sources, including transportation vehicles, and meteorological patterns. Models for the delivery of doses of toxic agents to the lung incorporate the physical characteristics of the agents, the structure of the lung, and patterns of breathing; such models are now widely available as software packages for personal computers.

How are models for environmental problems derived? Models may be empirical, based on observational data that describe how the response of interest occurs in relation to the environmental exposure of concern. Epidemiological data, derived from observation of populations, may be analyzed for this purpose. For example, the application of time-series methods for statistical analysis has related concentrations of air pollutants to the daily number of deaths, while controlling for weather (Dockery and Pope 1994; Samet et al. 1997). These analyses make assumptions as to the mathematical form of the relationship between air pollution and mortality—for example, linear (straight-line)—and the temporal relationships between concentration on a particular day and response on the same or subsequent days. The accuracy and precision of the estimates of the response based on the model reflect the extent and quality of the data available as well as the accuracy and precision of the assumptions in the model. Once developed, a model of the relationship between air pollution and mortality might be used to estimate risks for entire affected populations or to project the consequences of alternative control strategies.

Models may also be based on an understanding of the underlying relationship between exposure to an agent and the response. For ionizing radiation, for instance, physical models, well grounded in experiments, relate exposure to absorption, even at the level of specific organs, and the scientific community is moving toward a mechanistic understanding of the basis of carcinogenesis by some types of radiation (NRC 1998b). Statistical approaches for analyzing observational data may also rely on biologically based models of the underlying process (Moolgavkar 1994).

As a formal representation of the relationship between exposure and outcome, a model offers a framework for making underlying assumptions transparent and for identifying points of uncertainty that need to be judged by one or more assumptions. A model can thus be the basis for fully informed decision making, identifying clearly what is known and not known. A comprehensive approach to modeling risk, by clearly defining gaps in knowledge, can foster research with findings likely to be relevant to the development of policy. For example, the NRC has used the framework in Figure 4.1 as a basis for developing a policy-relevant agenda for research on airborne particulate matter (NRC 1998a). The schema in Figure 4.1 was used to identify key uncertainties that needed to be addressed through further research. Frameworks for identifying gaps in knowledge can also help to characterize the role of uncertainty in decision making (see under "Quantitative Risk Assessment," below).

Some of the assumptions of a model may carry significant implications for policy development. Among these is the mathematical form of the relationship between exposure and response (Fig. 4.4) and the presence of a *threshold,* which is a particular level of exposure that must be exceeded for a response to occur. A common assumption is a linear relationship between the amount of exposure to an agent and the risk of an adverse health effect. The assumption of linearity offers conceptual and computational simplicity as risk increases in a straight-line fashion with exposure; for most agents, the knowledge base is not sufficiently advanced to challenge this assumption. Most health models also assume that the response follows exposure without a threshold; these are the so-called nonthreshold models. This assumption is considered a "conservative" default with regard to public health, as any level of exposure is projected to carry risk.

Alternatives to linear exposure-response relationships may be found or assumed on a biological basis. Response may follow exposure in a curvilinear or other pattern; these curves would possibly indicate a risk of an adverse health effect substantially above or below that predicted by a linear relationship. For some exposures, like ionizing radiation, the response—risk of cancer—tends to flatten at higher exposures, as cells are actually killed by the radiation rather than undergoing a change that may lead to malignancy (NRC 1992). The finding of a threshold that must be exceeded may have substantial policy implications, as the threshold would set a boundary between levels of exposure carrying or not carrying risk.

Another key feature of a model may be its representation of the effects of the agent of interest as it acts in concert with other factors to affect risk. The other factors might include age, sex, genetically determined susceptibility, and other exogenous factors, such as cigarette smoking. In considering the risk associated with exposure to two factors, researchers might assume the combined effect to be additive; that is, the risk from the combined exposure is the sum of the risks from exposures to the two factors individually. Deviation from additivity may be observed: *antagonism* refers to a combined risk that is less than additive, and *synergism* to a combined risk greater than additive. The assumption of a model as to the pattern of combined effects might be based on observation or on a mechanistic understanding of how the two factors act together to produce the response.

The focus of this discussion has been on models for relating health responses to environmental agents. To date, the purposes of policymaking have been served successfully by relatively straightforward models. Typically, the concept of uncertainty applies to the risk of expected kinds of ad-

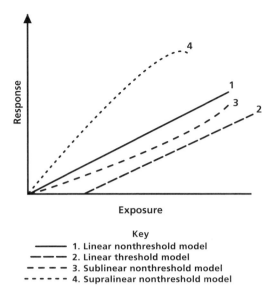

Key
1. Linear nonthreshold model
2. Linear threshold model
3. Sublinear nonthreshold model
4. Supralinear nonthreshold model

Figure 4.4   Examples of exposure-response models used for carcinogens.

verse effects. However, the consequences of ecosystem change at the global level may require more complex representations to capture consequences that extend from the local to the global level. The underlying relationships may be nonlinear, and modeling approaches based on chaos and complexity theory may prove valuable for representing the consequences of change in the global system. These types of models can more flexibly represent the seemingly unpredictable behavior of complex systems than can more restraining, alternative approaches. The analysis of models of global change requires the concept of "surprise" for the possibility of unexpected outcomes (*Schneider and Turner 1997*). See Chapter 5 for more on modeling of complex systems.

Quantitative Risk Assessment

One approach for using scientific information to develop policies on environmental problems is quantitative risk assessment. The application of quantitative risk assessment to environmental problems is increasing in the United States and elsewhere, in some instances with a statutory requirement (NRC 1983, 1994). Quantitative risk assessment is a framework for organizing information about risk and characterizing the risk in a fashion that is meaningful for decision making (Samet and Burke 1998).

Risk assessment is a term now used widely for a systematic approach to characterizing the risks posed to individuals and populations by environmental pollutants and other potentially adverse exposures. A seminal 1983 NRC report, *Risk Assessment in the Federal Government: Managing the Process* (often referred to as the "Red Book" because of its cover), defined

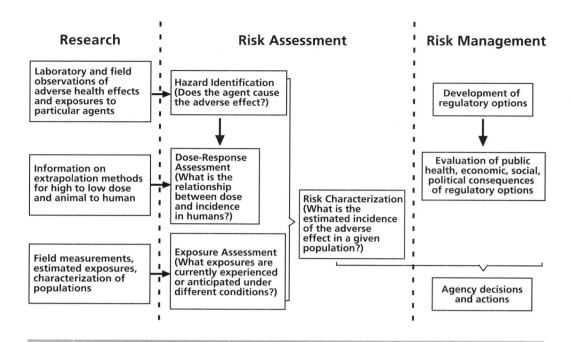

Figure 4.5 Elements of risk assessment and risk management. *Source:* Reproduced with permission from National Research Council 1983, Figure I.1.

risk assessment as "the use of the factual base to define the health effects of exposure of individuals or populations to hazardous materials and situations" (NRC 1983). This conceptualization of risk assessment is both qualitative and quantitative, although quantitative risk assessment should be considered as a component of risk assessment in its broadest context. The 1983 NRC report positioned risk assessment explicitly as a tool for translating the findings of research into science-based strategies for risk management (Fig. 4.5). Risk assessment evaluates and incorporates the findings of all relevant lines of investigation, from the molecular to the population level, through the application of a systematic process with four steps: hazard identification, dose-response assessment, exposure assessment, and risk characterization (Table 4.1). If there is no positive determination of the existence of a hazard, then the subsequent steps are not warranted.

Risk management follows and builds from risk assessment. Risk management involves the evaluation of alternative regulatory actions and the selection of the strategy to be applied. Risk communication is the transmission of the findings of risk assessments to the many stakeholders needing to know the results so they can participate in the policymaking process and to the general public. In this formalism, and in practice to some degree,

Table 4.1   The "Red Book" Paradigm: The Four Steps of Risk Assessment

1. Hazard Identification	A review of the relevant biological and chemical information bearing on whether an agent may pose a carcinogenic hazard and whether toxic effects in one setting will occur in other settings
2. Dose-Response Assessment	The process of quantifying a dosage and evaluating its relationship to the incidence of adverse health effects response
3. Exposure Assessment	The determination or estimation (qualitative or quantitative) of the magnitude, duration, and route of exposure
4. Risk Characterization	An integration and summary of hazard identification, dose-response assessment, and exposure assessment presented with assumptions and uncertainties. The final step provides an estimate of the risk to public health and a framework to define the significance of the risk.

Source: Reproduced with permission from Samet and Burke 1998, Table 5.2. © Oxford University Press.

those performing the risk assessment—risk assessors—and those managing the risks—risk managers—are separate groups of professionals, distinct from the researchers who develop the data used in risk assessments.

Two concepts central to the interpretation and application of a risk assessment are uncertainty and variability. Any assessment of risk involves the development of an underlying model with attendant assumptions that cover gaps in knowledge. Uncertainty refers to this lack of knowledge (NRC 1994). Examples of sources of uncertainty include extrapolation of findings from animals to humans, extrapolation from effects observable at high doses to the unobservable range of low doses, and use of models or assumptions to estimate the exposure of populations indirectly rather than using direct measurements. Analyses of uncertainty may be qualitative or quantitative. Qualitative analyses may involve expert judgments, whether accomplished informally or more formally using a systematic approach for achieving convergence among experts (National Council on Radiation Protection and Measurements 1996). Quantitative assessments of uncertainty may use sensitivity analyses (i.e., varying model assumptions and assessing the consequences) or model-based approaches that characterize the contributions of various sources of uncertainty to overall uncertainty.

Variability, although distinct from uncertainty, may also cloud the interpretation of a risk assessment; many sources of variability may affect a risk assessment (NRC 1994). These include variability in exposures and susceptibility; together, these two sources of variability could lead to a wide range of risk in a population. Overall or average estimates of risk, which do not address variation in risk across a population, may be misleading and may obscure the existence of a group at unacceptable risk. Recently, researchers have begun to place greater importance on understanding the vulnerability of some communities to a high rate of exposure from multiple environmental hazards simultaneously (NRC 1999).

Hazard identification is the first step of a risk assessment, addressing the question of whether the agent or factor poses a risk to human health. This step is inherently integrative, as it may draw evidence from structure-activity relationships for chemical agents, in vitro evidence of toxicity, animal bioassays, and epidemiological data (NRC 1983). Epidemiological data indicative of an adverse effect, when available, are strongly weighted in the evaluation of the weight of evidence to determine whether an agent presents a hazard. Data on humans provide direct evidence of a hazard without the need to extrapolate from knowledge of toxicity in analogous agents or from another species. In fact, as science has gained further understanding of the complexity of cross-species extrapolation from animal to human, such extrapolations are viewed with less certainty unless buttressed by an understanding of human and animal pathways of absorption and metabolism and of mechanisms of action.

Once a hazard has been identified, the second step—the dose-response assessment—is initiated to establish the quantitative relationship between dose and response. Dose—the quantity of material entering the exposed person—is not identical to exposure, which is defined as contact with a material at a potential portal of entry into the body: the skin, the respiratory tract, or the gastrointestinal tract (NRC 1991a). Typically, epidemiological studies characterize the relationship between exposure, or a surrogate for exposure, and response, expressed as the risk of a particular outcome; the dose-response relationship may be estimable if the relationship between exposure and dose can be established. For a risk assessment, description of the exposure-response relationship may be sufficient, as exposure can be linked directly to response. In combination with data on the distribution of exposure, the risk posed to a population by an agent can be estimated without moving to establish the dose-response relationship.

For the purpose of risk assessment, characterization of the exposure-response relationship in the range of exposures to humans is needed. For a few agents (e.g., environmental tobacco smoke), data on risks are available in the exposure range of interest, and risk to the general population can be estimated directly. For other agents, data from exposures above those received by the population may be available from worker groups (e.g., radon exposures of uranium miners), persons who have been accidentally exposed (e.g., radiation exposures of the population downwind of the Chernobyl nuclear plant), or persons who have been exposed during wartime (e.g., radiation exposures of the survivors of the atomic bomb blasts in Hiroshima and Nagasaki). For such exposures above the range of usual environmental levels, exposure-response relationships estimated at the higher exposures are extended downward.

For the purpose of risk assessment, information is needed on the full distribution of exposures in the population. Measures of central tendency (i.e., the mean or median) may be appropriate for estimating overall risk to the population, but reliance on central measures alone may hide the existence of more highly exposed persons with unacceptable levels of risk. Modern assessment of exposure is based on a conceptual framework that relates sources of pollutants to their effects, through the intermediaries of exposure and dose (see Fig. 4.1) (NRC 1991b, 1994). The concept of total personal exposure is central; that is, the assessment of health risks must consider exposures received by individuals from all sources and media. For some agents (e.g., lead), exposures may arise from multiple sources, media of exposure, and activities.

Risk characterization is the final step in risk assessment. During this phase, the data on exposure are combined with information on the exposure-response relationship to estimate the potential risk posed to exposed populations. The risk characterization becomes the fulcrum for decision making and the basis for communication with stakeholders (NRC 1996). Epidemiological studies may provide a direct risk characterization if the findings can be linked to a specific population and the exposure assessment component of the study is sufficient. More typically, however, the risk characterization is based on models.

The step of risk characterization puts the risk in perspective for risk managers and the public. The Red Book approach emphasized the presentation of the probability of harm from a biological perspective. However, a more recent NRC report has called for risk characterization to be broader and to consider social, economic, and political factors in describing risk and guiding options for risk management (NRC 1996). This report emphasized the need to engage stakeholders throughout the process of risk assessment to assure that the risk characterization addresses the full range of concerns (see also NRC 1999 and Chapter 5).

The consequences of the factor of concern may be characterized with a variety of types of measures. Some express the burden of disease associated with the factor. The number of occurrences of the event of concern may be estimated (e.g., the number of lung cancer deaths caused each year by passive smoking); a related measure, the population attributable risk, describes the proportion of the events of concern attributable to the factor. For example, in a comprehensive risk assessment of lung cancer associated with radon, an NRC committee estimated that about 18,000 lung cancer deaths annually—approximately 11 percent of lung cancer deaths in the United States—are attributable to indoor radon (NRC 1998b). Another measure is years of life lost because of exposure to a disease-causing agent. This quantity might be calculated as an average for exposed persons or for the entire population including exposed persons. A related measure of effect, the quality-adjusted years of life lost, takes into account the quality of the lost years. This measure was used for the Global Burden of Disease Study (Murray et al. 1994). However, different methods may assign different values to the quality of life (Dasbach and Teutsch 1996).

Estimates of cost may also be part of risk characterization (see also Chapter 5). The estimates would be based on the number of events and the associated costs of medical care and lost productivity. Deaths or years of life lost may be valued economically as well. Estimates of cost may be folded into a cost-benefit analysis in support of the implementation of environmental regulations or into a cost-effectiveness analysis for comparisons of different environmental policies.

Risk Assessment and Global Ecosystem Change

The paths by which global ecosystem change may adversely affect public health are diverse, involving direct and indirect mechanisms. How will these potential consequences of global ecosystem change be approached with models? The points of analogy with quantitative risk assessment provide a place to start. In the risk assessment paradigm, global ecosystem change can be viewed as the source of "exposure," leading directly or indirectly to changing population patterns of contact with physical and biological factors that affect health. The sources of change, like sources of air pollutants, correspond to human activities that underlie the changes in the global ecosystem. Models linking these activities to the pace of change in the ecosystem have already been developed: examples are the rise in the earth's temperature in relation to emissions and concentrations of greenhouse gases and the depletion of the stratospheric ozone layer in relation to the release of chlorinated fluorocarbons (see Chapter 7).

An essential first component in a model of global ecosystem changes and health is the linking of changes to the more proximate factors that actually affect health: exposure to ultraviolet radiation, contact with arthropod vectors of disease, and exposure to allergens, for example. Analogous to the exposure assessment component in a risk assessment, the overlay of contact of populations with the potentially injurious agent needs to be characterized. Information on the exposure-response relationship might be based on epidemiological evidence or perhaps on an experimental model. The requisite information would describe how risk changes as exposure varies. Complicating features in this paradigm include synergistic or antagonistic effects of multiple types of exposures and a concern for populations particularly vulnerable to the effects of such exposures.

However, a fuller picture must consider the global ecosystem as a dynamic system with many interacting components. As discussed under "Models and the Science/Policy Interface" (above), processes of global change may present outcomes that had been unanticipated previously. The context for making decisions and the use of scientific evidence are likely to be more complex than those in more traditional areas of environmental health. Chapter 5 develops these themes in more detail.

Conclusion

The scope and complexity of global ecosystem change are generating concerns about public health consequences of unquestionably unacceptable magnitude. In addressing more conventional problems in environmental health, science-based evidence has been used to convince decision makers of the need to act and to provide guidance among the policy options. One approach for developing estimates of harm to the public's health is risk assessment, a framework for identifying what is known about risks and where there are key gaps in scientific knowledge. Risk assessment seems to have ready applicability to characterizing some of the consequences of global ecosystem change, but the entire process of risk assessment and risk characterization must be embedded in a broader study of social, economic, and environmental systems and a greater appreciation for the context of policy development. These themes of integrated assessment are developed in Chapter 5.

SUGGESTED STUDY PROJECTS

Suggested study projects provide a set of options for individual or team projects that will enhance interactivity and communication among course participants (see Appendix A). The Resource Center (see Appendix B) and references in all of the chapters provide starting points for inquiries. The process of finding and evaluating sources of information should be based on the principles of information literacy applied to the Internet environment (see Appendix A).

The objective of this chapter's study projects is to develop an understanding of the relationship between scientific study and public policy on environmental health.

PROJECT 1: Environmental Health Issues and Public Policy

The objective of this study project is to deepen the understanding of the scope of environmental health issues that affect public policy.

Task 1. Identify an example of an environmental health issue that involves exposure of humans to a harmful substance in air, water, soil, or food.

Task 2. Identify a location for the environmental health issue selected in task 1 and describe how that issue affects public policy in that location.

Task 3. For the issue and location selected in task 2, do you think public policy should be changed? Justify your answer.

PROJECT 2: Quantitative Risk Assessment

The objective of this study project is to deepen the understanding of quantitative risk assessment as applied to environmental health.

Task 1. Identify an example of an environmental health issue that involves exposure of humans to a harmful substance in air, water, soil, or food.

Task 2. For the example in task 1, search the literature for information about the four steps of quantitative risk assessment for a particular population, as shown in Table 4.1. Consider how cofactors of an individual might affect risk and whether parts of the population are especially vulnerable.

Task 3. Summarize your results.

PROJECT 3: Risk Management

The objective of this study project is to deepen the understanding of risk management as applied to environmental health.

Task 1. Select an example of an environmental health issue that has affected public policy in some setting.

Task 2. For the setting selected in task 1, describe the strategy for risk management. Consider communication about risks to the public and participation by the public in the development of strategy.

Task 3. How might you evaluate the strategy for managing risk?

Task 4. Summarize your results.

References

Dasbach E, Teutsch SM. 1996. Cost-utility analysis. In *Prevention Effectiveness: A Guide to Decision Analysis and Economic Evaluation* (Haddix AC, Teutsch SM, Shaffer PA, Duñet DO, eds.). Oxford University Press, New York, pp. 130–42.

Dockery DW, Pope CA. 1994. Acute respiratory effects of particulate air pollution. *Annu Rev Public Health* 15: 107–32.

Last JM. 1993. Global change: Ozone depletion, greenhouse warming, and public health. *Annu Rev Public Health* 14: 115–36.

McClellan RO, Miller FJ. 1997. An overview of EPA's proposed revision of the particulate matter standard. *CIIT Activities* 17 (4): 1–23.

McMichael AJ. 1993. Global environmental change and human population health: A conceptual and scientific challenge for epidemiology. *Int J Epidemiol* 22: 1–8.

Moolgavkar SH. 1994. Biological models of carcinogenesis and quantitative cancer risk assessment. *Risk Anal* 14 (6): 879–82.

Morgan MG, Henrion M. 1990. *Uncertainty: A Guide to Dealing with Uncertainty in Quantitative Risk and Policy Analysis.* Cambridge University Press, Cambridge.

Murray CJ, Lopez AD, Jamison DT. 1994. The global burden of disease in 1990: Summary results, sensitivity analysis and future directions. *Bull World Health Organ* 72 (3): 495–509.

National Council on Radiation Protection and Measurements. 1996. *A Guide for Uncertainty Analysis in Dose and Risk Assessments Related to Environmental Contamination.* Report No. 14. NCRP, Bethesda, Md.

National Research Council. 1983. *Risk Assessment in the Federal Government: Managing the Process.* National Academy Press, Washington.

———. 1991a. *Human Exposure Assessment for Airborne Pollutants: Advances and Opportunities.* National Academy Press, Washington.

———. 1991b. *Frontiers in Assessing Human Exposures to Environmental Toxicants.* National Academy Press, Washington.

———. 1992. *Global Environmental Change: Understanding the Human Dimensions.* National Academy Press, Washington.

———. 1994. *Science and Judgment in Risk Assessment.* National Academy Press, Washington.

———. 1996. *Understanding Risk: Informing Decisions in a Democratic Society* (Stern PC, Fineberg HV, eds.). National Academy Press, Washington.

———. 1998a. *Research Priorities for Airborne Particulate Matter: No. 1. Immediate Priorities and a Long-Range Research Portfolio.* National Academy Press, Washington.

———. 1998b. *Health Effects of Exposure to Radon (BEIR VI).* National Academy Press, Washington.

———. 1999. *Toward Environmental Justice: Research, Education, and Health Policy Needs.* National Academy Press, Washington.

Samet JM, Burke TA. 1998. Epidemiology and risk assessment. In *Applied Epidemiology: Theory to Practice* (Brownson RC, Petitti D, eds.). Oxford University Press, New York, pp. 137–75.

Samet JM, Zeger SL, Kelsall JE, Xu J, Kalkstein LS. 1997. Air pollution, weather, and mortality in Philadelphia, 1973–1988. In *The Phase 1.B Report of the Particle Epidemiology Evaluation Project: Particulate Air Pollution and Daily Mortality. Analyses of the Effects of Weather and Multiple Pollutants.* Health Effects Institute Report. Health Effects Institute, Cambridge, Mass.

Electronic References

Environmental Protection Agency. 1999a. Plain English Guide to the Clean Air Act. http://www.epa.gov/oar/oaqps/peg_caa/pegcaain.html (Date Last Revised 11/30/1999).

———. 1999b. Taking Toxics out of the Air: Progress in Setting "Maximum Achievable Control Technology" Standards under the Clean Air Act. http://www.epa.gov/oar/oaqps/takingtoxics (Date Last Revised 12/16/1999).

Schneider SH, Turner BL. 1997. Surprise and Global Environmental Change. Report of the 1994 Aspen Global Change Institute Summer Session on Surprise and Global Environmental Change. http://sedac.ciesin.org/mva/iamcc.tg/articles/surprise.html (Date Last Revised 5/12/1997).

CHAPTER 5

Integrated Assessment

Joan L. Aron, Ph.D., J. Hugh Ellis, Ph.D.,
and Benjamin F. Hobbs, Ph.D.

Integrated assessment is a "structured process of dealing with complex issues, using knowledge from various scientific disciplines and/or stakeholders, such that integrated insights are made available to decision makers" (Rotmans 1998). The application of integrated assessment to studies of global ecosystem change and public health is somewhat analogous to the application of the methods of risk assessment to problems in environmental health and can profitably build on those experiences (see Chapter 4). Integrated assessment is distinctive, however, in its effort to link social, economic, and environmental factors as a basis for examining policies and decisions that affect diverse sectors. Because of this intent to link different fields and analyses, often for different purposes, integrated assessment can be a useful test of available information and its limits. Integrated assessment also extends beyond risk assessment to provide support for decisions to manage risk (see Fig. 4.5). In applications of integrated assessment, the context for decisions is more complex than regulations put forth by a single government agency (see Fig. 4.2). An integrated assessment may influence decisions made by multiple national governments, multiple agencies at different levels of government, and diverse interest groups.

Research on integrated assessment involves two very different approaches (Rotmans 1998). One group of researchers, usually trained in natural science, engineering, or economics, aims to develop mathematical models of the consequences of different policies. Another group of researchers, usually trained in qualitative methods of the social sciences, views integrated assessment itself as a participatory process. Neither approach alone is likely to succeed, and researchers are increasingly seeking new ways to join analytical studies with participatory methods (*International Human Dimensions Programme on Global Environmental Change*

2000). The effective use of scientific information requires the integration of multiple disciplinary perspectives.

An examination of the real-world consequences of a bad assessment should dispel any notion that integrated assessment is merely a technical exercise. Glantz (1982) documents high costs associated with an erroneous forecast of drought in 1977 for Yakima Valley in the state of Washington. The setting for the forecast was an area in which most agricultural activities depended on water that the U.S. Bureau of Reclamation provided for irrigation from April to October. After extremely dry conditions had prevailed in the Yakima River basin from October 1976 through January 1977, the Yakima Project superintendent of the U.S. Bureau of Reclamation issued an estimate of the total water supply available for 1977. Following the law in a 1945 consent decree, the formula for the total water supply available included natural flow, storage, and return flow, where return flow is water used for irrigation that returns to the stream for downstream users. The law also specified different classes of water rights so that the Yakima Project superintendent allocated the projected water supply to satisfy 98 percent of senior (higher priority) water rights and provide only 6 percent of normal flow for junior (lower priority) water rights. Farmers with junior water rights stood to lose perennial plants and trees that would require up to eight years to replace. This dire forecast of drought caused farmers to take actions costing millions of dollars:

1. The aggregate cost of digging irrigation wells was approximately $9,000,000.
2. Some farmers with perennial crops paid up to four times the usual price to lease water from farmers with annual crops who could afford to lose one year's production.
3. Some farmers transplanted mint crops (a perennial) at their own expense.

The drought forecast also led to expenditure of government funds:

1. A feasibility study of a project to divert water from the Columbia River basin cost $70,000.
2. A new pumping station to capture water below a dam's lowest outlet cost $3,000,000 and was ready by July 1977.
3. The state of Washington spent $400,000 to seed clouds with no effect proven scientifically.

However, by May 1977, the Yakima Project superintendent had sharply revised upward the estimate of the water supply to allow 50 percent of normal flow for junior water rights. By the end of the growing season, farmers with junior rights had received 70 percent of normal water allocations. The new pumping station that opened in July was never used. Farmers made legal claims against the U.S. Bureau of Reclamation to recover $20,000,000 lost by responding to the erroneous drought forecast in February. The U.S. Bureau of Reclamation suffered a severe loss of credibility.

An understanding of what went wrong with the forecast of drought in Yakima leads to general insights about problems in forecasts and how to avoid them. Surprisingly, despite uncertainties in weather, the estimate of return flow was the primary error in the forecast of extreme shortage of water. Implicit in the initial estimate of return flow was an assumption of flood control, which had been the main concern of water resource managers before October 1976. In balancing the estimate of return flow against diversions for flood control, the return flow was dropped from the initial estimate of the total supply of water available to irrigators in 1977. When forecasters corrected the technical error, the estimate of the total supply of water rose considerably. Retention of outmoded assumptions, termed *assumption drag*, is a widespread phenomenon that causes serious errors in forecasting (Ascher 1979, 202). Another instance of assumption drag was maintaining an assumption that low precipitation would persist even after normal monthly precipitation had returned in early spring. Forecasts for precipitation were strongly influenced by drought forecasts for the entire western United States. The contribution from stored water was also a problem, not because of inaccurate estimates of the quantity of water, but rather because of disputes over ownership. The Yakima Project superintendent decided to consider water stored in reservoirs as part of natural flow, for which senior water rights take priority, despite the fact that only water districts with junior water rights were paying for storage precisely because of their vulnerability in years with low natural flow. Combining these errors with worry about promising more water than could be delivered, the Yakima Project superintendent produced a very conservative forecast.

Poor communication within the government sector and with farmers contributed to the promulgation of an erroneous forecast. Even though the National Weather Service, the Soil Conservation Service, and the state of Washington's Department of Ecology were all producing water-related forecasts, these agencies provided no external checks on the development of forecasts of drought by the Yakima office of the U.S. Bureau of Reclamation. The Pacific Northwest Regional Office of the U.S. Bureau of Reclamation, which had an advisory role from its location in Boise, Idaho, disagreed with the Yakima forecast but had no clear authority to modify forecasts made at Yakima. The lack of open communication between forecasters and farmers was significant because too little attention was paid to the effects caused by announcing a forecast, particularly the consequences of offering less water than was actually available. In general, "efforts to shield expert research and decision making from public scrutiny and accountability invariably backfire, fueling distrust and counterproductive decisions" (Pielke et al. 1999, 311).

A salient feature underlying all of these difficulties is that the formula for allocating water in the 1945 consent decree had never been used before the period of water storage that began in October 1976. Consequently, the water resource managers had no direct experience on which to base their actions. It is intrinsically more difficult to manage without direct experience. Yet many pressing problems, such as preventing possible adverse effects of global climate change and containing nuclear waste in underground storage for 10,000 years, provide little or no direct experience. In general, "the less frequent, less observable, less spatially discrete, more gradual, more distant in the future, and more severe a predicted phenomenon, the more difficult it is to accumulate direct experience" (Pielke et al. 1999, 311). In grappling with integrated assessment, researchers of all disciplines need to be open to learning the use of a variety of tools.

This chapter presents an overview of major conceptual and methodological issues in the development of integrated assessment models (IAMs) and studies of the participatory process in assessment. One section devotes special attention to the multiple meanings of effectiveness when scientific assessments are applied to complex problems. The general discussion sets the stage for addressing the challenges of an integrated assessment of ecosystem change and public health on a global scale.

The History of Integrated Assessment Models

Quantitative approaches to integrated assessment attempt to link human actions to valued consequences by developing more complex computer-based models with interrelationships and feedbacks among system components (Dowlatabadi and Morgan 1993a; Shlyakhter et al. 1995; Parson and Fisher-Vanden 1997). The two basic functions of IAMs are

1. *system modeling:* simulation of the response of the physical, biological, and/or social system to changes in inputs, assumptions, and decisions; and
2. *decision evaluation:* comparison of alternative decisions in terms of their risks and performance on important objectives and application of value judgments by users to rank or screen alternatives.

In system models, several components are interconnected so that changes in one part of the system cause multiple indirect effects in other parts of the system, which may be classified by sector, geography, and other characteristics. Complex interconnections create time lags so that the effects of changes in one part of the system may not become apparent in other parts of the system until some time has passed. The state of a system may change abruptly as the value of a variable reaches a threshold level, such as the amount of acid rain that exceeds the ability of lakes to buffer its effects (see Chapter 7).

Selected elements of the Integrated Model to Assess the Greenhouse Effect (IMAGE) illustrate these concepts (Rotmans and Dowlatabadi 1998, 319). One important set of interrelationships in the IMAGE simulation determines how the force of climate change affects agricultural production, natural ecosystem health, and human health. The simulation also specifies how changes in terrestrial ecosystem health provide feedback into system components for atmospheric composition, evapotranspiration, and albedo, which are all part of atmospheric processes affecting climate (see Chapters 7, 8, and 9). Two possible decisions in the model are investments in infrastructure or the development of more resilient crops.

Many IAMs emphasize only system modeling and focus on assessing the effects of alternative scenarios or previously adopted policies. The ultimate purpose of such efforts is to provide policymakers with an understanding of how structures, assumptions, and policies produce system behavior and consequences (Gardiner and Ford 1980). This is sometimes termed the "intelligence" stage of decision making (Nazareth 1993) or, less formally, "embedding models in policy makers' heads" (Parson and Fisher-Vanden 1997). Although the term *integrated assessment* is of recent coinage, IAMs of this type go back to the 1960s. The most familiar early examples are urban and global systems dynamics models, such as the model used in the Club of Rome's "Limits to Growth" study (Forrester 1969; Meadows et al. 1972) and the assessment of the proposed supersonic transport (see also Chapter 7). More recent IAMs that emphasize systems modeling include the many systems developed to quantify the regional or global effects of possible climate change (e.g., Rotmans 1990; see the reviews by Dowlatabadi 1995, Parson and Fisher-Vanden 1997, Intergovernmental Panel on Climate Change 1996; Rotmans and Dowlatabadi 1998). Another is the Tracking and Analysis Framework, a system for evaluating the cost and environmental effects of the Title IV acid rain program of the 1990 U.S. Clean Air Act (Henrion et al. 1995).

Integrated assessment may be most useful to policymakers when assessments are explicitly linked to decision making (Bernabo and Eglinton 1992; National Acid Precipitation Assessment Program 1991). "Decision-focused" IAMs are those that build in formal capabilities for analyzing decisions. Alternatively, the developers and potential users of a model could ensure that it is designed to yield information on the performance and risks of policy options for use in decision processes taking place outside of the formal IAM. These can be termed *policy-oriented impact assessments* (Meo 1991). (See also "Evaluation of the Effectiveness of Integrated Assessment," below.)

Here, we look at decision-focused models, although methods for analyzing multiple objectives and aspects of risk are certainly also applicable to policy-oriented impact assessments. An advantage of incorporating the evaluation of decisions within a computer-based model is that users can receive immediate feedback on the implications of alternative value judgments. We have found that such feedback increases user confidence and satisfaction with the results (Hobbs and Meier 2000), consistent with general findings in the scientific literature on the importance of feedback in decision support systems (Foley and Van Dam 1982; Schneiderman 1992). More generally, some researchers have found that decision support systems that explicitly include the function of decision evaluation (e.g., an explicit analysis of the tradeoffs among multiple objectives) in addition to system modeling increase user satisfaction and do a better job of facilitating group discussion and compromise (e.g., Iz and Krajewski 1992; Nazareth 1993). Strategies to structure the development of models and generation of candidate policies as a collaborative process may be as important as the technology for supporting the analysis of decisions (Bonnicksen 1996).

An example of a decision-focused IAM involves the determination of "optimal" levels of sulfur dioxide (SO_2) reduction at large point sources in Canada so that prescribed environmental quality levels would be achieved at selected locations with sensitive receptors (Ellis et al. 1985). While originally conducted for the Ontario provincial environment ministry, the work expanded to a federal setting, with the result that the plant-by-plant SO_2 reductions arising from the analyses were eventually adopted by the provinces and signed into law. Other examples of decision-focused IAMs are (1) a Great Lakes climate change modeling system for assessing lake regulation, shore protection, and wetland policies under climate uncertainty (Hobbs et al. 1997); (2) the analyses of the RAINS model (Hordijk 1991) conducted in support of the sulfur and nitrogen protocols (roughly the European equivalent of Title IV of the 1990 U.S. Clean Air Act); and (3) the U.S. Environmental Protection Agency (EPA)'s (1995) Adaptation Strategy Evaluator, an explicitly multiobjective system that displays tradeoffs among adaptation strategies incorporating multiple objectives in diverse sectors—the coastal zone, water resources, and agriculture.

The Building of Decision-Focused Integrated Assessment Models

Several capabilities are key to the success of a decision-focused IAM:

1. Explicit representation of alternative policies
2. Credible physical models or the capability easily to modify the system to incorporate alternative model structures and parameters
3. Performance indicators, or *objectives,* that address the concerns of policymakers

4. Characterizations of the uncertainties associated with model outputs and the sources of those uncertainties
5. A user interface that encourages "play" (sensitivity analysis and exploration) while effectively communicating tradeoffs among objectives and uncertainties (possibly using advances in geographic information systems and visualization techniques)
6. The ability to quickly show the implications for policy choices of alternative value judgments concerning acceptable risks and tradeoffs among objectives

Table 5.1 provides additional elaboration by showing the tasks that must be performed for an IAM to execute successfully the functions of system modeling and decision evaluation. The two functions are shown as columns, with the tasks arranged in rows, beginning with formulation and ending with the justification and documentation of decisions. The function of decision evaluation is divided into the subfunctions of *evaluation of tradeoffs among objectives* and *evaluation of risks.* Several observers also stress the need to recognize the variety of perspectives and objectives involved in policymaking, explicitly to acknowledge and deal with ubiquitous uncertainties, and to foster greater communication among policymakers and developers of integrated assessments (see under "Integrated Assessment as a Participatory Process," below, and Dowlatabadi and Morgan 1993a; Dowlatabadi 1995; Shlyakhter et al. 1995).

In building models to support integrated assessments of global change, the key to success lies in formulating problems that have the potential to be *policy relevant.* That is, do the models have the capability of answering those questions that policymakers, stakeholders, or the general public deem important? While policy relevance as a laudable goal is hard to dispute, the best path to achieving that goal is far from obvious. Identification of the important questions can be very difficult, even with direct access to policymakers or whomever is the intended audience/user of the analyses (see under "Integrated Assessment as a Participatory Process," below). Practical experience has led to the development of three guidelines for modeling and analysis applied to integrated assessment.

Guideline 1. Interact early, interact often and repeatedly, and continually pose the question: What is the possible policy relevance of this model and this analysis? The discussion should be in terms that the users of the assessment, not just systems analysts, understand. Achieving this communication is difficult but central to conducting influential systems analyses. Without close contact with policymakers or other potential users, the likelihood of a systems analyst discovering the important questions is far from certain.

Guideline 2. Remember that real problems are multiobjective. All decisions involve tradeoffs among multiple objectives, and explicit consideration of those tradeoffs enhances greatly the policy relevance of systems modeling. This issue is explored in somewhat more depth under "Single versus Multiple Objectives" (below).

Guideline 3. Pay attention to the analysis and communication of uncertainty. The concerns here lie not so much in mathematics or technical sophistication but rather in communication. It is very easy for scientists to lose their audience and hence any chance for policy relevance with elegant stochastic analyses that seem to make perfect technical sense—at least to the analyst. Some practitioners of risk analysis even counsel that uncertainty should not be communicated routinely because analyses of uncertainty are difficult to understand, and policy-oriented presentations of many types of data (e.g., economic statistics) typically avoid the subject (Goldstein 1995). However, open communication is essential for participatory decision making (Pielke et al. 1999). Some stochastic (probabilistic) programming techniques hold much more promise than others in terms of their ability to facilitate communication (see "Deterministic versus Stochastic Approaches," below).

A Taxonomy of Approaches to Integrated Assessment Modeling

The selection of an approach to modeling should be driven entirely by the characteristics of the problem at hand. If the modeling approach should lead to a model that cannot be solved, the question changes: Can a related model requiring an altered statement of the problem still hold promise for policy relevance? Dealing with problems in this fashion therefore requires that the scientists performing the analysis have a facility with a variety of approaches. The following brief overview describes the attributes of some of the more important approaches to modeling: simulation versus optimization, linearity versus nonlinearity (including the important subdistinction of convex and nonconvex programming problems), single-objective versus multiple-objective approaches, deterministic versus stochastic approaches, and static versus dynamic approaches. A different kind of distinction is the technological approach for visualization and human interaction with the model (see under "Visualization and System-User Dynamics," below).

Simulation versus Optimization

The categorization of models as based on simulation or optimization represents the most basic modeling distinction. A good working definition is simply that simulation models help answer the question: What happens if I do this? Optimization models, on the other hand, help answer the question: What should I do? Simulation models are commonly termed *de-*

Table 5.1   Tasks in Constructing Decision-Focused Integrated Assessment Models

Tasks	System Modeling Function	Decision Evaluation Function	
		Concerning Objectives: What are the objectives, and what tradeoffs are acceptable?	*Concerning Risk:* What risk dimensions are we concerned about, and what risks are we willing to accept?
Formulation	Choices of system components to include Linking of components Scale (from detailed, spatially distributed process models to aggregate metamodels) Alternative judgments about system structure should be able to be accommodated in model.	Alternative formulation of objectives; forms of value functions/preference statements	Dimensions of risk of interest to user (regret thresholds; acceptable risk thresholds; utility functions)
Initial parameter estimation	Can accommodate both empirically based and subjective estimates, along with characterizations of their uncertainty	Initial characterization of multiattribute utility functions should be based on transparent methods that are easy to use and yet are likely to yield parameters (e.g., attribute weights) that are valid representations of preferences. Alternatively, can capture initial, partial statements of preferences.	
Computations and screening of alternatives	Efficient simulation, optimization methods	Rapid generation of noninferior (non-dominated, Pareto optimal) solutions; ability to generate linkages between value judgments, solutions in objective space, and solutions in decision space	Rapid generation of efficient solutions in terms of risk (e.g., stochastic non-dominance)

Information presentation	Structure, parameters of model should be readily apparent (graphical/object orientation).	Noninferior or subsets of options; use of more than one way to portray tradeoffs, including Cartesian plots, value paths, and tabulations	Stochastically nondominated solutions; use of more than one way to portray risks, including probability distributions (or their moments), conditional expectations, ranges (e.g., tornado diagrams), regret, decision trees, influence diagrams, correlations between uncertain parameters and outputs
System-user dynamics	*Exploratory modeling:* Users should be able readily to explore and alter system parameters and even structure.	User-directed sensitivity analyses of effect of different multiobjective value judgments. Multiple interactive approaches available (including some based on partial information on preferences), so that users can choose method most appropriate for their decision-making style and view the problem from different perspectives.	User-directed sensitivity analyses of effect of different risk attitudes upon decisions; variety of interactive approaches available
Communication among stakeholders	Presentation of individual models and their implications should be easily accomplished.	Group processes for eliciting and discussing judgments on objectives definition and priorities. Exchange views on priorities, acceptable tradeoffs, thresholds, focusing on implications. (Do differences among people matter?)	Group processes for eliciting and discussion of judgments on risk definitions and attitudes. Exchange views on risk attitudes, etc., focusing on implications. (Do differences among people matter?)
Justification/ documentation of decisions	Documentation of system structure and parameters should be easily generated.	Precise judgments (e.g., quantitative objectives, weights); imprecise judgments and arguments based on robustness of decision	Precise judgments (e.g., risk attitudes); imprecise judgments and arguments based on robustness of decision

scriptive, in contrast to optimization models, which are *prescriptive*. An example of a simulation-based integrated assessment involves halocarbon production and climate change (Holmes and Ellis 1996, 1997). Analyses proceed basically by prescribing input (including what is happening in terms of production and the state of the climate), then running the models (linking halocarbon production and climate change), and finally assessing the resulting output (e.g., global mean averaged chlorine loading concentration in the atmosphere). Contrast this model of operation with an optimization-based approach that might, for example, set a target level for maximum allowable chlorine loading concentration and then identify the least-cost strategy of reducing emissions that can achieve that goal. For this example, as is commonly the case, simulation models are often embedded within optimization models. This example also serves to point up some of the forces that come into play in deciding which approach to pursue. Holmes and Ellis (1996, 1997) selected a strictly simulation-based approach because they thought it was too difficult at the time to "wrap" an optimization model around a framework based on a simulation model that was already complicated.

Linearity versus Nonlinearity

This modeling distinction refers to a graphic representation of a quantitative relationship between two variables: the relationship can be represented as a straight line (*linear*) or not (*nonlinear*). Linear problems are easier to analyze, but the real world is generally nonlinear. Investigators from fields as diverse as biology, chemistry, climatology, and economics are exploring the effects of nonlinearity in generating complex outcomes, such as chaotic behavior, in simulation models (e.g., Zimmer 1999; Whitesides and Ismagilov 1999; Rind 1999; Arthur 1999). Nonlinearities are also fundamental to the construction of quantitative objectives in optimization models. The state of computationally tractable solution techniques is so advanced that there simply is no need to replace a nonlinear problem/model with a linearized approximation, except perhaps for linear programs of truly prodigious size, and even that exception is arguable.

In optimization, nonlinear problems are classified according to convexity. The distinction between convexity and nonconvexity refers to the geometric shape of the representation of how a variable is mathematically related to other variables. For simplicity, consider height above a flat surface as the mathematical function. If the shape is convex, then height is somewhat like the roof of a dome over a sports stadium, and a straight line from one point on the roof to any other remains under or at the roof. If the shape is not convex, then height might be more like the turrets of a medieval castle, and a straight line from one point on the roof to any other may well be above the roof itself. For an optimization model, convexity guarantees relative ease in the calculation of global optima, which is the general term for points higher than all others or points lower than all others. With convexity, the computer algorithms for searching for optima do not depend on where the search begins. However, the optima are not necessarily unique, in that alternative decisions may be able to reach the same objective in a different way, much as two different turrets on a castle may be able to provide access to the same height.

The presence of nonconvexity, as distinct from nonlinearity, most assuredly does not signal the need to formulate and solve a related but different convex programming problem. It does, however, require that additional steps be taken to ensure that good solutions at least are identified, for in most cases, this is the best that one can do. The suite of traditional and newer (heuristic) solvers readily available today renders the chances of solving even the most recalcitrant nonconvex problems very good. The presence of nonlinearity and nonconvexity also strongly suggests the need for a broad view of the problem and definitions of its relevant features.

Single versus Multiple Objectives

The need for multiobjective approaches has been cited above and is reaffirmed here with the assertion that every public policy–related management problem, certainly including global change and public health problems, is multiobjective. The nature of the multiple objectives, of course, varies greatly from application to application. Typically, the objectives are in the categories of cost minimization; the achievement of environmental standards (if they exist) of many kinds, including standards explicitly linked to human health; a wide variety of stochastic considerations (e.g., stochasticity manifested in the context of robustness—solutions that perform acceptably well if certain input assumptions prove incorrect); objectives directly or indirectly representing measures of equity and fairness (e.g., allocating the costs of emission control among polluters); and objectives that reflect aesthetic considerations, to name but a few.

Multiobjectivity is a key to engaging decision makers and producing analyses that are policy relevant. As Dowlatabadi (1995, 294) dryly notes, "policy decision makers do not seem to have made their decisions on the basis of cost-effectiveness or cost-benefit analyses in the past." Gardiner and Ford (1980) argue that early systems dynamics models, such as the "Limits to Growth" model, failed to focus on objectives and tradeoffs and that this failure diminished their effectiveness. The increasing involvement of diverse stakeholders in integrated assessment (e.g., Cohen 1994) will only increase the need to be explicitly multiobjective and to identify explicitly for all parties how tradeoffs are measured and evaluated in comparison with other possibilities.

The explicit recognition of the multiobjective nature of a problem helps to create a modeling environment in which decision makers and analysts are able to maintain appropriate roles and in which information essential to effective decision making can be generated and conveyed. It is the fundamental nature of multiobjective problems that there is no single optimal solution. Rather, the focus is the tradeoffs among the objectives implied by choosing one solution over another. One solution is said to *dominate* another if it is better in every respect, such as one medication that is more effective, less expensive, and safer than another medication; in this highly unrealistic situation, making a decision involves no tradeoffs among objectives. Therefore, the interest lies in analysis of the set of *nondominated solutions* (i.e., those that are not clearly dominated by any other). Differing perceptions of the relative importance of different objectives are a major source of policy disagreements; analysis of these tradeoffs is essential if decision support systems are effectively to promote discussion among interested parties (Thiessen and Loucks 1992; for an example of how multiobjective methods can foster consensus, see Brown 1984).

Operationally, the way in which tradeoffs are generated distinguishes the two basic methods of multiobjective analysis: *generating methods* and *preference-oriented techniques* (Cohon 1978; Goicoechea et al. 1982; Chankong and Haimes 1983). The aim of generating methods is to create either an approximation or an exact representation of the set of nondominated solutions, forming the basis for exploration of the tradeoffs among objectives. There is no attempt made to incorporate decision makers' preferences in any formal or explicit manner. Preference-oriented methods, by contrast, use explicit quantitative statements of decision makers' preferences to identify a preferred solution. Preferences can be elicited in one "shot," resulting in a functional representation of preferences that can be used to rank alternatives (*prior articulation methods*). Or preferences can be elicited iteratively as users react to solutions, with the users' reactions guiding the generation of the next trial solutions (*progressive articulation methods*).

Though preference-oriented techniques can help policymakers to understand the implications of preferences and preference conflicts for decision making, many of them suffer from several disadvantages. Most are rigid in the way that preferences must be stated, and they tend to reveal sparse information about the set of nondominated solutions, thus limiting the insight gained from analysis. Also, the presence of multiple decision makers can cause complications that defeat most of the preference-based methods. Arrow's classic work demonstrated that a single rule cannot represent individual preferences across all options; moreover, social norms play an important role that is distinct from an aggregate of individual preferences (Cantor and Yohe 1998, 62–63). In practice, some combination of methods will often work best. A generating technique might be emphasized first to develop an appreciation of the range of choice and the tradeoffs. In reacting to these results, decision makers may be able to articulate preferences, perhaps a particular portion of the nondominated set worthy of further, detailed exploration. The process could continue until a single or a small collection of solutions is identified. More formal preference-oriented methods could then be used, but some results indicate that this is best done only after the nondominated set is explored with a generating method.

Experiments involving diverse groups of stakeholders applying several multiobjective methods to the same problem confirm the importance of using a combination of methods (e.g., Hobbs et al. 1992). Different multiobjective methods applied by the same person can yield very different rankings (see also Hobbs 1986; von Winterfeldt and Edwards 1986). Further, no single approach was trusted or valid for all participants, and they preferred to use more than one method to gain insight. This conclusion was reinforced by energy planning studies involving large groups of stakeholders in British Columbia (e.g., Meier and LeClair 1995). Nondominated set generation was applied along with multiple methods to quantify preferences (e.g., both Analytic Hierarchy Process weights and tradeoff-based weights). After applying the methods, individual stakeholders reviewed the results, resolved conflicts among the methods, and made initial recommendations, which were then the basis of group discussions. The stakeholders stated that the multimethod process was essential to building insight and confidence. The processes resulted in sets of recommendations that the power utilities largely accepted. In contrast, previous attempts by one of the utilities to apply a single preference-based method in isolation were unsuccessful, in large measure because the stakeholders did not trust the method or results.

A large and rich literature on multiobjective theory and applications has developed since the 1970s, but it is surprising how infrequently formal multiobjective methods have actually been used in integrated assessment. There are several possible reasons for this apparent lack of acceptance and success:

1. It is difficult to portray tradeoffs among nondominated solutions when there are four or more objectives.

2. There is a tendency among many multiobjective specialists to apply a single "one size fits all" method for the elicitation of values in a given situation and to use it in a manner that assumes that the user's preferences are completely formed, coherent, readily expressed in quantitative terms, and consistent over time. In reality, psychologists "view preferences as adaptable and constructed as needed in particular contexts" (Cantor and Yohe 1998, 61), so that users do not have fully formed *utility functions*. A utility function is a quantitative expression of an individual's preferences

that can take into account attitudes toward risk, such as a risk-averse preference for certain possession of one thousand dollars over the opportunity to toss a coin to win two thousand dollars. As noted earlier, different methods work better for some people than for others. Users are frustrated by the arbitrariness and lack of control involved in using a "black box"-like method to translate value judgments (such as numerical weights of different objectives) into ranks of alternatives. As a result, multiobjective methods are often used inappropriately—to supplant rather than support hard thinking, insight, and careful articulation of reasons for supporting certain alternatives (Parkin 1992).

3. Finally, there has been lack of careful evaluation by real decision makers in realistic settings of what works and what does not (see "Evaluation of the Effectiveness of Integrated Assessment," below). Convincing comparisons of methods in actual problem settings, including integrated assessment, are few (see surveys of experimental evaluations by von Winterfeldt and Edwards 1986; Hobbs 1986; Weber and Borcherding 1993). Symmetrically, there is almost no effort to evaluate most public decisions retrospectively.

Consequently, there is a gap between a theoretical literature full of elegant but untested multiobjective methods—each with its own advocates—and potential users who adopt and sometimes (unintentionally) misuse simple methods rather than attempt to sort out the conflicting claims. Corner and Corner (1995) and Stewart (1992) note that most published decision analyses use additive value functions. And in most cases, users choose weights using simple methods (such as rating importance on a 0–100 scale or stating the ratio of importance of two attributes) that cannot be trusted to yield weights that actually represent their willingness to trade off objectives. As has been shown in theory (Schoemaker 1981) and in several experiments (e.g., Hobbs et al. 1992), weights based on some vague notion of "importance" can differ drastically from weights inferred from explicit tradeoff judgments.

These problems have several undesirable results. One is a research community that often seems out of touch with the needs of potential users. Another is the application of methods in a manner that obscures rather than illuminates the tradeoffs and distorts preferences by quantifying them in inappropriate ways. Consequently, possible users of multiobjective methods will be wary of them—and the potential of the multiobjective approach to promote understanding of the problem and dialogue among stakeholders will not be realized.

Deterministic versus Stochastic Approaches

It is difficult to imagine a realistic integrated assessment of global change and public health that does not contain important stochastic elements. Often the most difficult issue in contemplating a stochastic approach when building models and designing analyses lies in assessing how uncertain information will be presented to decision makers and whether there is policy-relevant "value added" in adopting a stochastic approach. These are easy questions to pose; their resolution is problematic, especially with regard to the notion of assessing added value. No single approach works for all problems, but a reasonable philosophy is to attempt to build the (stochastically) simplest model first and then continually subject it to guideline 1 described under "The Building of Decision-Focused Integrated Assessment Models" (above)—ask whether the analyses are possibly policy relevant. Communicating stochastic information is a complex subject unto itself, and much can be learned from experiences in risk assessment (see Chapter 4). But there is a fourth guideline to follow.

Guideline 4. Begin simply. Refrain from focusing extensively on those stochastically feasible outcomes that promise little additional insight in terms of policy relevance (admittedly easier said than done), and communicate in terms familiar to decision makers (see guideline 1, which encourages interaction between analysts and decision makers).

In principle, one might expect that explicit consideration of uncertainty, like consideration of multiple objectives, should be a powerful means of engaging decision makers and enhancing the policy relevance of integrated assessment, but it often does not. Advocates of multiobjective approaches might further venture to say that, when an integrated assessment has had both multiobjective and stochastic elements and has "worked," it has done so *because* of multiobjectivity and *in spite of* stochasticity. Rather than a mechanism to enhance realism and policy relevance, explicit consideration of uncertainty all too often tends to alienate decision makers. Decision makers routinely process forecasts without estimates of errors associated with them (Goldstein 1995). Probabilistic concepts are generally not readily understood, even by reasonably technically literate people.

In workshops held to evaluate an IAM for assessing the effect of climate change upon Great Lakes management (Chao et al. 1999), U.S. and Canadian water resource managers tested both scenario-based methods and more sophisticated approaches based on decision trees, which are explicit representations of the probabilities of multiple possible outcomes of a decision, and Bayesian analysis, which incorporates prior personal experience in assigning probabilities to possible outcomes. The managers preferred deterministic scenarios. Despite recognizing the need to consider uncertainty explicitly, they were distinctly uncomfortable with using subjective probabilities. This points to the critical need to enhance the communicability of probabilistic approaches in integrated assessment—that is, to cast stochastic problems in terms accessible to users. The involvement of users throughout the entire process of building models is likely to aid in communication between developers and users of an assessment (Bonnicksen 1996).

Another limitation involves the basic paradigms that attempt to describe and prescribe how people process information and make decisions under uncertainty. One approach is to presume *risk neutrality* and use expected values of the stochastic outcomes (i.e., the averages) of the model, although policymakers are manifestly *risk averse,* a technical term that describes individuals very concerned about avoiding the worst possible outcomes. *Expected utility maximization* is a reasonable and popular alternative that can capture an individual's attitudes toward risk, but it, like any other model that exists today, is limited in its explanatory power. Furthermore, people's attitudes toward risk differ and are labile (i.e., they are not set in stone), and their expression can depend on many variables, such as the context and phrasing of the question (Fischoff et al. 1979).

Regret theory is an alternative approach to the theory of choice under uncertainty that has several attractive attributes (see Loomes and Sugden 1982, 1987; Keasey 1984; Sugden 1986, 1993; Loomes et al. 1992). It originated through the recognition that commonly observed patterns of choice violated conventional (von Neumann and Morgenstern 1947; Savage 1954) expected utility axioms (see Kahneman and Tversky 1979; Machina 1982). Perhaps its most compelling attribute is its recognition of certain aspects of human behavior, described by Loomes and Sugden (1982, 820) as follows: "Regret theory rests on two fundamental assumptions: first, that many people experience the sensations we call regret and rejoicing; and second, that in making decisions under uncertainty, they try to anticipate and take account of those sensations." Regret theory has been implemented in several large-scale modeling contexts, for example, "no regrets" strategies for reducing greenhouse gas emissions (Mills et al. 1991) and strategy development for the control of acid rain (Ellis 1988).

Another approach is to avoid the need to specify a particular utility function (and thus attitude toward risk) altogether. This can be done by using the concept of *stochastic dominance* to eliminate alternatives that would not be found attractive under any utility function within a given general class (e.g., risk-averse utility functions). In application, the method has been an effective screening tool for eliminating many, if not most, of the alternatives considered without requiring a utility function to be specified (Levy 1992; Mosler 1984; Mosler and Holz 1994). A related technique is to use partial information on risk preferences to limit the range of possible utility functions (in the spirit of Sarin 1977). Then the technique determines which alternatives would be found inferior (i.e., worse than some other alternative in every respect) for all utility functions within that range.

For large systems, traditional uncertainty analysis can be successful in the important task of sensitizing users to the extent of uncertainty (e.g., Reilly et al. 1987; Morgan et al. 1990; Hope et al. 1993; Dowlatabadi and Morgan 1993b). Psychologists have shown that people, even long-term experts in the field, can grossly underestimate the range of uncertainty in estimates (Tversky and Kahneman 1974; Dowlatabadi 1995; Shlyakhter et al. 1995). This can have disastrous consequences if a "brittle" policy is adopted that performs poorly under circumstances other than the narrow ones assumed in its development. But error propagation can nonetheless fall prey to the generation of such broad ranges of ostensibly statistically feasible outcomes so as to potentially lose policy relevance. In the context of integrated assessments of climate change, Klemes (1994) calls these results "information pollution." This spread can be reduced by Monte Carlo Bayesian analysis, in which the initial results of error propagation studies are combined with empirical observations of outputs, yielding *posterior* distributions of model parameters and model outputs that may be significantly more precise than the original distributions. This approach has been used, for example, to refine models of rise in sea level (Patwardhan and Small 1992) and to update beliefs about the effects of climate change in an analysis of alternatives for managing the levels of Lake Erie (Hobbs et al. 1997). Yet the resulting distributions can still be very wide, perhaps too much to distinguish meaningfully among policy alternatives.

Regret-based approaches avoid (at a cost) many of the shortcomings of conventional error propagation approaches (e.g., explosive growth in model uncertainty), but regret-based methods are no panacea for the problems of error propagation. All else being equal, they are conservative and sometimes excessively (perhaps unaffordably) so. Yet a potentially significant advantage of regret-based methods is that they, by design, focus on possible effects of uncertainty in a more easily communicable and, perhaps, policy-relevant manner. Such methods, for example, often address uncertainty from a *minmax* (Savage 1954) perspective (developing policy scenarios that minimize the worst effects that could occur—economic, environmental, etc.—given predetermined joint probability laws for the random components in the system). In that sense, regret-based methods use the same information as traditional error propagation but intentionally refrain from propagating the entire range of probabilistic outcomes. Practitioners may still prefer, however, to view the entire probability distribution of possible events (Parson and Fisher-Vanden 1997). In that case, stochastic dominance can be useful for analyzing the results of error propagation. Even though the distribution of outcomes for different policies may overlap considerably, stochastic dominance will generally show that some policies will always be preferable, given very weak conditions that constrain risk preferences.

Static versus Dynamic Approaches

Most of the history of integrated assessment involves static analyses (i.e., not changing over time), not because these applications are inherently static, but for the critically important reason that static problems are typically computationally tractable—at least much more so than their dynamic counterparts. Here is an issue where tractability plays a dominant role, for the transition from static to dynamic models and analyses represents a quantum jump in complexity, analytical effort, model solution, and, often, requirements for input data. The gradations of dynamic analysis vary greatly from simpler repeated static analysis to explicit modeling efforts involving, say, closed loop feedback optimal control, which places a system in a desired state in some optimal fashion (e.g., as quickly as possible) using feedback to make adjustments over time. In addition to issues involving optimal control, dynamics play a critically important role when one attempts realistically to model the financial aspects of pollution management and the effects of those aspects on optimal control strategies that may change with time. In problems of complex ecosystem change, a framework of sequential decision making permits the recognition that new knowledge will be gained over time (Funtowicz and Ravetz 1991, 157–60).

Visualization and System-User Dynamics

Visualization in the sense intended here involves much more than simply developing enhanced graphical user interfaces. But even that task is far from simple and, as Loucks (1995) notes, it is crucial to the success of decision support systems. The advances that need to be made touch upon the fundamental means through which people assimilate, interpret, and interact with complex, highly dimensioned information. And in many respects, advances in visualization will require that researchers in the integrated assessment community with expertise in more traditional quantitative methods must also draw from nontraditional systems disciplines, notably cognitive science. There exists a growing body of exciting work in cognitive science and its relationship with visualization, and a synthesis of that work with multiobjective and risk analysis methodology may yield improved models and methods for integrated assessment.

The use of abstract visual metaphors to show data did not begin until the late eighteenth century, with the work of Crome and Playfair. Playfair developed the bar chart in his "Commercial and Political Axis" (1786) and the pie chart and circle graph in "The Statistical Breviary" (1801). Subsequent developments involved such famous names as Bessel (the graphic table) and Fourier, who introduced the cumulative distribution curve (Feinberg 1979). Modern-day visualizations allow the viewer to analyze and compare multivariate data, to recognize both local and global properties of data, and to study the reality of time. Structure in data that would otherwise be beyond our processing abilities can be revealed with today's high-resolution interactive graphic capabilities. Defanti et al. (1989, 13) contend that "the information . . . undergoes a qualitative change because it brings the eye-brain system, with its great pattern-recognition capabilities, into play in a way that is impossible with purely numeric data." Tufte (1983, 9) notes that "graphics are instruments for reasoning about quantitative information."

As a case in point, visualization techniques were applied to the EPA's Regional Oxidant Model, which is an air quality model designed to examine the transport and deposition of airborne pollutants. The Regional Oxidant Model simulates most of the significant chemical and physical processes responsible for the production of ozone over a 100-km domain for episodes lasting 15 days. Separate visualizations were used to examine each chemical species in each of the three atmospheric layers associated with every run of the model. The researchers also created images of the model input vectors. An animated visualization of wind over the domain resulted in the improvement of this parameter's specifications (Rhyne et al. 1993).

In operations research, visualization allows managers to experiment with alternative scenarios generated by decision models by interactively varying visual representations of the model parameters. The goals of this approach are to help managers gain confidence in the use of the models and to equip those managers to use their own experience and judgment of the original problem domain to help analysts improve those models. The majority of the early visual interactive modeling applications used simulation as the modeling mechanism (Hurrion 1986), but applications are emerging in such areas as network modeling (Anghern and Luthi 1990) and dynamic programming (Lemberski and Chi 1984). Some important applications for these methods, such as water resources planning (Johnson and Loucks 1980; Loucks 1995), have emerged. Related developments involve the integration of process models and sophisticated visualization techniques into geographic information systems (see Chapter 3 and *Mitasova and Mitas 1998; Taylor 1997*).

A promising area of research is the use of these evolving methods of visualization to represent multiobjective tradeoffs and risks. An important rationale for improved visualization methods is that many errors in decision making under multiple objectives can be attributed to difficulties in processing multidimensional data. Experimental investigations have uncovered systematic errors of several sorts. First, decision makers generally pay attention to a single attribute and are less willing to make tradeoffs on other objectives than they should be (Gardiner and Edwards 1975). They may not even consider alternatives below some threshold that they set on that objective (Tversky 1972). Second, decision makers tend to localize their searches to those regions in criteria space in which their initial preferences lie. This behavior is referred to as *anchoring* (Tversky and Kahneman 1974). Therefore, an effective visual model should ideally include all of the alternatives, or at least a representative set, and all of the objectives. Carswell and Wickens (1987) and Goettl et al. (1991) argue that the integration of variables in displays helps the viewer to make any judgments that depend on all of these variables. Additionally, the model might use color or some other visual property to highlight those alternatives or attributes that the decision maker seems to be overlooking. Bennett et al. (1993) discuss different ways in which images can be highlighted in integrated displays. Of course, all of these features need to be integrated in the visual model without violating some important principles in graphic design—not the least of which is to avoid clutter (Tufte 1983, 1990, 1997). In the development of applications for public health, studies of the interpretability of epidemiological maps will also be useful (see Chapter 3).

Integrated Assessment as a Participatory Process

In the development of policies to address complex problems, technical expertise may serve as a barrier to participation by interested constituencies (deLeon 1997). An approach to integrated assessment as a participatory process attempts to counter this tendency by using a suite of participatory methods often referred to as interactive, deliberative, or communicative methods (Rotmans 1998). Participatory methods have long been used in other contexts. For example, potential consumers gather in focus groups to discuss the appeal of new products, professionals meet in consensus conferences to assess scientific and technological research, and organizations in the corporate and military sectors conduct games for strategic planning. In integrated assessment, participatory methods aim for the active involvement of nonscientists, who are usually government officials, stakeholders, or members of the general public. The hope is that the use of participatory methods will ensure the relevance of the integrated assessment process and provide public support necessary in democratic societies.

Some participatory methods seek to elicit information from nonscientists (Rotmans 1998). In the *dialogue method,* scientists who are designing an integrated assessment query the intended users of the assessment. *Policy exercises* set up participants in a negotiation process structured around a simplified representation of a complex system. As participants play roles during a policy exercise, they provide information about human behavior and preferences for particular policies. The participants are not necessarily the users of the assessment. Both dialogues and policy exercises are often conducted in conjunction with the development of a formal computer-based IAM.

The participatory methods of *mutual learning* bring together scientists and nonscientists as coproducers of knowledge (Rotmans 1998). In an approach using focus groups, scientists may act as facilitators by providing scientific information to various groups of stakeholders. In a more interactive approach, scientists and stakeholders work together to define problems and propose solutions. Methods of mutual learning, as with other participatory methods, are often, but not necessarily, linked to IAMs.

"Mutual construction" of science and policy as scientists and policymakers work closely together also imposes limitations on integrated assessment that may not be obvious to the participants (Shackley and Wynne 1995; Rayner and Malone 1998). Consensus on decisions is sometimes achieved for pragmatic reasons, "though such decisions are frequently presented subsequently as having been purely scientific in character" (Jäger 1998, 144). "Mutual construction" may generate "regulatory science," in which an assessment is driven by pressure to find answers quickly in response to a specific problem arising in a political or legal context (Jäger 1998, 144). The procedures for evaluation and assessment in regulatory science incorporate advice and publications from scientific experts but stand in contrast to "new research" of a more basic nature.

The distinction between basic science and regulatory science has important implications for the process of integrated assessment. The pressures of time may demand the utilization of existing models and data. It simply may not be possible to consider and select from an entire array of approaches (see "A Taxonomy of Approaches to Integrated Assessment Modeling," above). These conditions increase the dangers of assumption drag, the retention of outmoded assumptions (Ascher 1979, 203). Pulling in models and data across disciplines also increases the likelihood of assumption drag because it is difficult to maintain currency in many areas.

Integrated assessment, of necessity an interdisciplinary endeavor, often lacks credibility in the basic component disciplines (Rotmans and Dowlatabadi 1998, 301). The problems are especially acute in the characterization of social and economic phenomena in IAMs that attempt to address fundamental questions about changes in behavior in response to environmental changes, processes of technological innovation and diffusion, and transitions in demographic and epidemiological characteristics, including urbanization and migration (Rotmans 1998, 167). However, not all processes at all scales are relevant to the outcomes in IAMs; the establishment of IAM credibility should focus on whether component processes can be safely ignored or simply parameterized rather than whether the catalogue of processes embedded in the model is apparently comprehensive (Schneider 1997, 234).

The translation of information between the domains of science and policy requires multiple ongoing dialogues rather than a single consensus. The process itself may involve multiple models, since "intelligent decision-making requires the appropriate use of many different models designed for specific purposes—not reliance on a single, comprehensive model of the world" (Sterman 1991). In the case of global climate change, Root and Schneider (1995) propose a "strategic cyclical scaling" of communication between physical scientists studying global processes and biological scientists studying impacts at a local or regional scale. Henderson-Sellers (1996) envisions an analogous "cyclic social co-operative" between basic climate research and the needs of policymakers. Any system or systems for coordinating this information must handle great complexity. Henderson-Sellers (1996) decries the "scenario bazaar" of inadequately explained models of global climate change but also recognizes that "universally agreed scenarios" might close off useful diversity of perspectives.

Another challenge to the "mutual construction" of science and policy is the incorporation of more and more diverse sets of actors into the decision-making process. Traditional frameworks either implicitly or explicitly assume a single decision maker who weighs the costs and benefits of various decisions in the "rational actor paradigm" (Jaeger et al. 1998). Indeed, the common expression of "embedding models in policymakers' heads" creates a naive image of the political process in which one person in authority makes each policy decision (Parson and Fisher-Vanden 1997). One IAM addresses this problem by incorporating a classification of three different perspectives: (1) risk-accepting "hierarchists" who believe in rational allocation of resources, (2) risk-averse "egalitarians" who believe in strategies to reduce needs, and (3) risk-seeking "individualists" who believe in expanding the resource base (Rotmans and Dowlatabadi 1998, 361). Multiple parallel projects on integrated assessment modify the process itself to bring in more perspectives (Parson and Fisher-Vanden 1997; Schneider 1997). More people involved in more complex decisions only add complications to an already difficult problem of communication. Structuring activities to involve participants in the development of models can aid in developing communication and consensus (Bonnicksen 1996).

In attempting to understand how to conduct a successful integrated assessment, it is useful to consider problems according to the level of uncertainty about the system to be managed and the magnitude of the stakes (i.e., consequences) in the decision (Funtowicz and Ravetz 1991, 145; Jaeger et al. 1998, 195). *Applied science* refers to situations with relatively little uncertainty and low stakes in the outcome. In these situations, scientific experts generally agree on appropriate methods and can utilize large and reliable databases. When system uncertainty and the magnitude of the consequences are considerable, the situation of applied science becomes one of *professional consultancy,* for which quantitative analysis of environmental risks must be supplemented explicitly by experienced qualitative judgments. When system uncertainty and the magnitude of the consequences are high, the situation of professional consultancy becomes one of *postnormal science,* which is permeated by qualitative judgments. Dialogue and advocacy become more central to the process of inquiry. Extended peer communities are needed because "the essential function of quality assurance can no longer be performed by a restricted corps of insiders" (Funtowicz and Ravetz 1991, 149). Integrated assessment for global issues is most closely related to the concept of postnormal science.

The solutions to complex problems require an institutional structure for an open and participatory process (Pielke et al. 1999). Identifying stakeholders in the solution and ensuring their participation are essential for success. The process should include clear expression of the limitations and uncertainties of models that are used and explicit consideration of adaptive strategies that are not highly dependent on specific predictions (see "Evaluation of the Effectiveness of Integrated Assessment," below). Although participants in the process are unlikely to comprehend every aspect of a model, they should be able to ask questions directly pertinent to the interpretation of results (Table 5.2). It is also important to understand the political and ethical context surrounding an integrated assessment (Pielke et al. 1999). Those who provide analyses have professional and organizational linkages that influence their perspectives and possibly raise conflicts of interest (Pielke et al. 1999; Ascher 1979, 12–13; Levins 1998). Responsible formal analysis is transparent, which may defeat many interests. Attempts to shield the process from scrutiny will ultimately prove counterproductive.

Evaluation of the Effectiveness of Integrated Assessment

Ideally, standard criteria for evaluating the effectiveness of integrated assessment should permit a clear understanding of the strengths and weaknesses of various approaches. That is, how does one assess an assessment? The answer to this question is not straightforward. The interplay of science, technology, and the decision process requires a multifaceted discussion.

Studies of successful interactions between science and policy do not always fit prevalent beliefs about integrated assessments. A notable example is an assessment of stratospheric ozone depletion (see Chapters 6 and 7) that was highly influential in shaping environmental policy in 1985 and yet was mostly technical atmospheric science without estimates of impacts (Jäger 1998, 151). Nevertheless, the report successfully conveyed a strong message that human activities were threatening the earth's protective layer of stratospheric ozone. It is possible that the political effect of scientific consensus would have been harmed by divisive controversy about the pro-

---

Table 5.2   Checklist for the Model Consumer

1. What is the problem at hand? What is the problem addressed by the model?
2. What is the boundary of the model? What factors are endogenous? Exogenous?[a] Excluded? Are soft variables included?[b] Are feedback effects properly taken into account? Does the model capture possible side effects, both harmful and beneficial?
3. What is the time horizon relevant to the problem? Does the model include as endogenous components those factors that may change significantly over the time horizon?
4. Are people assumed to act rationally and to optimize their performance? Does the model take noneconomic behavior (organizational realities, noneconomic motives, political factors, cognitive limitations) into account?
5. Does the model assume that people have perfect information about the future and about the way the system works, or does it take into account the limitations, delays, and errors in acquiring information that plague decision makers in the real world?
6. Are appropriate time delays, constraints, and possible bottlenecks taken into account?
7. Is the model robust in the face of extreme variation in input assumptions?
8. Are the policy recommendations derived from the model sensitive to plausible variations in its assumptions?
9. Are the results of the model reproducible? Or are they adjusted[c] by the model builder?
10. Is the model currently operated by the team that built it? How long does it take for the model team to evacuate a new situation, modify the model, and incorporate new data?
11. Is the model documented? Is the documentation publicly available? Can third parties use the model and run their own analyses with it?

Source: Data from Sterman 1991, Checklist for the Model Consumer.

[a]Endogenous variables are calculated by the model. Exogenous variables are provided from sources external to the model.

[b]Soft variables are qualitative and difficult to quantify.

[c]Forecasts produced by models are often modified using the judgment of the model builder about what is reasonable.

jections of impacts that depend on assumptions about the future emissions of ozone-depleting gases (Jäger 1998, 151). Scholars do not agree about the relative importance of various factors that contributed to the successful development of policy for stratospheric ozone (Pielke 1997, 261).

Another complication is that the salient effects of environmental hazards may not be the same in all populations. For example, Britain and Germany had different experiences in the development of policies to remove lead from gasoline (Haigh 1998). In Britain, public sentiment expressed in

the Campaign against Lead in Petrol focused on the adverse effects on human health of exposure to lead. In Germany, concern about the destruc-
tive effect of nitrogen oxides on forests resulted in support for catalytic
converters in automobiles. Since catalytic converters do not function properly with leaded gasoline, the concern about forests in Germany translated
into strong support for removing lead from gasoline.

One aspect of a general evaluation of integrated assessment is the influence of technical information that has been provided to people making
decisions. Three dimensions of this influence are the attention received,
the explicitness of use, and the decisiveness of the influence on the choice
of decision outcomes (Ascher 1979, 17). As noted above, assessment of the
problem of stratospheric ozone depletion in 1985 certainly gained attention, although which factors played decisive roles on the choice of decision
outcomes are not entirely clear (Pielke 1997, 261). Examples under "The
History of Integrated Assessment Models" (above) describe the explicitness of use of technical tools in terms of the satisfaction of users or the fa-
cilitation of group consensus. Bonnicksen (1996) points out how difficult
it is to ascribe success to a particular approach because real-world assessment problems cannot be replicated in a laboratory setting; nevertheless,
he uses the achievement of group consensus on a particular environmental problem as a measure of success. The section on the history of IAMs
also presents an example of direct influence on decision outcomes of the
Canadian government, which adopted model-derived recommendations
on reduction of SO_2 emissions at large point sources in Canada to achieve
prescribed levels of environmental quality (Ellis et al. 1985).

The measures of influence above do not address the accuracy of the
technical information provided in an assessment. Indeed, accuracy of information is not a prerequisite for influence. The "Limits to Growth"
model illustrates this point (Ascher 1979, 34–36). When the model was
published in 1972, it received a great deal of publicity and respectability for
a position that no growth was optimal for maintaining the world's standard of living. The argument was by no means original, but the use of computer simulation modeling was relatively novel and apparently objective.
Moreover, the sponsorship of the study by the Club of Rome, a business
group, gave the impression to the public that the bias of the study would
have been in favor of the opposite conclusion. In reality, many members of
the Club of Rome were predisposed to the position of no growth. The scientific community severely criticized the model's assumptions and methods, demonstrating that "the capacity to attract attention and the ability to
persuade are quite different" (Ascher 1979, 36).

Many integrated assessments provide technical information in the
form of an explicit description of what will happen in the future—they
make predictions. Models of the integrative earth sciences (i.e., sciences of

the solid earth, oceans, and atmosphere) demonstrate the general problems faced by researchers attempting to validate the predictions generated by complex system models used for integrated assessment; such predictions are not in the form of invariant physical laws but rather are highly dependent on time and place (Sarewitz and Pielke 1999). Observations of characteristics of populations and ecosystems are inherently statistical and so have intrinsic uncertainty—it is not meaningful to state an exact count for the entire population of a country or the world (Cohen 1995, 18–21). Many data on complex global phenomena are subject to serious systematic errors that make it difficult to draw reliable conclusions (Funtowicz and Ravetz 1991, 140–41).

For those with a critical understanding of statistical variation and bias (see Chapter 2 under "Potential Sources of Error in Epidemiological Studies"), the ability of a model to reproduce behavior observed in the past and present provides some criteria for testing validity. The approach of reasoning by analogy avoids formal systems models and instead makes models from direct analogies with historical events, such as changes in climate and the collapse of ancient civilizations (Meyer et al. 1998). However, real-world systems are dynamic and subject to change in unexpected ways. For example, forecasting of food prices based on detailed information on crop yields can be stymied by unpredictable political events, such as an oil embargo that raises energy prices (Schneider 1997). These shocks to the system are called "surprises" in the literature on integrated assessment (*Schneider and Turner 1997*). Consistency of a model with data from the past or present provides no guarantee that the model will work in the future. One can never be certain about what lessons to draw from historical analogies.

Techniques from the "harder" physical sciences invariably must be integrated with the social sciences. As an example, variation in social vulnerability seems to be more important than variation in physical environmental factors in accounting for the societal consequences of extreme weather and climate events (Kunkel et al. 1999). To make matters even more difficult, "predictions themselves are events that cause impacts on society" (Pielke et al. 1999, 313). Even if a prediction works fairly well in describing certain natural phenomena during a period of basic scientific research, individual and societal responses when the prediction is announced to the public can and do alter the associated effects. In fact, changing behavior to reduce adverse societal consequences is typically the purpose of making such announcements. Behavioral responses to predictions may also generate unintended societal losses, as shown in the example of Yakima in the introduction to this chapter.

The very notion of using models for the evaluation of alternative policies or actions means that forecasts are conditional. "Such forecasts are not appraisable in terms of accuracy" (Ascher 1989, 470), because real-world policies are never identical to idealized policies in models. Evaluation should focus on a clear exposition of the method and theory at the core of the model. It is difficult to determine where errors lie in complex models (Ascher 1981, 253). "The problem is that the methodology is often unclear, particularly when formal modeling is involved" (Ascher 1989, 473). The core of the model should also be open to question and debate. "Yet methodological hard cores often seem especially difficult to change, partly because they are so intimately associated with specific professional training" (Ascher 1989, 473).

Difficulties in determining the accuracy of a model lead to disagreements that prevent the adoption of policies to address important problems. "Even agreement across models is no indication of validity; they could all be wrong" (Ascher 1981, 253). However, "accurate prediction of phenomena may not be necessary to respond effectively to political or socioeconomic problems created by the phenomena" (Pielke et al. 1999, 313). Building scenarios about the future may provide insight into the implications of different policies (Sarewitz and Pielke 1999, 130). It may be very useful to identify potential surprises, even if they are not likely; these efforts should be "clearly labeled as 'dangerous possibilities' rather than as likelihoods" (Ascher 1979, 212). Some investigators wish to avoid the pitfalls of quantitative predictions by using the terminology of "projection" instead. It is not always clear in practice how strongly investigators distinguish between projections and forecasts; certainly, the distinction between projection and forecast is not recognized in standard English usage (Henderson-Sellers 1996). Building scenarios without claiming predictive capability does not entirely eliminate the problem of accuracy. An essential dilemma is that, although qualitative insights are more robust than quantitative forecasts, quantitative outputs are used to develop confidence in a model (*Kandlikar and Risbey 1996*).

With uncertainties in any system, monitoring and adaptive approaches to management should be considered in an integrated assessment. For example, although prediction of the exact time and location of an earthquake is not technically feasible, advances in earthquake preparedness have reduced the vulnerability of populations to earthquake damage (Pielke et al. 1999). More generally, diversification of economic and technological infrastructures, strengthening of responses to disasters, and management that emphasizes flexibility and lifelong learning can create adaptive societies that should be able to reduce social vulnerability.

Perhaps the most important message about the evaluation of integrated assessment comes from a retrospective analysis of a major U.S. assessment of the acid rain problem, which has been praised for the quality of its science and criticized for its poor linkage with policy (Herrick and

Jamieson 1995). A proposed solution to address the anticipated needs of policymakers is not adequate because of implausible assumptions "that policy makers will define the problem in the same terms; that their 'needs' are transparent and remain constant as time, circumstances, and political fortune change; and that they agree on the implicit assumptions, structural biases, and other technical elements of the assessment framework" (Herrick and Jamieson 1995, 110). Instead, the process of integrated assessment must grapple with questions of values up front. The role of the scientific community should be to enable the establishment of a community of understanding. For effective decisions to be made, "there must be widespread agreement on what questions are being asked, why they are important, what counts as answers to them and what the social use of these answers might be" (Herrick and Jamieson 1995, 111–12).

The Integrated Assessment of Ecosystem Change and Public Health

The Building of a Framework

Public health is affected by changes in the global ecosystem, broadly defined to include natural resources, population growth, and patterns of economic development; all of these components interact with each other as well (Fig. 5.1). The case studies in Chapters 11, 12, 13, and 14, along with numerous examples in other chapters, demonstrate the diversity and complexity of the effects of global change on public health. The development of a framework for understanding these effects must therefore extend beyond the traditional approach of characterizing risk in terms of exposure to a specific health hazard (see Chapter 4).

Since global change results from the interaction of natural and anthropogenic dynamics in the global ecosystem, an integrated assessment must synthesize information from many disciplines (Chapter 1). An analysis of natural dynamics should address air and climate as well as water, land, and energy; an analysis of anthropogenic dynamics should address migration, urbanization, industrialization, and agricultural development (see Fig. 1.1). The linkage between information about global change and information about public health may take many forms. Empirical epidemiological studies, including analyses that take advantage of geographic information systems, provide evidence about relationships and can support assumptions used as a basis for modeling (see Chapters 2 and 3). Models of global change can be linked with more traditional models for epidemiology and disease control. For example, a key parameter for transmission of disease might be dependent on specific environmental and social parameters, such as temperature or quality of housing. In the area of infectious diseases, typical applications of models include the immunization of children and the control of mosquitoes that transmit malaria (see

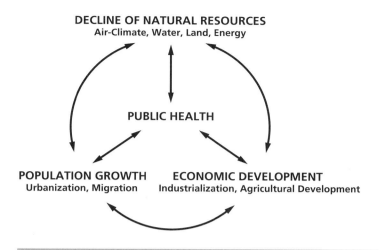

DECLINE OF NATURAL RESOURCES
Air-Climate, Water, Land, Energy

PUBLIC HEALTH

POPULATION GROWTH   **ECONOMIC DEVELOPMENT**
Urbanization, Migration   Industrialization, Agricultural Development

Figure 5.1 Integrated assessment of ecosystem change and public health.

Chapter 10 and Aron 2000; Anderson and May 1991). Other models have been developed for the assessment of risk from exposures to chemicals and radiation (see Chapter 4).

A very general framework must be expanded to address particular issues. For example, an integrated assessment framework proposed for climate change and infectious diseases focuses on diseases carried by vectors, which are animals that can transmit disease to humans (see Chapter 10; Chan et al. 1999). Of the major vector-borne diseases considered, most are transmitted by bites from flying insects, such as the mosquitoes that transmit malaria; in the case of schistosomiasis, the vectors are snails that shed infective stages into freshwater habitats. The assessment framework adds more structure to the description of natural resources by using components for nutrient cycle changes, community relocation, and biodiversity loss, which are all related to assemblages of biological species and the nutrients in their environment. The anthropogenic dynamics include control/adaptation, nutrition, sanitation, population/economy, and migration/travel. The components of the biology of transmission are vector physiology and behavior, vector migration, vector breeding, pathogen reproduction, pathogen transmissibility, and pathogen virulence. All of these elements in the framework are linked to describe the epidemiological effect of climate change, which in this assessment includes both natural climatic variability and anthropogenically induced global climate change. Such a framework can be useful for integrating information from a variety of disciplines and identifying important gaps in research.

Major uncertainties are inevitable (see "Evaluation of the Effectiveness of Integrated Assessment," above). It is difficult even to obtain consensus on the nature of the uncertainties involved in an integrated assessment. The experience with handling the issue of evidential adequacy in an assessment of acid rain led to the realization that "what is regarded as unknown or uncertain depends profoundly on what one needs to know" (Herrick and Jamieson 1995, 110). Environmental hazards to human health occur at a wide range of scales—global, regional, community, household, and individual (see Table 2.2). Multiple disciplines and subdisciplines bring different perspectives on spatial and temporal scales, statistical testing, and approaches to integrated assessment modeling. Different disciplines also bring very different attitudes toward environmental hazards, as demonstrated in a study of beliefs about the effect of climatic warming on the global economy (Schneider 1997, 242). Classical economists in the study did not believe that warming even as high as six degrees Celsius by the year 2100 would have a catastrophic economic impact despite the fact that such a change would be equivalent to the transition from an ice age to an interglacial epoch. Natural scientists in the study were more concerned about catastrophic consequences and provided estimates of damage that were larger than those of the economists by an order of magnitude. Such variation reflects profound differences in beliefs about the ability of human ingenuity to compensate for the loss of functions in the natural world. Views about the interdependence between human society and the natural environment are central to debates about global change (see Chapter 6).

Consequently, one of the thorniest issues to be addressed is the effect of technological and behavioral adaptation to change. In reviewing debates about the impact of global climate change on agriculture, Schneider (1997) contrasts extreme assumptions about "the dumb farmer" and "the genius farmer." According to the "dumb" assumption, farmers do not change their practices and simply have to accept losses caused by changes in climate; according to the "genius" assumption, farmers can readily adapt to changes in pests, crops, weather, technology, policy, and climate. The debate about adaptation in public health is similar. One common scenario is the spread of an infectious disease into new areas because of changes in climate that provide more favorable conditions for transmission. On the one hand, estimates of the impact of global climatic change focus on its sole influence on the conditions of transmission (e.g., Lindsay and Birley 1996). On the other hand, public health officials point to the capacity to reduce or even eliminate the impact were an infectious agent to be introduced under those altered conditions. Differences in assumptions about adaptation lead to tremendous variation in the expected losses due to global change. One trend in IAMs has been the incorporation of assumptions about adaptation and technological change that lie between the "dumb" and "genius" extremes (Schneider 1997).

There are no simple answers to complex problems affecting ecosystem change and public health. However, other experiences in integrated assessment demonstrate that the establishment of an open and participatory process is essential (Pielke et al. 1999; Herrick and Jamieson 1995; Shrader-Frechette 1991). In developing an integrated assessment, scientists should not be isolated from its policy context when determining the right questions to ask and the criteria for evidence.

Multiple Benefits and Costs

Decisions in the joint management of health issues and ecosystems involve complex tradeoffs (Wolman 1995; Graham and Wiener 1995). Multiple benefits and costs generate multiple objectives to be considered in decisions (see "Single versus Multiple Objectives," above). The benefit of reducing exposure to mosquitoes that can transmit infectious agents must be weighed against the damage to natural resources and human health caused by the application of insecticides or the drainage of wetlands to control populations of mosquitoes. The benefits of the reduction of acid rain might seem at first to be less complicated, since acid rain is harmful to humans, forests, freshwater fish, and even structures made of limestone and marble (see Chapters 7 and 13). However, the fact that the impact of acid rain is felt far away from its source complicates the geographic structure of who pays for and who benefits from the cleanup. Acid rain is only one of many pollutants that cross national boundaries and require international environmental negotiations (see Chapters 6 and 13; Commission for Environmental Cooperation 1997).

A single environmental change may have both positive and negative aspects. For example, a drought in West Africa that contributed to reductions in the transmission of malaria fostered the spread of tick-borne relapsing fever (Trape et al. 1996). Heavy rainfall in Brazil is hazardous in urban areas, where it causes mudslides and outbreaks of leptospirosis, an infectious disease transmitted by rats; yet the regular flooding of the Amazon River is important for maintaining its ecological health and sustaining human populations that depend on it (see Chapter 14). More generally, climate plays a dual role as a hazard and as a beneficial natural resource (Meyer et al. 1998).

The Placement of Value on Benefits and Costs

One approach to making decisions compares costs and benefits. Analytically, it is useful to consider economic costs and benefits that literally reduce the elements of the decision to a common currency. This approach is difficult to apply to global ecosystem change and public health because many of the benefits provided by ecosystems and human life are not in a market system of goods and services. Although the possibility of catastrophic damage in the future has led many to advocate the use of a precautionary principle to commit resources to prevent such damage, the direct application of this principle to making decisions is not altogether clear (Funtowicz and Ravetz 1991, 154; Foster et al. 2000). However, experience in public health and environmental management has generated insights about methodology for valuation in the context of making decisions.

The economic impact of disease is usually expressed as a cost of illness (World Bank 1997, 8; Haddix et al. 1996). The cost of illness includes the direct cost of medical treatment and the indirect costs of lost earnings, lost time, or lost work. In practice, it is difficult to measure these indirect costs for people outside of the work force. Estimating the cost of disease is especially controversial when applied to mortality in an attempt to place a value on human life (World Bank 1998, 6). Other limitations of such estimates are the omission of pain and suffering and the omission of broader, more diffuse effects on the economy and society. An alternative approach is a method of contingent valuation called willingness-to-pay, which asks people what they would pay to avoid disease or reduce the risk of disease (World Bank 1997, 8; Haddix et al. 1996).

A study of the economic impact of an outbreak of foodborne illness illustrates the basic method of determining the cost of illness. In 1992, an outbreak of hepatitis A virus associated with a catering facility in Denver, Colorado, resulted in 43 cases of hepatitis and the potential exposure of 5,000 people to whom the catering facility had served food (Dalton et al. 1996). The total costs were estimated to be $809,706. Standard public health practice was to provide immune globulin injections to prevent illness in those who might have been exposed. The cost of controlling the possible spread of disease, which included the immune globulin injections and the time of health department personnel, accounted for about 85 percent of the total. The remainder of the cost combined the direct cost of medical treatment, loss of work, and loss to the catering business from discarding food that was potentially contaminated. When possible contamination of food occurs earlier in the chain of distribution, the potential effects on human health and business are even greater. For example, the aquaculture industry has growing concerns about possible microbial contamination of food derived from aquatic animals grown in culture (Greenlees et al. 1998).

Placing a value on ecosystems raises difficulties similar to those in placing a value on human life and health. One area of research is estimating the economic value of goods and services provided by ecosystems (Costanza et al. 1991). For example, the World Bank considered the economic benefits of establishing a national park in Haiti in terms of reduced damage to irrigation systems downstream (World Bank 1998, 7). Protection of the watershed in the park was therefore linked to enhanced agricultural productivity due to irrigation and to avoidance of costs of replacing damaged infrastructure. The willingness-to-pay method is also used. A study in Croatia asked what tourists would pay for reforestation of coastal areas damaged in war (World Bank 1998, 10). Another method of contingent valuation is willingness-to-accept. In Madagascar, people living near a new national park were asked what level of compensation they would accept so that they would agree not to enter the forests in the park; international tourists were also asked what they would be willing to pay for access to the park (World Bank 1998, 10).

Comparisons of benefits and costs raise additional problems. Even if values can be assigned at one point in time, it is necessary to aggregate measures across time periods to account for processes that unfold over many years, decades, and centuries. Standard economic analysis of investments discounts the future; as a benefit occurs further off in time, it has less and less value. This kind of discounting is not appropriate for considering the transfer of resources between generations (Norgaard and Howarth 1991). Aggregate measures of cost and benefit also ignore the distribution of resources within society (Jaeger et al. 1998, 165). For many complex problems of ecosystem change and public health, it simply may not be acceptable to assign economic costs and benefits. Cost-effectiveness is an alternative strategy that determines the best way of achieving a specified goal; a cost-effectiveness analysis does not reveal whether the benefits justify the resources invested (World Bank 1998, 4). As explained under "The History of Integrated Assessment Models" (above), the decision on how best to reduce emissions of SO_2 in Canada was based on an analysis requiring a specified goal for acid deposition to be met.

The Integrated Management of Ecosystems and Public Health

A better understanding of the dynamic linkages between ecosystems and public health is leading to new and diverse opportunities for interventions earlier in processes that could become direct threats to public health. Improvements in climate forecasting, such as advance warnings for El Niño and La Niña climatic anomalies (see Chapter 8), are becoming part of the public health community's planning for response to natural disasters and epidemics (Kovats et al. 1999). For example, groups that coordinate the control of malaria in southern Africa are paying attention to climate forecasts for that region, where rainfall is associated with La Niña (see Chapter 12). Organizations working to manage and conserve natural resources can join with organizations that provide community-based family planning services and reproductive health care (Engelman 1997). In Honduras, food security has emerged as a central theme in discussions of family planning (Vogel and Engelman 1999, 36). Programs that control the emissions and formation of toxic hazards, such as acid rain, can be evaluated jointly on the basis of environmental impacts and health effects (see under "Multiple Benefits and Costs," above). Actions in the public health sector require cross-sectoral communication, since ecosystem changes have broad social, economic, and ecological consequences (see, e.g., Glantz 1996).

Indicators for integrated assessment of ecosystem change and public health are necessary to support the management of integrated programs. The tremendous diversity in processes generates a considerable challenge. Within public health, for example, infectious and noninfectious diseases have long been quite separate areas of specialization (see Box 2.3). The management issues for wilderness areas are different from issues in agriculture, although ultimately the preservation of species is linked to the preservation of agriculture (Vandermeer 1997). A process of integrated assessment can aid communication within the public health community, between the public health community and other sectors, and more broadly between scientists and decision makers.

The example of malaria demonstrates how the public health sector perceives the strengths and weaknesses of indicators. A simplified classification of malaria uses indicators for different types of malaria, such as forest, urban, and coastal (Aron and Silverman 1994). These indicators may incorporate information about rainfall, temperature, vegetation, elevation, landforms, and the distribution of species of anopheline mosquitoes that transmit malaria (Rubio-Palis and Zimmerman 1997). Indicators can assist the public health sector in understanding how variation in the transmission of malaria should influence strategies for the control of malaria. However, the application of indicators generates methodological problems. Where malaria is most common, cases of malaria may be recorded based on the presence of fever without confirmation of the diagnosis by examination of parasites in the blood. In a study in Niger, hardly any of the malaria cases reported during the dry season had parasites (Buck and Gratz 1990). Differences between the microclimate experienced by mosquitoes and standard observations made by a weather service can make it difficult to interpret the effects of changes in climate on mosquitoes. In this context, *microclimate* refers to the fact that adult biting mosquitoes are exposed to temperature and humidity at the surface of a leaf or on the interior wall of a human dwelling; in addition, the survival of mosquito larvae depends on microclimatic conditions in aquatic habitats. Moreover, the relationship between malaria and the environment is not static because the characteristics of biological populations change over time (see Chapter 10). Over the past decades, parasites have developed resistance to antimalarial drugs while mosquitoes have developed resistance to insecticides. Therefore, strategies for the control of disease must change over time as well. Currently, the mosquitoes that transmit malaria are unable to breed in polluted water; conditions for urban malaria epidemics would be enhanced considerably if some mosquitoes adapted to polluted water sources in growing urban areas (Buck and Gratz 1990). The development of indicators has generated concerns that many operational decisions for the control of malaria require more detail (Teklehaimanot 1991).

A broad examination of the health of entire ecosystems requires biophysical and socioeconomic indicators as well as health-related aspects (Rapport 1998). Indices of ecosystem health may be grouped into vigor, organization, and resilience (Costanza et al. 1998). Ecological vigor is related to function and productivity, such as the net primary productivity of plant life (see Chapter 8). Ecological organization means structure and biodiversity, which might be measured as an index of species diversity. Ecological resilience refers to the capacity of an ecosystem to absorb disturbances, which might be measured as the inverse of the time required for the size of a particular biological population to return to its predisturbance level (i.e., shorter times to recover mean greater resilience).

The application of risk assessment to ecosystems is not new. Ecological risk assessments have been conducted with structures that combine ecosystem dynamics with a more traditional style of risk assessment for hazards (see Chapter 4 and Cairns 1998). An ecological risk assessment might consider the consequences of changing land use and industrial discharges on water quality in a river basin or the effect of practices in forestry on the survival of an endangered species (Hunsaker et al. 1990). The selection of endpoints for assessment for the U.S. EPA used to be done within the agency, but interested parties now participate in the process (*EPA 1999*). The experience of ecological risk assessment is an important component of integrated assessment, whose aims are a broad examination of many sectors and explicit consideration of decisions to manage risk. Integrated assessment of global ecosystem change also requires a larger spatial scale than the local and regional scales where ecological risk assessments have been applied.

Methods for expressing linkage between ecosystem change and public health must build upon the experience of the field of environmental health, which has a long tradition of using indicators as "an expression of the link between environment and health, targeted at an issue of specific policy or management concern and presented in a form which facilitates interpretation for effective decision making" (Corvalán et al. 1996, 25). An environmental health indicator must embody a linkage between environment and health and, as such, is more than an environmental indicator or a health indicator. For example, an indicator of water pollution and an indicator of life expectancy demonstrate environmental exposures and health effects without an explicit connection. An environmental health indicator expresses linkage between environment and health in one of two ways. An exposure-based indicator projects from information about an environmental hazard to give an estimated measure of risk to human health, while an effect-based

---

Table 5.3   Criteria for Environmental Health Indicators

---

1. Based on a known linkage between environment and health
2. Sensitive to changes in the conditions of interest
3. Directly related to a specific question of environmental health concern
4. Related to environmental and/or health conditions that are amenable to action
5. Consistent and comparable over time and space
6. Robust and unaffected by minor changes in methodology/scale used for their construction
7. Unbiased and representative of the conditions of concern
8. Scientifically credible, so that they cannot be easily challenged in terms of their reliability or validity
9. Easily understood and applicable by potential users
10. Available soon after the event or period to which they relate (so that policy decisions are not delayed)
11. Based on data that are available at an acceptable cost-benefit ratio
12. Based on data of a known and acceptable quality
13. Selective, so that they help to prioritize key issues in need of action
14. Acceptable to the stakeholders

Source: Reproduced with permission from Corvalán et al. 1996, Table 2.3.

indicator projects from health outcomes to environmental causes (Corvalán et al. 1996, 26). Environmental health indicators should meet the majority of criteria in Table 5.3, especially the first four criteria.

Expressing the linkage between ecosystem change and public health involves many problems at many different spatial and temporal scales. The protection of the tropical rainforest serves as an example. At the local level, conservation of a tropical rainforest provides resources to the local population, such as products that can be harvested for their economic value. The preservation of natural resources for future generations might generate interest in services for family planning that would limit the size of local populations and the pressure on local resources. At a regional level, sustainable development in rural areas could reduce migration from rural to urban areas, thereby limiting the pressure on urban infrastructure. The growing percentage of the world's population in urban areas increases demand for urban employment as well as services in health, education, transportation, electrical power, and water and sanitation (see Chapter 6). At a global level, the maintenance of tropical forests plays an important role in the carbon cycle, which is part of the system of global climate (see Chapter 8). The resources from the tropical forest might not only provide local benefit but also eventually lead to the development of pharmaceuticals. The benefits of the biodiversity of the forest could then be enjoyed by pharmaceutical companies and by people undergoing medical treatment (Grifo et al.

1997), although achieving an equitable distribution of those benefits requires considerable planning and forethought (Mays et al. 1997). Issues in the management of ecosystems and the conservation of biological diversity have generated a considerable literature (e.g., Wilson and Perlman 2000; Gibson et al. 1998; Pritchard et al. 1998).

There are many potential candidates for indicators. Some indicators might focus on ecosystem change, such as increasing temperature. Such indicators could include physical measurements of air temperature around the world and of the thickness of polar ice. Knowledge of biological species whose ranges are sensitive to climatic factors can supplement physical measurements. For example, fossil assemblages from the edges of ecosystems show that populations of insects responded rapidly to warming in North America and Europe during the last deglaciation (Epstein 1997, 72). Other biological indicators would be appropriate for measuring direct threats to human health. For example, since rodent populations carry many agents pathogenic to humans (see Chapter 10), monitoring of rodent populations can contribute to an early warning system for public health authorities to take action (Epstein 1997, 72–73).

Indicators need to be evaluated in terms of scientific credibility and utility in the decision-making process for particular programs, following guidelines developed for more traditional environmental health indicators (see Table 5.3). The research process involves consideration of many possible factors in different disciplines (see Chapter 1), the application of epidemiological tools to develop and test associations between ecosystem change and human health (see Chapter 2), the integration of geographic information systems and remote-sensing information into epidemiological studies (see Chapter 3), and the development of integrated assessment models to support the analysis of decisions to manage complex problems (see Chapter 4 and the rest of this chapter). For the results of scientific research to lead to sound decisions, the research process must occur within a larger context of participatory decision making (see under "Integrated Assessment as a Participatory Process," above).

Conclusion

Integrated assessment attempts to analyze social, economic, and environmental factors of complex issues in public policy. The focus on decisions in a policy context is a hallmark of this effort. The process of integrated assessment includes a variety of methods, loosely aggregated as quantitative and qualitative approaches. However, the scientific study of global change is a work in progress. The scientific challenge is to understand the linkages among complex ecosystem phenomena that operate at multiple spatial and temporal scales and their implications for public health in the future. The challenge for integrated assessment is establishing a process that uses sci-

entific knowledge and involves a broad array of people in making fundamental decisions of public policy. A multifaceted strategy for research
comprises all five chapters in Part I with the aim of developing better connections between empirical studies and the management of complex
ecosystems and public health. Even though much is already known, the
only certain expectation is surprise.

SUGGESTED STUDY PROJECTS

Suggested study projects provide a set of options for individual or team
projects that will enhance interactivity and communication among course
participants (see Appendix A). The Resource Center (see Appendix B) and
references in all of the chapters provide starting points for inquiries. The
process of finding and evaluating sources of information should be based
on the principles of information literacy applied to the Internet environment (see Appendix A).

The objective of this chapter's study projects is to deepen understanding of the application of integrated assessment to complex environmental
problems.

PROJECT 1: Multiple Objectives in Decision Making

The objective of this study project is to deepen understanding of how decisions related to complex environmental problems involve multiple objectives.

Task 1. Select an example of an ecosystem issue that has affected public policy in some setting.

Task 2. For the problem and setting in task 1, identify the stakeholders.

Task 3. For the problem and setting in task 1, identify multiple objectives that a decision must address and the tradeoffs between objectives.

Task 4. Summarize your results.

PROJECT 2: Participatory Decision Making

The objective of this study project is to deepen understanding of how decisions are made with public participation.

Task 1. Select an example of an ecosystem issue that has affected public policy in some setting.

Task 2. For the issue in task 1, describe how the public was involved in
the determination of policy, considering onset, duration, and type of involvement.

Task 3. Was the policy generally accepted, or was there widespread dis-
satisfaction?

Task 4. How might you evaluate the effectiveness of public participation?

Task 5. Summarize your results.

PROJECT 3: Integrated Assessment for Ecosystem Change
and Public Health

The objective of this study project is to deepen understanding of the complex process of integrated assessment of ecosystem change and public health.

Task 1. Select an example of an issue of ecosystem change and public health in some setting.

Task 2. For the example selected in task 1, describe the major interactions between public health and ecosystems following the schema in Figure 5.1.

Task 3. Identify the stakeholders in a decision.

Task 4. Identify multiple objectives and tradeoffs between objectives.

Task 5. Compare two strategies for managing the problem.

Task 6. Summarize your results.

Acknowledgments

We thank John Wiener, Mark Meo, Jim Hammitt, and Roger Pulwarty for helpful comments on this chapter. We also thank Mickey Glantz and Jan Stewart for assistance in locating sources of information. Support for Ellis and Hobbs was provided by the National Science Foundation under grant SBR9634336.

References

Anderson RM, May RM. 1991. *Infectious Diseases of Humans: Dynamics and Control.* Oxford University Press, New York.

Anghern A, Luthi H. 1990. Intelligent decision support systems: A visual interactive approach. *Interfaces* 29 (6): 17–28.

Aron JL. 2000. Mathematical modeling: The dynamics of infection. In *Infectious Disease Epidemiology: Theory and Practice* (Nelson KE, Williams CM, Graham NMH, eds.). Aspen Publishers, Gaithersburg, Md., Chap. 6.

Aron JL, Silverman BA. 1994. Models and public health applications. In *Parasitic and Infectious Diseases: Epidemiology and Ecology* (Scott ME, Smith G, eds.). Academic Press, San Diego, pp. 73–81.

Arthur WB. 1999. Complexity and the economy. *Science* 284 (5411): 107–9.

Ascher W. 1979. *Forecasting: An Appraisal for Policy-makers and Planners.* Johns Hopkins University Press, Baltimore.

———. 1981. The forecasting potential of complex models. *Policy Sci* 13:247–67.

———. 1989. Beyond accuracy. *Int J Forecasting* 5:469–84.

Bennett KB, Toms ML, Woods DD. 1993. Emergent factors and effective configural displays. *Hum Factors* 35:71–97.

Bernabo JC, Eglinton PD. 1992. *Joint Climate Project to Address Decision Makers' Uncertainties.* Prepared for U.S. Environmental Protection Agency and Electric Power Research Institute, Science and Policy Association, Washington (May).

Bonnicksen TM. 1996. Reaching consensus on environmental issues: The use of throwaway computer models. *Politics Life Sci* 15 (1): 23–34.

Brown CA. 1984. The central Arizona water control study: A success story for multiobjective planning and public involvement. *Water Resources Bull* 20:331–38.

Buck AA, Gratz NG. 1990. *Niger: Assessment of Malaria Control. Niamey, January 24–Feb-*

ruary 15, 1990. Report for U.S. Agency for International Development Vector Biology and Control Project, Arlington, Va.

Cairns J Jr. 1998. Ecological risk assessment: A predictive approach to assessing ecosystem health. In *Ecosystem Health* (Rapport D, Costanza R, Epstein PR, Gaudet C, Levins R, eds.). Blackwell Science, Malden, Mass., Chap. 14.

Cantor R, Yohe G. 1998. Economic analysis. In *Human Choice and Climate Change*, Vol. 3, *Tools for Policy Analysis* (Rayner R, Malone E, eds.). Battelle Press, Columbus, Ohio, Chap. 1.

Carswell CM, Wickens CD. 1987. Information integration and the object display—an interaction of task demands and display superiority. *Ergonomics* 30:511–27.

Chan NY, Ebi KL, Smith F, Wilson TF, Smith AE. 1999. An integrated assessment framework for climate change and infectious diseases. *Environ Health Perspect* 107 (5): 329–37.

Chankong V, Haimes YY. 1983. *Multiobjective Decision Making: Theory and Methodology.* North-Holland Series in System Science and Engineering, Vol. 8. North-Holland, New York.

Chao PT, Hobbs BF, Venkatesh BN. 1999. How should climate uncertainty be included in Great Lakes management? Modeling Workshop results. *J Am Water Resources Assoc* 35 (6): 1485–94.

Cohen JE. 1995. *How Many People Can the Earth Support?* W. W. Norton & Co., New York.

Cohen SJ, ed. 1994. *Interim Report No. 2, Mackenzie Basin Impact Study: Report of the Midstudy Workshop.* Canadian Climate Center, Atmospheric Environment Service, Toronto.

Cohon JL. 1978. *Multiobjective Programming and Planning.* Academic Press, New York.

Commission for Environmental Cooperation. 1997. *Continental Pollutant Pathways.* CEC, Montreal.

Corner JL, Corner PD. 1995. Characteristics of decisions in decision analysis practice. *J Operational Res Soc* 46:304–14.

Corvalán C, Briggs D, Kjellström T. 1996. Development of environmental health indicators. In *Linkage Methods for Environment and Health Analysis: General Guidelines* (Briggs D, Corvalán C, Nurminen M, eds.). Office of Global and Integrated Environmental Health, World Health Organization, Geneva, Chap. 2.

Costanza R, Daly HE, Bartholomew JA. 1991. Goals, agenda, and policy recommendations for ecological economics. In *Ecological Economics: The Science and Management of Sustainability* (Costanza R, ed.). Columbia University Press, New York, Chap. 1.

Costanza R, Mageau M, Norton B, Patten BC. 1998. Predictors of ecosystem health. In *Ecosystem Health* (Rapport D, Costanza R, Epstein PR, Gaudet C, Levins R, eds.). Blackwell Science, Malden, Mass., Chap. 16.

Dalton CB, Haddix A, Hoffman RE, Mast EE. 1996. The cost of a food-borne outbreak of hepatitis A in Denver, Colorado. *Arch Intern Med* 156 (9): 1013–16.

Defanti TA, McCormick BH, Brown MD. 1989. Visualization: Expanding scientific and engineering research opportunities. *Computer* 22 (8): 12–25.

deLeon P. 1997. *Democracy and the Policy Sciences.* State University of New York Press, Albany.

Dowlatabadi H. 1995. Integrated assessment of climate change: An incomplete overview. *Energy Policy* 23 (4/5): 289–96.

Dowlatabadi H, Morgan MG. 1993a. Integrated assessment of climate change. *Science* 259:1813–14.

———. 1993b. A model framework for integrated studies of the climate problem. *Energy Policy* 21 (3): 209–21.

Ellis JH. 1988. Acid rain control strategies: Options exist despite scientific uncertainties. *Environ Sci Technol* 22:1248–55.

Ellis JH, McBean EA, Farquhar GJ. 1985. Deterministic linear programming model for acid rain abatement. *ASCE J Environ Engineering Div* 111 (2): 119–39.

Engelman R. 1997. Earthly dominion: Population growth, biodiversity, and health. In *Biodiversity and Human Health* (Grifo F, Rosenthal J, eds.). Island Press, Washington, D.C., Chap. 2.

Environmental Protection Agency. 1995. *Documentation of Adaptation Strategy Evaluator Systems.* Climate Change Division, Washington.

Epstein P. 1997. Biodiversity and emerging infectious diseases: Integrating health and ecosystem monitoring. In *Biodiversity and Human Health* (Grifo F, Rosenthal J, eds.). Island Press, Washington, D.C., Chap. 3.

Feinberg SE. 1979. Graphical methods in statistics. *Am Statistician* 33 (4): 165–78.

Fischoff B, Slovic P, Lichtenstein S. 1979. Knowing what you want: Measuring labile values. In *Cognitive Processes in Choice and Decision Behavior* (Wallsten T, ed.). Erlbaum, Hillsdale, N.J.

Foley JD, Van Dam A. 1982. *Fundamentals of Interactive Computer Graphics.* Addison-Wesley, Reading, Mass.

Forrester JW. 1969. *Urban Dynamics.* MIT Press, Cambridge.

Foster KB, Vecchia P, Repacholi MH. 2000. Science and the precautionary principle. *Science* 288 (5468, May 12): 979, 981.

Funtowicz SO, Ravetz JR. 1991. A new scientific methodology for global environmental issues. In *Ecological Economics: The Science and Management of Sustainability* (Costanza R, ed.). Columbia University Press, New York, Chap. 10.

Gardiner PC, Edwards W. 1975. Public values: Multiattribute utility measurement for social decision-making. In *Human Judgment and Decision Processes* (Kaplan MF, Schwartz S, eds.). Academic Press, New York, pp. 1–37.

Gardiner PC, Ford A. 1980. Which policy run is best, and who says so? *TIMS Studies Management* 14:241–57.

Gibson C, Ostrom E, Ahn T-K. 1998. Scaling Issues in the Social Sciences. IHDP Working Paper No. 1 (May). A Report for the International Human Dimensions Programme on Global Environmental Change, Bonn, Germany.

Glantz MH. 1982. Consequences and responsibilities in drought forecasting: The case of Yakima, 1977. *Water Resources Res* 18 (1): 3–13.

———. 1996. *Currents of Change: El Niño's Impact on Climate and Society.* Cambridge University Press, Cambridge.

Goettl BP, Wickens CD, Kramer AF. 1991. Integrated displays and perception of graphical data. *Ergonomics* 34:1047–63.

Goicoechea A, Hanson DR, Duckstein L. 1982. *Multiobjective Decision Analysis with Engineering and Business Applications.* J. Wiley & Sons, New York.

Goldstein B. 1995. Routine uncertainty analysis: Certainly not. *Risk Policy Rep* 2 (8): 32, 34–35.

Graham JD, Wiener JB, eds. 1995. *Risk versus Risk: Tradeoffs in Protecting Health and the Environment.* Harvard University Press, Cambridge.

Greenlees KJ, Machado J, Bell T, Sundlof SF. 1998. Food borne microbial pathogens of cultured aquatic species. *Vet Clin North Am Food Anim Pract* 14 (1): 101–12.

Grifo F, Newman D, Fairfield AS, Bhattacharya B, Grupenhoff JT. 1997. The origins of prescription drugs. In *Biodiversity and Human Health* (Grifo F, Rosenthal J, eds.). Island Press, Washington, D.C., Chap. 6.

Haddix AC, Teutsch SM, Shaffer PA, Duñet DO, eds. 1996. *Prevention Effectiveness: A Guide to Decision Analysis and Economic Evaluation.* Oxford University Press, New York.

Haigh N. 1998. Roundtable 4: Challenges and opportunities for IEA—science-policy interactions from a policy perspective. *Environ Modeling Assessment* 3:135–42.

Henderson-Sellers A. 1996. Can we integrate climatic modelling and assessment? *Environ Modeling Assessment* 1:59–70.

Henrion M, Sonnenblick R, Hoo KS. 1995. *The Tracking and Analysis Framework: The Collaborative Development of a Tool for Integrated Assessment.* Prepared for the National Acid Precipitation Assessment Project, Lumina Decision Systems, Los Altos, Calif.

Herrick C, Jamieson D. 1995. The social construction of acid rain: Some implications for science/policy assessment. *Global Environ Change* 5 (2): 105–12.

Hobbs BF. 1986. What can we learn from experiments in multiobjective decision analysis? *IEEE Trans Systems Man Cybernetics* SMC-16:384–94.

Hobbs BF, Chankong V, Hamadeh W, Stakhiv EZ. 1992. Does choice of multicriteria method matter? An experiment in water resources planning. *Water Resources Res* 28 (7): 1767–80.

Hobbs BF, Chao PT, Venkatesh BN. 1997. Using decision analysis to include climate change in water resources decision making. *Climate Change* 37 (1): 177–202.

Hobbs BF, Meier P. 2000. *Energy Decisions and the Environment: A Guide to the Use of Multicriteria Decision Methods.* Kluwer Academic Press, Norwell, Mass.

Holmes KJ, Ellis JH. 1996. Potential environmental impacts of future halocarbon emissions. *Environ Sci Technol* 30 (8): 348–55.

———. 1997. Simulation of halocarbon production and emissions and effects on ozone depletion. *Environ Management* 21 (5): 669–85.

Hope C, Anderson J, Wenman P. 1993. Policy analysis of the greenhouse effect: An application of the PAGE model. *Energy Policy* 21:327–38.

Hordijk L. 1991. Use of the RAINS model. *Environ Sci Technol* 25 (4): 596–603.

Hunsaker CT, Graham RL, Suter GW II, O'Neill RV, Barnthouse LW, Gardner RH. 1990. Assessing ecological risk on a regional scale. *Environ Management* 14 (3): 325–32.

Hurrion RD. 1986. Visual interactive modeling. *Eur J Operational Res* 23:281–87.

Intergovernmental Panel on Climate Change. 1996. Integrated assessment of climate change: An overview and comparison of approaches and results. In *Climate Change 1995: The Supplementary Report to the IPCC Scientific Assessment.* Cambridge University Press, Cambridge.

Iz P, Krajewski L. 1992. Comparative evaluation of three interactive multiobjective programming techniques as group decision support tools. *INFOR* 30 (4): 349–63.

Jaeger CC, Renn O, Rosa EA, Webler T. 1998. Decision analysis and rational action. In *Human Choice and Climate Change,* Vol. 3, *Tools for Policy Analysis* (Rayner R, Malone E, eds.). Battelle Press, Columbus, Ohio, Chap. 3.

Jäger J. 1998. Current thinking on using scientific findings in environmental policy making. *Environ Modeling Assessment* 3:143–53.

Johnson LE, Loucks DP. 1980. Interactive multiobjective planning using computer graphics. *Computers Operations Res* 7:89–97.

Kahneman D, Tversky A. 1979. Prospect theory: An analysis of decision under risk. *Econometrica* 47:263–91.

Keasey K. 1984. Regret theory and information: A note. *Econ J* 94:645–48.

Klemes V. 1994. Design implications of climate change. In *Proceedings of the First National Conference on Climate Change and Water Resources Management* (Ballentine TM, Stakhiv EZ, eds.). Institute for Water Resources, U.S. Army Corps of Engineers, Ft. Belvoir, Va., III-9–19.

Kovats RS, Bouma M, Haines A. 1999. *El Niño and Health.* WHO/SDE/PHE/99.4. World Health Organization, Geneva.

Kunkel KE, Pielke RA Jr, Changnon SA. 1999. Temporal fluctuations in weather and climate extremes that cause economic and human health impacts: A review. *Bull Am Meteorol Soc* 80 (6): 1077–98.

Lemberski MR, Chi UH. 1984. Decision simulators speed implementation and improve operations. *Interfaces* 14:1–15.

Levins R. 1998. Environmental assessment: By whom, for whom, and to what ends? In *Ecosystem Health* (Rapport D, Costanza R, Epstein PR, Gaudet C, Levins R, eds.). Blackwell Science, Malden, Mass., Chap. 5.

Levy H. 1992. Stochastic dominance and expected utility: Survey and analysis. *Management Sci* 38:555–93.

Lindsay SW, Birley MH. 1996. Climate change and malaria transmission. *Ann Trop Med Parasitol* 90 (6): 573–88.

Loomes G, Starmer C, Sugden R. 1992. Are preferences monotonic? Testing some predictions of regret theory. *Economica* 59 (233): 17–34.

Loomes G, Sugden R. 1982. Regret theory: An alternative theory of rational choice under uncertainty. *Econ J* 92:805–24.

———. 1987. Some implications of a more general form of regret theory. *J Econ Theory* 41 (2): 270–87.

Loucks DP. 1995. Developing and implementing decision support systems: A critique and a challenge. *Water Resources Bull* 31 (4): 571–82.

Machina MJ. 1982. Expected utility analysis without the independence axiom. *Econometrica* 50:277–323.

Mays TD, Duffy-Mazan K, Cragg G, Boyd M. 1997. A paradigm for the equitable sharing of benefits resulting from biodiversity research and development. In *Biodiversity and Human Health* (Grifo F, Rosenthal J, eds.). Island Press, Washington, D.C., Chap. 12.

Meadows DH, Meadows DL, Randers J, Beherens WW. 1972. *The Limits to Growth: A Report of the Club of Rome's Project on the Predicament of Mankind.* Universe Books, New York.

Meier P, LeClair D. 1995. *Resource Trade-off Decision Analysis for B.C. Hydro's 1995 IRP.* Prepared for BC Hydro, Vancouver. IDEA, Washington, D.C.

Meo M. 1991. Policy-oriented climate impact assessment. *Global Environ Change* 1:124–38.

Meyer WB, Butzer KW, Downing TE, Turner BL II, Wenzel GW, Wescoat JL. 1998. Reasoning by analogy. In *Human Choice and Climate Change,* Vol. 3, *Tools for Policy Analysis* (Rayner R, Malone E, eds.). Battelle Press, Columbus, Ohio, Chap. 4.

Mills E, Wilson D, Johansson TB. 1991. Getting started: No regrets strategies for reducing greenhouse gas emissions. *Energy Policy* 19 (6): 526–42.

Morgan MG, Henrion M, Small M. 1990. *Uncertainty: A Guide to Dealing with Uncertainty in Quantitative Risk and Policy Analysis.* Cambridge University Press, New York.

Mosler KC. 1984. Stochastic dominance decision rules when the attributes are utility independent. *Management Sci* 30:1311–22.

Mosler KC, Holz H. 1994. An interactive decision procedure with multiple attributes under risk. *Ann Operations Res* 52:151–70.

National Acid Precipitation Assessment Program. 1991. *The Experience and Legacy of NAPAP: Report to the Joint Chairs Council of the Interagency Task Force on Acidic Deposition.* NAPAP Oversight Review Board, Washington, D.C.

Nazareth DL. 1993. Integrating MCDM and DSS: Barriers and counter strategies. *INFOR* 31 (1): 1–15.

Norgaard RB, Howarth RB. 1991. Sustainability and discounting the future. In *Ecological Economics: The Science and Management of Sustainability* (Costanza R, ed.). Columbia University Press, New York, Chap. 7.

Parkin J. 1992. A philosophy for multiattribute evaluation in environmental impact assessment. *Geoforum* 23 (4): 467–75.

Parson EA, Fisher-Vanden K. 1997. Integrated assessment models of global climate change. *Annu Rev Energy Environ* 22:589–628.

Patwardhan A, Small M. 1992. Bayesian methods for model uncertainty analysis with application to future sea level rise. *Risk Analysis* 12:513–25.

Pielke RA Jr. 1997. Asking the right questions: Atmospheric sciences research and societal needs. *Bull Am Meteorol Soc* 78 (2): 255–64.

Pielke RA Jr, Sarewitz D, Byerly R Jr, Jamieson D. 1999. Prediction in the earth sciences and environmental policy making. *Eos* 80 (28): 311–13.

Pritchard L Jr, Colding J, Berkes F, Svedin U, Folke C. 1998. The Problem of Fit between Ecosystems and Institutions. IHDP Working Paper No. 2 (May). A Report for the International Human Dimensions Programme on Global Environmental Change, Bonn, Germany.

Rapport D. 1998. Dimensions of ecosystem health. In *Ecosystem Health* (Rapport D, Costanza R, Epstein PR, Gaudet C, Levins R, eds.). Blackwell Science, Malden, Mass., Chap. 3.

Rayner R, Malone E. 1998. Why study human choice and climate change? In *Human Choice and Climate Change*, Vol. 3, *Tools for Policy Analysis* (Rayner R, Malone E, eds.). Battelle Press, Columbus, Ohio, Introduction.

Reilly JM, Edmonds JA, Gardner RH, Brenkert AL. 1987. Uncertainty analysis of the IEA/ORAU CO_2 emissions model. *Energy J* 8 (3): 1–29.

Rhyne T, Bolstad M, Rheingans P. 1993. Visualizing environmental data at the EPA. *IEEE Computer Graphics Applications* 13 (2): 34–38.

Rind D. 1999. Complexity and climate. *Science* 284 (5411): 105–7.

Root T, Schneider SH. 1995. Ecology and climate: Research strategies and implications. *Science* 269:334–41.

Rotmans J. 1990. *IMAGE: An Integrated Model to Assess the Greenhouse Effect.* Kluwer Publishers, Dordrecht.

———. 1998. Methods for IA: The challenges and opportunities ahead. *Environ Modeling Assessment* 3:155–79.

Rotmans J, Dowlatabadi H. 1998. Integrated assessment modeling. In *Human Choice and Climate Change*, Vol. 3, *Tools for Policy Analysis* (Rayner R, Malone E, eds.). Battelle Press, Columbus, Ohio, Chap. 5.

Rubio-Palis Y, Zimmerman RH. 1997. Ecoregional classification of malaria vectors in the neotropics. *J Med Entomol* 34 (5): 499–510.

Sarewitz D, Pielke R Jr. 1999. Prediction in science and policy. *Technol Soc* 21:121–33.

Sarin RK. 1977. Screening of multiattributed alternatives. *Omega* 5 (4): 481–89.

Savage J. 1954. *The Foundations of Statistics.* J. Wiley & Sons, New York.

Schneider SH. 1997. Integrated assessment modeling of global climate change: Transparent rational tool for policy making or opaque screen hiding value-laden assumptions? *Environ Modeling Assessment* 2:229–49.

Schneiderman B. 1992. *Designing the User Interface: Strategies for Effective Human-Computer Interaction.* Addison-Wesley, Reading, Mass.

Schoemaker PJH. 1981. Behavioral issues in multiattribute utility modeling and decision analysis. In *Organizations: Multiple Agents with Multiple Criteria* (Morse JN, ed.). Springer-Verlag, New York.

Shackley S, Wynne B. 1995. Global climate change: The mutual construction of an emergent science-policy domain. *Sci Public Policy* 22 (4): 218–30.

Shlyakhter A, Valverde ALJ, Wilson R. 1995. Integrated risk analysis of global climate change. *Chemosphere* 30 (8): 1585–1618.

Shrader-Frechette KS. 1991. *Risk and Rationality: Philosophical Foundations for Populist Reforms.* University of California Press, Berkeley and Los Angeles.

Sterman JD. 1991. A skeptic's guide to computer models. In *Managing a Nation: The Microcomputer Software Catalog* (Barney GO, Kreutzer WB, Garrett MJ, eds.). Westview Press, Boulder, Colo., pp. 209–29.

Stewart TJ. 1992. A critical survey on the status of multiple criteria decision making theory and practice. *Omega* 20:569–86.

Sugden R. 1986. New developments in the theory of choice under uncertainty. *Bull Econ Res* 38 (1): 1–24.

———. 1993. An axiomatic foundation for regret theory. *J Econ Theory* 60 (1): 159–80.

Teklehaimanot A. 1991. Dissenting opinion. In *Malaria: Obstacles and Opportunities* (Oaks SJ, Mitchell VS, Pearson GW, Carpenter CCJ, eds.). Institute of Medicine, National Academy Press, Washington, App. B.

Thiessen EM, Loucks DP. 1992. Computer assisted negotiation of multiobjective water resources conflicts. *Water Resources Bull* 28 (1): 163–77.

Trape JF, Godeluck B, Diatta G, Rogier C, Legros F, Albergel J, Pepin Y, Duplantier JM. 1996. The spread of tick-borne borreliosis in West Africa and its relationship to sub-Saharan drought. *Am J Trop Med Hyg* 54:289–93.

Tufte ER. 1983. *The Visual Display of Quantitative Information.* Graphics Press, Cheshire, Conn.

———. 1990. *Envisioning Information.* Graphics Press, Cheshire, Conn.

———. 1997. *Visual Explanations: Images and Quantities, Evidence and Narrative.* Graphics Press, Cheshire, Conn.

Tversky A. 1972. Elimination by aspects: A theory of choice. *Psychol Rev* 79 (4): 281–99.

Tversky A, Kahneman D. 1974. Judgment under uncertainty: Heuristics and biases. *Science* 211:453–58.

Vandermeer J. 1997. Biodiversity loss in and around agroecosystems. In *Biodiversity and Human Health* (Grifo F, Rosenthal J, eds.). Island Press, Washington, D.C., Chap. 5.

Vogel CG, Engelman R. 1999. *Forging the Link: Emerging Accounts of Population and Environment Work in Communities.* Population Action International, Washington, D.C.

von Neumann J, Morgenstern O. 1947. *Theory of Games and Economic Behaviour.* Princeton University Press, Princeton.

von Winterfeldt D, Edwards W. 1986. *Decision Analysis and Behavioral Research.* Cambridge University Press, New York.

Weber M, Borcherding K. 1993. Behavioural influences on weight judgments in multiattribute decision-making. *Eur J Operational Res* 67:1–12.

Whitesides GM, Ismagilov RF. 1999. Complexity in chemistry. *Science* 284 (5411): 89–92.

Wilson EO, Perlman DL. 2000. *Conserving Earth's Biodiversity, with E. O. Wilson* (CD-ROM). Island Press, Washington, D.C.

Wolman MG. 1995. Human and ecosystem health: Management despite some incompatibility. *Ecosystem Health* 1:35–40.

World Bank. 1997. *Health Aspects of Environmental Assessment.* Environmental Assessment Sourcebook Update, No. 18. Environment Department, World Bank, Washington.

———. 1998. *Economic Analysis and Environmental Assessment.* Environmental Assessment Sourcebook Update, No. 23. Environment Department, World Bank, Washington.

Zimmer C. 1999. Life after chaos. *Science* 284 (5411): 83–86.

Electronic References

Environmental Protection Agency. 1999. Ecological Risk Guidelines. National Center for Environmental Assessment. http://www.epa.gov/nceawww1/ecorsk.htm (Date Last Revised 5/3/1999).

International Human Dimensions Programme on Global Environmental Change. 2000. IHDP Home Page. http://www.uni-bonn.de/IHDP (Date Last Revised 2/28/2000).

Kandlikar M, Risbey J. 1996. Uses and Limitations of Insights from Integrated Assessment Modeling. http://sedac.ciesin.org/mva/MVAUG/uginsights.html (Date Last Revised 1996).

Mitasova H, Mitas L. 1998. Process Modeling and Simulations. NCGIA Core Curriculum in GIScience. http://www.ncgia.ucsb.edu/giscc/units/u130/u130.html (Date Last Revised 12/2/1998).

Schneider SH, Turner BL. 1997. Surprise and Global Environmental Change: Report of the 1994 Aspen Global Change Institute Summer Session on Surprise and Global Environmental Change. http://sedac.ciesin.org/mva/iamcc.tg/articles/surprise.html (Date Last Revised 5/12/1997).

Taylor G. 1997. Multimedia and Virtual Reality. NCGIA Core Curriculum in GIScience. http://www.ncgia.ucsb.edu/giscc/units/u131/u131_f.html (Date Last Revised 12/17/1997).

ENVIRONMENTAL CHANGES

Part II presents a selection of vital issues on global change, with special emphasis on atmospheric changes and the hydrological cycle. Part II begins with a chapter that explores the relationship between human populations and the environment (Chapter 6). This chapter has three sections: a broad overview of historical relationships between demographic, technological, and economic changes and human health; a survey of contemporary changes, such as urbanization, migration, population growth, and aging, that are likely to affect human health in the future; and an analysis of recent attempts to reach international agreement on issues of environmental degradation on a global scale, especially stratospheric ozone depletion and global climate change. The next chapter describes four main types of large-scale atmospheric degradation—stratospheric ozone depletion, acid rain, urban smog (ozone), and enhanced global warming due to anthropogenic emissions of greenhouse gases (Chapter 7). The perspective of a chemist elucidates both the sources and consequences of these atmospheric changes. The chapter that follows provides an ecological perspective on health through an explanation of the interdependence of cycles of energy, water, and carbon and how they shape the earth on which we live (Chapter 8). Of particular note is the use of the climatic fluctuations of El Niño as an example of interactions in the earth's dynamics with direct implications for public health. A chapter on water resources provides an overview of the use of water during the twentieth century and the growing importance of management of this most precious resource in sustainable development (Chapter 9). Topics include the quantity and quality of water as well as the effects of agriculture, forestry, population growth, and urbanization and their potential interactions with global climate change. A chapter on ecology and infectious disease lays out cycles of transmission of infectious agents and how they are affected by specific environmental changes (Chapter 10). This foundation provides a framework for the study of emerging infectious diseases that examines the contribution of humans, animals, and abiotic factors in the spread of disease.

See Part I for approaches to research. See Part III for case studies on global ecosystem change and public health.

CHAPTER 6

Human Populations
in the Shared Environment

Dennis C. Pirages, Ph.D., Paul J. Runci, M.A.L.D.,
and Robert H. Sprinkle, M.D., Ph.D.

While other chapters of this book focus directly on human health effects likely to result from various ecosystem changes, this chapter places such concerns in a broader perspective, examining health-related consequences of long-term changes in human social organization and behavior. Changes in the ways people live, interact, work, behave, organize, and travel all have important implications for sensitive relationships between human beings and the environment.

This chapter has three sections. The first focuses on some broader historical relationships between large-scale demographic, technological, and economic changes and human health. The second section concentrates on specific contemporary changes in the human environment, such as those related to urbanization, migration, population growth, and aging, that are likely to have a significant effect on future human health and the future incidence of disease. The last section explores the potential for international cooperation to protect public health in all countries of the world by analyzing recent attempts to reach international agreement on issues of environmental degradation on a global scale.

Social Evolution, Progress, and the Environment

The many human populations that collectively compose the human race have been constantly evolving, both biologically and socioculturally, within the constraints of changing physical environments shared with a vast array of other species (see Durham 1991). While the size of these individual human populations has varied over time, the total human population of the world has increased only very slowly for much of history (Durham 1991; Ponting 1991, 18–36). For the most part, a quasi equilibrium has been maintained between *Homo sapiens,* the sustaining capabilities of natural systems, and pathogenic microorganisms. The history of human progress has, however, been punctuated by large-scale outbreaks of disease that have substantially trimmed human populations and played a significant role in shaping human social evolution. In China alone, for example, some 291 major epidemics were recorded between the years 243 B.C. and 1911 A.D. (McNeill 1977, 259–69). Even earlier, during the development of agricultural society, rapid population growth was accompanied by an increase in infectious diseases and nutritional deficiencies (Armelagos et al. 1991). Cultural patterns have influenced and been influenced by the prevalence of infectious diseases throughout human history (Armelagos and Dewey 1970; Brown 1981).

The continued well-being of a burgeoning world population depends upon maintaining a balance between human beings and pathogenic microorganisms. The accepted traditional view of human progress is that the industrial revolution and its associated improvements in living standards will continue to spread worldwide. After a short interval of "transformational dislocation," the advantages of new technologies would be reflected in improved medical care and longer life spans. Thus, Murray and Lopez, in their mainstream assessment of future disease burdens, project that by the year 2020 the "diseases of modernity"—heart disease, depression, traffic accidents, cerebrovascular disease, and pulmonary obstructions—will be the leading causes of death worldwide (Murray and Lopez 1996, 375).

From another perspective, however, the future of infectious diseases is less certain. The large-scale increases in the size and density of the global human population, changes in human values and behavior, increased economic activity and worldwide industrial growth, and related changes in the natural environment all create new opportunities for pathogenic microorganisms to make inroads into human populations. There is now growing concern that an era of major scientific and technological breakthroughs in biology and medicine, often portrayed as a "war" on pests and disease, might be giving way to a period in which innovations in fighting infectious diseases no longer come as relatively easily as they have in previous decades and in which emerging and resurgent diseases are once again becoming a serious threat to human prosperity and health (Garrett 1994, 618–20). The issue of emerging infectious diseases has engaged the public health community in the United States both domestically and in the arena of foreign relations (*Centers for Disease Control and Prevention 2000; National Science and Technology Council Committee on International Science, Engineering and Technology 1997*).

The Effects of Industrialization

The recent rapid growth of the world's total human population has been facilitated by a spreading industrial revolution, which began to take hold

in Europe some five hundred years ago. Around that time, the feudal societies of the continent slowly began to change, and a class of merchants and
capitalists emerged in urban centers. This economic transition was subsequently facilitated by the advent of fossil fuel–based technologies that
stimulated myriad changes in human social organization and behavior,
many of which have had major implications for human health. For instance, the growth of industrial economies fostered the emergence of large
and polluted cities as the primary centers of industrial activity, which in
turn created the denser human populations and unsanitary conditions
that facilitated the spread of infectious diseases. The rise of urban industrial economies also brought changes in land use patterns far beyond the
cities themselves. Growing urban populations as well as growing industrial
countries created an expanding "ecological footprint," as much larger
quantities of food and other raw materials were imported from outlying
rural areas (Wackernagel and Rees 1996, 51–57).

These changes in the distribution of human populations and their
production activities were eventually accompanied by major shifts in the
way that "industrial man" viewed relationships with nature. First, because
an increasing fraction of the population had less direct contact with the
land and natural systems, the notion of humanity as separate from—and
even dominant over—nature gained broad acceptance. This manner of
thinking about the world, epitomized by the Enlightenment, emphasized
the ability of the human mind to understand and eventually master nature
through the application of reason. It produced a set of cultural values—a
"dominant social paradigm"—based on rationalism, positivism (the be-
lief that values and facts can be kept in separate spheres), and reduction-
ism (the attempt to study individual empirical phenomena in isolation
from the broader context of cause and effect). The concepts of balance between human beings and nature and of limits to the earth's resources mer-
ited less and less consideration as human industries and empires spread
out across a planet offering humans seemingly endless frontiers and challenges (see Norgaard 1994; Dunlap 1980).

The industrial revolution and the associated modern world view that
grew out of the Enlightenment has gradually spread outward from Europe
and now continues to transform other regions in Asia, Africa, and Latin
America. It brings with it changes in economic activity, transportation, human settlement patterns, wealth distributions, education, social behavior,
political activity, and resource consumption. In general, people in the industrialized countries of North America, East Asia, and Europe are more
economically affluent, educated, urbanized, and healthy, while those in
countries in the earlier stages of development are struggling to transform
preindustrial economies and to improve living standards. Countries that
have already undergone industrialization now have a lower incidence of

Table 6.1   Causes of Death as Percentages of Total Deaths in Selected Regions, 1990

Disease	OECD[a]	Latin America	Asia[b]	Sub-Saharan Africa
Tuberculosis	0.00	2.5	5.7	4.7
Human immunodeficiency virus (HIV)	0.50	0.90	—[c]	2.91
Diarrheal disease	0.00	5.08	7.17	11.58
Nutritional deficiencies	0.20	2.52	1.42	1.80
Malaria	0.00	0.46	1.40	8.92
Road traffic accidents	1.84	3.62	2.40	1.89
Self-inflicted injuries	1.57	0.73	1.21	0.20
Malignant neoplasms	24.74	11.47	11.56	5.23
Perinatal conditions	0.65	6.51	5.98	6.13
Respiratory infections	3.8	5.9	9.97	12.47
Cardiovascular diseases	43.85	26.22	24.37	9.9
Other	22.85	34.09	28.82	34.27

Source: Data from Murray and Lopez 1996, 434–60.

[a]The OECD (Organization for Economic Cooperation and Development) is an international organization whose membership consists of the world's wealthiest nations, mainly those of North America, western and central Europe, and Australasia.

[b]Does not include India and China.

[c]Not available.

disease and suffer from a different set of maladies—afflictions of affluence more than the infectious diseases associated with poverty (see Table 6.1). But the affluence of the industrial world combined with the increasing resource demands of industrializing countries such as China, India, and Brazil are putting an ever-greater strain on the physical environment. While affluence and improved living standards have proven, in many respects, the key to better overall human health, a related global intensification of resource consumption could ultimately undercut the environmental foundations of improved health. Although some business leaders and social scientists may still maintain that the earth offers endless frontiers and inexhaustible bounty to which human ingenuity holds the key, many others hold that the current state of the earth's natural systems suggests that humanity is increasingly pressing against environmental limits to sustain the spread of an energy- and resource-intensive way of life (for arguments on both sides of this debate see, e.g., Meadows et al. 1992; McMichael 1993; Cohen 1995; Simon 1981; Simon and Kahn 1984).

A key element of the sustainability question is the extent to which changes in the human environment increase the exposure and vulnerability of various populations to disease. At present, the continuing global spread of industrial activity provides an ever-changing context within which the interactions between people and pathogenic microorganisms take place. This transformation of the planet involves massive changes in land use, such as the conversion of forests to agricultural land, the growth of cities, and the extraction and use of larger quantities of renewable and nonrenewable resources (e.g., water, fossil fuels, wood, and mineral ores). Since all of these activities are commonly deemed essential to the modernization and development processes, their potentially detrimental health effects are often afforded inadequate consideration (see Dunlap 1980; Norgaard 1994, 1–10).

An Emerging Global System

The accelerating spread of the industrial revolution and technological advancement are core elements of the process referred to as *globalization*. Facilitated by innovations in transportation and telecommunications, the world is being transformed into a "global village" within which there is much more frequent contact between rich and poor neighborhoods than ever before. Thus, microorganisms that may emerge in remote geographic locations can spread more rapidly and easily to other areas by virtue of the increased speed and frequency of international travel. For example, international air travel has grown severalfold over the last three decades and the corresponding movement of bacteria and viruses has increased apace (see, e.g., Shell 1997, 45–46). And the rising volume of trade among countries globally has permitted various kinds of pests to hitchhike around the world, often carrying infectious diseases with them (Culotta 1991). The increasing progress of the globalization process, which entails economic liberalization and integration among countries, and the reduction or removal of state controls on the international movement of people and goods suggest that the global transport of disease-causing organisms may encounter even fewer obstacles in the future.

Seemingly insignificant individual behaviors within states, such as driving automobiles, using air conditioners, growing rice, and raising cattle, now have unintended additive consequences. In aggregate, these activities contribute to the emissions of pollutants that threaten to alter the earth's environmental services (e.g., those provided by the earth's atmosphere and oceans) on which all life depends (Orians 1995, 12). While the aggregation of human activities is now significantly affecting the atmospheric commons, the world is divided into nearly 200 nation-states, each of which acts individually to safeguard its own sovereignty and independence of action.

Global and regional environmental changes thus pose challenges to the way that human populations are organized politically. These large-scale environmental problems do not respect existing political boundaries and often span multiple jurisdictions, creating needs for coordinated management. Some scholars go as far as to suggest that the emergence of global environmental problems is fundamentally altering the nature of international relations by undermining the principle of sovereignty on which the international system is based. The sovereignty principle recognizes the right of nation-states to manage affairs within their own borders independently to the extent that their actions do not have harmful consequences beyond their own territories. But recognition that there are global environmental and health effects flowing from individual or national actions can potentially undermine or constrain the legal independence of nations (see Conca 1994; Zacher 1992).

While health consequences of environmental change have implications for the future nature of international organization, economic and political changes within states also have implications for human health—as the major transformation that swept through Eastern Europe and the Soviet Union between 1989 and 1991 has shown. The early years of transition from communism to democratic government and market economies have entailed major disruptions in the social and economic institutions on which the people in those countries have depended over the past several decades. For many, this has meant lower living standards and quality of life and poorer nutrition and health care services; these and other factors—among them the region's severely degraded environmental conditions, which resulted from decades of inefficient economic production—have contributed to a resurgence of infectious diseases and a legacy of other adverse health effects due to environmental exposures. For example, thousands in the former Soviet Union have suffered from a massive epidemic of vaccine-preventable diphtheria (see Table 6.2), while the toll of thyroid cancer and other radiation-induced illnesses continues to be monitored in the aftermath of the Chernobyl nuclear accident (see Chapter 2).

Anthropogenic Changes and Future Human Health

The effects of industrial modernization on interactions between human populations, between humans and the natural environment, and between humans and disease-causing microorganisms have a wide variety of health-related ramifications for the future. Assuming that current means of production remain largely unaltered, continued growth in resource demands and consumption will place increasing strains on the earth, both as a mine for raw materials and as a sink for waste disposal (see Sachs 1993).

Facilitated by scientific and technological progress, sheer growth in human numbers, and growing consumer appetites, the enormous annual throughput of materials in the world economy pushes natural systems ever closer to—and in some instances beyond—their limits at all levels, from

Table 6.2   Diphtheria Cases in the Former Soviet Union, 1991–1996

Region	Population (millions)	No. Cases					
		1991	1992	1993	1994	1995	1996
Russia	149.90	1,869	3,897	15,211	39,582	35,652	13,604
Central Asia[a]	55.17	58	114	908	2,984	6,978	2,571
Caucasus[b]	16.69	43	58	170	1,171	1,331	471
Western states[c]	67.16	1,143	1,655	3,142	3,596	6,020	3,432
Baltics[d]	7.99	13	20	31	295	431	137

Source: Data from Vitek and Wharton 1998, Table 1.

[a]Kazakhstan, Kyrgyzstan, Tajikistan, Turkmenistan, Uzbekistan.

[b]Armenia, Azerbaijan, Georgia.

[c]Ukraine, Moldova, Belarus.

[d]Estonia, Latvia, Lithuania.

local to global. Continuation of these trends will present serious future challenges to human health, particularly in the poorer countries that have not yet industrialized.

A major qualitative difference between the consumption power of wealthy and poor populations that has direct health implications is the ability of the wealthy to distance themselves in time and space from the ecological consequences of their consumption decisions. Wealthier populations, within and among nations, generally develop ecological dependencies on surrounding regions, have a wider range of consumption options, and have a greater ability to "export" waste streams and environmental externalities, including health risks. They achieve this in many instances by disposing of wastes via global common property resources such as the atmosphere and oceans, thereby dispersing the costs of pollution over the planet's systems and species and into the future across generational boundaries. The phenomenon of global climate change provides a good example. Many argue that wealthy nations' use of fossil fuels and venting of waste gases to the atmosphere will result in all of humanity—and, indeed, the biosphere as a whole—bearing related environmental and health consequences for generations to come.

Poorer populations in any country, on the other hand, often bear directly a larger portion of the environmental and epidemiological consequences of their resource consumption decisions; since they have fewer options, they typically extract resources and dispose of wastes in their local surroundings. In many parts of the rural developing world, for instance, the immediate needs for fuel wood and arable land often combine with poor sanitation infrastructure and distorted land use and resource access policies to promote local land and water degradation, deforestation, and soil erosion (see Paarlberg 1994; Ribot 1993).

Food Production and Health in the Less-Industrialized World

Future patterns of environmental change and human health will be driven in large part by demographic trends that are now clearly visible. The rapid growth of human populations in many regions of the developing world has necessitated a quest for new sources of water, food, and other resources and is forcing migration into marginal and previously remote areas. In parts of Africa, Latin America, and Asia, for instance, land use change is often characterized by the steady encroachment of human populations on previously remote, forested areas, which might be converted to lands for grazing, agriculture, and industry. During the 1980s alone, because of these processes, the planet as a whole lost 8 percent of its forest cover; over that period, estimated rates of deforestation were highest in the tropical forests of Asia (11%), Africa (7.4%), and Latin America (7.5%) (World Resources Institute [WRI] 1996, 201–3).

While there are numerous local and global environmental consequences of deforestation, one major health effect stems from the fact that clearing of forests has historically brought human beings into contact with uncommon disease-causing microorganisms (see, e.g., Ponting 1991, 224–28). Scientists have estimated that there are millions of unknown organisms in these habitats and that newly encountered viruses and bacteria will continue to infiltrate human populations with unknown effects as forests fall (see Chapter 10). Recent outbreaks of the Ebola virus in Zaire and hemorrhagic fever in Bolivia, for example, have been linked to human encroachment on forests (see Garrett 1994, chap. 1; Gibbons 1993). Loss of biodiversity may also have broader health consequences through the loss of genetic resources for pharmaceutical development or agricultural systems (Grifo and Rosenthal 1997).

Growing populations and food resource requirements in the developing world have necessitated more intense cultivation of existing agricultural lands. The use of monocultures in agriculture may also directly accelerate the loss of biodiversity and increase vulnerability to the spread of pests, adversely affecting production (Grifo and Rosenthal 1997). Modern cultivation practices have resulted in an increasingly heavy use of agrochemicals to control pests and to increase yields. Ironically, agrochemicals have contributed further to land degradation and increased human health risks. In the past forty years, chemical fertilizer use in developing countries has increased 9-fold, while pesticide use has grown more than 30-fold. Many of the compounds in widespread use have long been banned in the United States and Europe (e.g., DDT, alachlor) and have been linked to cancer, liver and kidney toxicity, endocrine disruption, and possibly immunosuppressive effects that could increase the risks of infectious diseases and cancer (Commission for Environmental Cooperation 1997; Repetto and Baliga 1996).

Table 6.3   Undernourished Population and Dietary Energy Supply, 1994–1996

Region	Undernourished Population		Daily Dietary Energy Supply per Capita (calories)
	millions	%	
Sub-Saharan Africa	210	39	2,150
Near East/North Africa	42	12	2,990
East and Southeast Asia	258	15	2,740
South Asia	254	21	2,360
Latin America and the Caribbean	63	13	2,780
Transition economies[a]	—[b]	—[b]	2,850
Industrialized countries	—[b]	—[b]	3,340
World	828[b,c]	19[b]	2,720

Source: Data from Food and Agriculture Organization (FAO) 1998a, Table 1, and FAO 1998b.

[a]Countries in Eastern Europe and the former Soviet Union.

[b]The transition economies and the industrialized countries are not included.

[c]The discrepancy in the total is due to rounding and the fact that the total includes Oceania.

One of the cruel ironies of agrochemical use is that it has at times precluded from entry into foreign markets the export crops of developing nations and thus has blocked their access to much-needed foreign currency. Between 1984 and 1994, the U.S. Food and Drug Administration detained shipments of fruits and vegetables from Guatemala worth nearly $18 million because of excessive pesticide levels (see Repetto and Baliga 1996). The growth in pesticide markets of the developing world, estimated to be rising between 2.5 percent and 3.5 percent annually, is especially troubling, since inadequate health infrastructure, widespread malnutrition, and lack of access to clean water and sanitation already place the health of large segments of the populations of developing countries at great risk.

Despite the more intensive use of agrochemicals and the overall increases in agricultural yields in the developing world, hunger and food insecurity will continue to present major problems to human populations in some regions (Table 6.3). Over the past two decades, agricultural production has kept pace with population growth rates in most regions, with the exception of sub-Saharan Africa, where population growth has been especially rapid. Although the question of environmental limits and carrying capacity is important in relation to the ability of the world to continue to sustain rapidly growing human populations, the more immediate challenges are those of economics, logistics, and distribution. Even in areas where food supply has kept pace with population growth, malnutrition persists.

Advances in information and agricultural technologies have opened up the possibilities of higher levels of food production and better nutrition for millions. The availability of these technologies to poor countries where yields are inadequate, however, is a more challenging and complex question. Similarly, while potential world agricultural yields reflect the substantial agricultural growth potential in Europe and North America, the demand for food will grow the fastest in regions such as the Middle East and Africa, where population growth is most rapid. Continued increases in food production depend on some combination of increasing the productivity of land already under cultivation and increasing the amount of arable land. Irrigation is important in both activities but may also generate harmful environmental effects (see Chapter 9). Conversion of land to agriculture often involves deforestation, which can affect water resources, enhance the spread of diseases, and increase the emissions of greenhouse gases that may lead to global climate change (see Taylor 1997 and Chapters 7, 9, 10, and 13). Reconciling food supply and demand will continue to be a great challenge for the foreseeable future (see Dyson 1996, chap. 4).

Human Population Movements

Since the end of the cold war, there has been an increase in population movements generated by a resurgence of ethnic conflict. The breakup of empires and spheres of influence has created large, dislocated, insecure minorities. Differentials in physical security, income, and economic opportunities provide the pushes and pulls that motivate migrations and refugee flows (Weiner 1992, 91–95). Several recent studies have also focused on the potential role of environmental degradation as a driver of migration and refugee flows. Although environmental degradation and resource limitation may not be the only causes of mass population movements, evidence suggests that they are often among a handful of primary or proximate causes (see, e.g., Jacobson 1988; Suhrke 1991; Hazarika 1991; Homer-Dixon 1991).

While media attention often focuses on migration-related issues in Western nations, most international refugee and migration flows actually occur among developing countries. The largest flows have occurred within Africa, Asia, and the Middle East. In South Asia alone, some 35–40 million refugees have crossed international borders within the region in recent years (Weiner 1992, 93–94). Comparable numbers of refugees and migrants are estimated to have been on the move in Africa since the late 1980s (WRI 1994, 31). Recent ethnic conflicts in the central African nations of Rwanda, Zaire, and Burundi have intensified refugee flows in that region and taken a health toll on affected populations through malnutrition, poor sanitary conditions, and exposure to disease in densely populated refugee camps.

Contemporary large-scale population movement is obviously a factor changing the balance between human beings and microorganisms. Migrants and refugees frequently live in marginal and unsanitary areas and have poor access to clean water, sufficient food, or health services. Crowded, unsanitary refugee camps are ideal locations for the propagation and spread of diseases such as cholera and typhoid (see Pirages 1995, 7, and Chapters 2, 10, 11, and 14). Thus, future large population movements present serious health risks affecting both migrant populations themselves and recipient countries.

Urbanization and the Rise of Megacities

The United Nations has estimated that, by the year 2015, there will be 26 urban areas of 10 million inhabitants or more (known as *megacities*), all but 4 of which will be in less-industrialized nations (*United Nations Centre for Human Settlements [UNCHS] 1997*). The dramatic demographic shifts projected are illustrated by a comparison between Osaka and Lagos, with Osaka larger in 1995 but experiencing no growth while Lagos becomes the third largest urban agglomeration in 2015 (Table 6.4). While urban growth might not seem to present a problem in itself, the fact that the most rapid growth is occurring in the cities of the less-industrialized world raises concerns, given their more limited resources, infrastructure, and formal institutional structures (Ezcurra and Mazari-Hiriart 1996).

Ecological factors in rural areas of poorer countries are among the major contributors to the growing global trend of rural-to-urban migration. Degradation of the agricultural resource base resulting from desertification, deforestation, drought, and salination of soils from irrigation has increased the number of rural landless in some regions of Africa and Asia and has led to declines in agricultural productivity, particularly in resource-poor areas (Brown 1999).

Compounding matters is the fact that, throughout the developing world, growth rates of human populations in rural areas are especially high and often exceed the ability of local environmental systems to sustain them. Under these circumstances, marginalized rural residents frequently move to urban centers, where economic opportunities and the overall quality of life are perceived to be better. The "push" of deteriorating rural conditions combined with the "pull" of industrial growth in major cities of the developing world, such as Jakarta, Shanghai, Bangkok, and Kuala Lumpur, are complementary dynamics (World Commission on Environment and Development 1987, 95–102). A potential silver lining to Asia's ongoing economic crises is the situation's possible dampening effect on the region's urban population growth rates.

The prevalence of poverty, pollution from municipal and industrial wastes, and the inadequacy of access to infrastructure and services (particularly sanitation and sewer) in urban areas mean that the urban poor of

Table 6.4   Projected Population and Growth Rates for the World's Megacities, 1995–2015

City	Population (thousands)		Avg. Ann. Growth Rate (%)
	1995	2015	
Tokyo	26,959	28,887	0.35
Bombay	15,138	26,218	2.78
Lagos	10,287	24,640	4.46
São Paulo	16,533	20,320	1.04
Dhaka	8,545	19,486	4.21
Karachi	9,733	19,377	3.50
Mexico City	16,562	19,180	0.74
Shanghai	13,584	17,969	1.41
New York City	16,332	17,602	0.38
Calcutta	11,929	17,305	1.88
New Delhi	9,948	16,860	2.67
Beijing	11,299	15,572	1.62
Manila	9,286	14,657	2.31
Cairo	9,690	14,418	2.01
Los Angeles	12,410	14,217	0.68
Jakarta	8,621	13,923	2.43
Buenos Aires	11,802	13,856	0.81
Tianjin	9,415	13,530	1.83
Seoul	11,609	12,980	0.56
Istanbul	7,911	12,328	2.24
Rio de Janeiro	10,181	11,860	0.77
Hangzhou	4,207	11,407	5.11
Osaka	10,609	10,609	0.00

Source: Data from *United Nations Centre for Human Settlements 1997*, Table 10.

megacities in developing countries suffer the "worst of both worlds." They experience the problems that characterize underdevelopment (e.g., more deaths from infectious disease, higher infant mortality rates) and those that characterize industrial populations (e.g., higher rates of death from heart disease, neoplasms, and accidents). In addition, the urban poor have shorter life expectancies and often are worse off nutritionally than are rural dwellers (World Bank 1991, 51). Nonetheless, many migrants in the developing world still regard themselves as better off in cities than in the countryside (Parnwell 1993, 18–24).

A development that makes the urbanization trend especially troubling is the recent increase in the incidence of infectious waterborne diseases (such as cholera, typhoid, hepatitis A and E) and other diseases that thrive around stagnant water. Megacities, because of their size and the insufficiency of their infrastructures, present favorable "laboratories" in which such organisms will be able to flourish; the ever-increasing numbers and densities of the urban poor also might present favorable conditions for the spread of disease.

Demographic Change and Health in the Industrialized World

Major population changes are by no means limited to the less-industrialized world. While populations are growing rapidly and becoming younger on average in many less-industrialized countries, the general trend in most industrialized nations like Japan, Germany, Italy, Hungary, and Spain is one of either slow or negative growth. By the year 2030, as life expectancy grows longer and birthrates decline, the percentage of the population over the age of 60 in countries in the Organization for Economic Cooperation and Development will have doubled to 32 percent from 1990 levels of 16 percent (World Bank 1994, 26).

Some of the implications of population aging are clear. For instance, individuals' health care needs and costs frequently rise with age. The care required by older people often involves more expensive technology, hospitalization, and long-term nursing care. Consequently, countries spend more on health care as their populations grow older. Since spending on health and spending on pensions rise together, pressure on a nation's health care resources and government budgets grows exponentially as populations age (World Bank 1994, 2).

In the United States, for instance, where the "very old" (those over the age of 75) is the fastest-growing segment of the elderly population, the sustainability of the current health care and social security infrastructures is at risk. This fraction of the population in the United States will grow to nearly 8 percent by the year 2020 (American Association of Retired Persons, Resources for the Future 1993). Health problems that researchers believe will be associated with global environmental degradation—for example, increased incidence of skin cancers due to higher levels of ultraviolet radiation from ozone depletion (see Chapter 7) and respiratory problems due to higher average ambient temperatures in many areas (see Chapter 13)— could also have particularly serious effects on the elderly.

Even though economically developed nations represent a declining fraction of the world's population, their economies still account for a majority of global economic activity and pollution that contribute to the increasing likelihood of global climate change. For example, while the industrialized nations of the north account for approximately 20 percent of the world's population, they account for nearly 50 percent of global annual consumption of fossil fuels (WRI 1994, 3–8).

International Cooperation for Global Public Health

A major dilemma stems from the fact that dealing with many emerging health issues will require the cooperation and coordination of numerous countries. The ecology of disease depends on global trends in population, development, and environmental degradation (Pedersen 1996). Effective international relations are therefore essential for crafting responses. A review of international negotiations surrounding the global environmental issues of stratospheric ozone depletion and climate change provides lessons on the process by which the global community can act to protect health or is prevented from doing so.

The ozone and climate change problems have been characterized by high degrees of uncertainty, which greatly magnifies the challenge of negotiating international responses. In fact, the early phases of the domestic and international debates centered around arguments over the extent to which a problem meriting action actually even existed. As the stratospheric ozone problem illustrates, the availability of good scientific evidence of a problem is a necessary, but by no means sufficient, condition for international action. By the late 1970s, a relatively small set of widely used industrial chemicals (e.g., chlorofluorocarbons, carbon tetrachloride, and halons) had been linked to destruction of the earth's ozone layer. Moreover, given the long-lived presence of these chemicals in the stratosphere, the scientific community predicted that the thinning of the ozone layer was likely to continue, posing an increasing risk to the health of plants and animals around the world.

As events showed over the next several years, demonstrated environmental and health risks alone were inadequate in catalyzing international action. Major chemical firms that manufactured ozone-depleting chemicals and depended on global sales of the chemicals, such as DuPont in the United States and ICI in France, continued to deny the problem and to lobby their governments vigorously against participation in an international ban. In the absence of viable substitutes for these common chemicals, they warned that the economic effects of a ban would be devastating—and would be likely to yield no environmental benefit. Only in the early 1980s, as it began to appear that substitute chemicals could be commercially available in the near future, did industry alter its position, thereby enabling the governments of the United States and other major industrialized countries to do the same. And only in 1985, with the conclusion of the Vienna Convention for the Protection of the Ozone Layer, did the international community adopt a treaty formally acknowledging the existence of a stratospheric ozone problem. That same year, the discovery of the ozone hole over Antarctica, the so-called smoking gun, accelerated the timetable (see Box 6.1). Two years later, in the Montreal Protocol on

Scientific Discovery and the History of Ozone Diplomacy

The history of the scientific understanding of stratospheric ozone depletion is often compressed into a simple story of a chemical theory that was dramatically confirmed in 1985 by the discovery of the Antarctic ozone hole, the so-called smoking gun (see Fig. 7.1). Similarly, the history of ozone diplomacy is often compressed into a simple story of an international community prompted to act by reports of the Antarctic ozone hole. However, the development of the theory of stratospheric ozone depletion was considerably more circuitous, and the speed of the international response was influenced by changes in technology and international law.

The original theory of stratospheric ozone depletion catalyzed by halogens (chlorine, bromine) was proposed in the early 1970s by Rowland and Molina to account for a gradual thinning in the middle stratosphere at about 40 kilometers (see mechanism III in Chapter 7). In contrast, the ozone hole over Antarctica affects the lower stratosphere at about 14–22 kilometers (see Fig. 7.2) and develops much more rapidly than could be explained by the original theory. Therefore, a new chemical or meteorological mechanism had to be found to explain the dramatic, rapid ozone loss at the South Pole. It was soon clear that the likely explanation had to take into account the unique conditions of the polar springtime, that is, exceedingly cold temperatures (colder than anywhere else on the globe in the lower atmosphere), the existence of surfaces provided by stratospheric clouds (upon which heterogeneous chemical reactions can take place), and unusual twilight solar illumination conditions (a rapid switch between the long polar night and the long polar day near the spring equinox). Molina proposed yet another catalytic cycle involving oxides of chlorine (different from the one used to explain the loss of ozone in the middle stratosphere), which actually proceeds faster at the lower temperatures encountered in the Antarctic lower atmosphere (see mechanism IV in Chapter 7) and quantitatively explains the rapid ozone loss. As a result of this body of work, Rowland and Molina were awarded the Nobel Prize in Chemistry in 1995.

The international diplomatic community was able to respond quickly to reports of accelerated stratospheric ozone depletion in 1985 because they had already established an innovative framework convention, a process permitting review and adaptation that has become the model for international environmental negotiations. Advances in engineering also played a role through the development of satisfactory substitutes for chlorofluorocarbons, chemicals whose production is being phased out to protect the ozone layer. Innovations in science, technology, and diplomacy all contributed to the rapid international response to protect the ozone layer.

Substances that Deplete the Ozone Layer (1987), the international community took concrete measures to phase out and ban ozone-depleting chemicals.

Scholars of international relations and environmental politics consider the Vienna Convention and the Montreal Protocol a success story. These agreements marked the first time that the international community had come together to address an environmental issue of truly global scope.

Also, many observers have been impressed with the speed with which countries responded; the 13 years that elapsed between the identification of the threat to the ozone layer and the conclusion of the Montreal Protocol was, in their minds, a blink of an eye compared with the pace of other multilateral negotiations, such as the Law of the Sea and the General Agreement on Tariffs and Trade.

From the perspective of international law and diplomacy, however, the greatest successes of the Vienna Convention and Montreal Protocol might lie in the procedural innovations and precedents they created for future global environmental negotiations. The utility of this model became apparent almost immediately as a new global environmental challenge, global climate change, began to emerge in the late 1980s.

A major innovation in the negotiations surrounding the stratospheric ozone problem was the use of a treaty instrument known as a "framework convention." Unlike most treaties, which commit participating countries to a concrete set of rules and behaviors, a framework convention is a more open-ended agreement under which countries agree to continue to work with one another, to meet at specified intervals, and to take action as necessary on the problem at hand. This structure provided countries with a means of opening and maintaining a formal discourse on a problem without having to extend themselves beyond their comfort levels, given the state of scientific knowledge and the surrounding uncertainties. As the science improved and the uncertainties grew smaller, most notably due to the discovery of the ozone hole in 1985, a forum for the adoption of subsequent protocols—amendments to the framework agreement—already existed.

As environmental groups, scientists, and some policymakers began to urge international action (see Box 6.2) to combat a threat that they perceived to be looming on the horizon—global climate change—the ozone negotiations offered some useful insights. Unlike ozone depletion, which concerned a relatively small group of chemicals and industries in a few countries, global climate change concerned a variety of behaviors and practices, associated mainly with fossil fuel use and agriculture, that were dispersed across the globe. Moreover, where scientists had demonstrated that ozone-depleting chemicals made their way into the stratosphere and began depleting ozone within about seven years, the potentially disastrous climatic effects of carbon dioxide and other gaseous emissions remained unproven and were unlikely to become manifest until well into the next century. In short, the range of human behaviors concerned, the range of environmental impacts, the costs and benefits of international response, and the levels of uncertainty were exponentially larger in the case of climate change than in that of ozone depletion.

Nevertheless, only four years elapsed between the emergence of the cli-

Nongovernmental Organizations in Global Environment and Health Issues

Nongovernmental organizations (NGOs) play an important role in global environment and health issues, often working in partnership with national governments, specialized agencies of the United Nations system, and other international organizations. In the arena of biological conservation, the World Resources Institute (WRI) collaborated with the United Nations Environment Programme (UNEP) and the World Conservation Union (IUCN) to protect the world's biodiversity (*WRI 2000*). This group gathered scientists, community leaders, and representatives of governments, NGOs, development assistance agencies, and industry during the preparation of a major publication on a strategy for global biodiversity that appeared during negotiations for the Convention on Biological Diversity. The Pew Center on Global Climate Change, as well as WRI and other NGOs, have developed initiatives for climate change (*Pew 2000; WRI 2000*). Partnerships among national governments, U.N. agencies, and other international organizations have been formed to foster the use of satellite-based remote sensing technology for research in global change (see Table 3.4). The ability of the World Health Organization (WHO) to promote health depends on a network of partnerships with national governments, agencies of the U.N. system, development banks, intergovernmental organizations, NGOs, academia, and private sector groups (*WHO 1999*). The activities of NGOs complement the formal negotiations of nation-states.

---

mate change problem as a matter of public concern in the summer of 1988 and the opening for signature of the Framework Convention on Climate Change (FCCC) at Rio de Janeiro in June 1992 (*United Nations Environment Programme 2000*). In the spirit of the Vienna Convention, the FCCC pledged participants to work together and to act according to a "precautionary principle" to avoid irreversible damage to the global environment. The business of establishing binding country commitments to the reduction of greenhouse gas emissions was left to subsequent conferences of the parties to the FCCC, at which protocols to the agreement would be negotiated.

As history has shown in the years since 1992, creating the framework agreement was the easy part of the international negotiations on climate change. The process of adopting protocols to the framework has proven slow and politically contentious for a host of reasons. First, since the largest likely culprit with regard to climate change is the global use of fossil fuels (i.e., coal, oil, and natural gas), addressing the problem requires a major transition in the world's energy and transportation infrastructures and technologies. In distinct contrast to ozone-depleting chemicals, no substitutes for fossil fuels are currently available on a sufficient scale or at acceptable cost.

Second, equity issues associated with climate change seem in many ways intractable. For example, developing countries are reluctant to curtail the growth in energy consumption that will enable them to expand their economies and improve living standards for their citizens. This, after all, was the pathway by which the world's wealthy countries achieved the affluence they enjoy today. Is it equitable that poorer countries be penalized for a problem that, for the most part, they have not created? If they are to develop along a non–fossil energy path, they argue, the industrialized countries have an obligation to provide them with the capital and technology by which they might do so.

Finally, while uncertainty regarding the science of climate change has grown smaller and scientific consensus is growing stronger, the "smoking gun" linking carbon emissions to climate change has yet to be found. While higher concentrations of carbon dioxide in the earth's atmosphere do trap more incoming solar radiation, leading to global warming, the linkages between carbon emissions and changes in climate are still not well understood (see Chapters 7 and 8). Consequently, many politicians and industries remain steadfastly opposed to any early action to address the problem.

In 1997, at a conference of the parties to the FCCC in Kyoto, Japan, a protocol was introduced that committed industrialized countries to ambitious emissions reductions between the years 1998 and 2012. Fewer than 50 of the 150 nations attending the conference signed the Kyoto Protocol at the meeting. The United States, the world's largest emitter of greenhouse gases, was not among the signatories. The U.S. Senate, for its part, promised to veto any climate treaty sent to it for ratification unless and until developing countries also agreed to reductions in emissions. On the scientific front, many felt that the agreement was unrealistic in its targets, would entail devastating economic costs, and would also be ineffective in addressing the problem. The following year, when the parties assembled in Buenos Aires, Argentina, to continue negotiations, little progress was made. Considering the conflicting interests of industries and environmentalists, developing and industrialized countries, and present and future generations that are intertwined in the climate change issue, many observers are concerned that the international community may attempt to act in earnest only in the future, after the impacts of global climate change are demonstrated. Then, of course, humanity's only course of action will be to attempt to adapt to changes as they unfold. The time for preventive action will have passed long before.

The problems of global climate change and stratospheric ozone may be partly inspirational and partly cautionary tales for those concerned with the international response to emerging global health issues. On the one hand, the international community has demonstrated the ability to move relatively quickly and to devise innovative political responses to complex and uncertain problems. On the other hand, these events indicate that effective international action may be likely only after the danger becomes clear and present. Politicians and executives who might be willing to pay today to avert *possible* disaster in the more distant future are unlikely to be rewarded by voters and shareholders.

Conclusion

Throughout history, human populations and their respective environments have enjoyed highly dynamic and interactive relationships. Just as human actions have continually transformed the earth, environmental conditions and constraints have played inestimable roles in shaping human organization and culture and in providing a constant impetus for technological innovation.

Until the advent of the modern era some five centuries ago, the interactions of human populations with environmental systems occurred mainly on a localized scale; the size and technological capabilities of human populations effectively limited their capacities to alter natural systems on a global level. But the acceleration of technological process since the dawn of the industrial age has greatly enhanced humanity's reach—to the extent that the human race may now be both the greatest perpetrator and victim of global environmental changes.

In many respects, the urgent environmental and health issues confronting humanity on the frontier of the twenty-first century are the same as those it has always faced: epidemic disease, resource limitations, poverty, affluence, and the unintended consequences of new technologies. It is principally the scale and complexities of the problems that have grown and changed across time.

Although there is substantial cause for concern with regard to humanity's ability to break out of its historic pattern of interaction with the natural environment—a cycle of seeking, reaching, and eventually exceeding frontiers—there is also cause for optimism. As the cases of the Montreal Protocol and the Framework Convention on Climate Change show, recent innovations in human organization and cooperation hold the promise of mitigating, reversing, and perhaps even preventing dangerous anthropogenic changes to the environment, in effect dampening the worst consequences of technological innovation. A rising global awareness of the linkages between human and Earth systems, and of the uncertainties surrounding those linkages, offers the hope and possibility that precaution might be coming to play a larger role in the future of all human activity as it relates to the shared environment.

SUGGESTED STUDY PROJECTS

Suggested study projects provide a set of options for individual or team projects that will enhance interactivity and communication among course participants (see Appendix A). The Resource Center (see Appendix B) and references in all of the chapters provide starting points for inquiries. The process of finding and evaluating sources of information should be based on the principles of information literacy applied to the Internet environment (see Appendix A).

The objective of this chapter's study projects is to gain deeper insights and detailed knowledge of anthropogenic activities affecting the environment and subsequently human health.

PROJECT 1: Population Growth, Industrialization, and Human Health

The objective of this project is to create a keener awareness and understanding of the global effects of population growth, the ensuing industrialization process in developing countries, and their effects on human health.

Task 1. Identify one to three examples from Asia, Africa, or Latin America and describe patterns of population growth, the industrialization process, and their effects on human health.

Task 2. Develop one of your examples in depth and create a presentation. Use one of the following options:

a. a written paper;
b. a Power Point presentation (lecture);
c. a visually oriented presentation (creation of charts, statistics, photographs or video).

PROJECT 2: Ecological Consequences of Consumption Patterns

The objective of this project is to deepen the understanding of the negative impact of developed countries and wealthy societies on the pool of available resources caused by high rates of consumption.

Task 1. Study the consumption of energy and water in three different wealthy regions. Report your findings supported by statistics.

Task 2. Pick one wealthy region and study its food consumption in depth. Include transportation and waste of food in your report.

Task 3. Pick one wealthy region (not necessarily the same as in Task 2) and report on sanitation and waste removal.

Task 4. On the basis of the reports from tasks 1–3, write a paper on the ecological consequences of the patterns of consumption by wealthy societies.

PROJECT 3: Environmental Treaties

The objective of this project is to create a more detailed knowledge of international efforts to counteract ecological and environmental decline.

Task 1. Search for environmental treaties, establish criteria for selecting a subset, and present them in an annotated list.

Task 2. Study the Montreal Protocol and give an in-depth report.

Task 3. Study the Framework Convention on Climate Change and give an in-depth report.

References

American Association of Retired Persons, Resources for the Future. 1993. *Aging of the U.S. Population: Economic and Environmental Implications.* American Association of Retired Persons, Washington.

Armelagos GJ, Dewey JR. 1970. Evolutionary response to human infectious diseases. *Bioscience* 20 (5): 271–75.

Armelagos GJ, Goodman AH, Jacobs KH. 1991. The origins of agriculture: Population growth during a period of declining health. *Popul Environ* 13 (1): 9–22.

Brown LR. 1999. Feeding nine billion. In *State of the World, 1999* (Brown LR, Flavin C, French H, eds.). W. W. Norton & Co., New York.

Brown PJ. 1981. Cultural adaptations to endemic malaria in Sardinia. *Med Anthropol* 5 (3): 313–37.

Cohen JE. 1995. *How Many People Can the Earth Support?* W. W. Norton & Co., New York.

Commission for Environmental Cooperation. 1997. *Continental Pollutant Pathways.* CEC, Montreal.

Conca K. 1994. Rethinking the ecology-sovereignty debate. *Millennium J Int Studies* 23 (3): 701–11.

Culotta E. 1991. Biological immigrants under fire. *Science* 254 (December 6): 1444–47.

Dunlap R. 1980. Paradigmatic change in social science: From human exemptionalism to an ecological paradigm. *Am Behav Sci* 24 (1): 5–14.

Durham WH. 1991. *Coevolution: Genes, Culture, and Human Diversity.* Stanford University Press, Stanford.

Dyson T. 1996. *Population and Food: Global Trends and Future Prospects.* Routledge, New York.

Ezcurra E, Mazari-Hiriart M. 1996. Are megacities viable? *Environment* 38 (1): 6–35.

Food and Agriculture Organization. 1998a. *The State of Food and Agriculture, 1998.* FAO, Rome.

———. 1998b. *Mapping Nutrition and Malnutrition: Dietary Energy Supply (1994–1996).* FAO, Rome.

Garrett L. 1994. *The Coming Plague: Newly Emerging Diseases in a World out of Balance.* Penguin Books, New York.

Gibbons A. 1993. Where are new diseases born? *Science* 261 (August 6): 680–81.

Grifo F, Rosenthal J, ed. 1997. *Biodiversity and Human Health.* Island Press, Washington, D.C.

Hazarika S. 1991. *Bangladesh and Assam: Land Pressures, Migration, and Ethnic Conflict.* Occasional Paper of the Project on Environmental Change and Acute Conflict. American Academy of Arts & Sciences, Cambridge, Mass.

Homer-Dixon TF. 1991. On the threshold: Environmental changes as causes of acute conflict. *Int Security* 16 (2): 76–116.

Jacobson JL. 1988. *Environmental Refugees: A Yardstick of Habitability.* Worldwatch Paper No. 86. Worldwatch Institute, Washington.

McMichael AJ. 1993. *Planetary Overload: Global Environmental Change and the Health of the Human Species.* Cambridge University Press, Cambridge.

McNeill WH. 1977. *Plagues and Peoples.* Doubleday, New York.

Meadows DH, Meadows D, Randers J. 1992. *Beyond the Limits.* Green Publishing Co., Post Mills, Vt.

Murray CJL, Lopez AD, eds. 1996. *The Global Burden of Disease.* Harvard University Press, Cambridge.

Norgaard R. 1994. *Development Betrayed: The End of Progress and a Coevolutionary Revisioning of the Future.* Routledge, London.

Orians GH. 1995. Thought for the morrow: Cumulative threats to the environment. *Environment* 37 (7): 6–36.

Paarlberg RL. 1994. The politics of agricultural resource abuse. *Environment* 36 (8): 6–9, 33–42.

Parnwell M. 1993. *Population Movements and the Third World.* Routledge, London.

Pedersen D. 1996. Disease ecology at a crossroads: Man-made environments, human rights and perpetual development utopias. *Soc Sci Med* 43 (5): 745–58.

Pirages D. 1995. Microsecurity: Disease organisms and human well-being. *Washington Q* 18 (4): 5–12.

Ponting C. 1991. *A Green History of the World.* Penguin Books, London.

Repetto R, Baliga SS. 1996. *Pesticides and the Immune System: The Public Health Risks.* World Resources Institute, Washington.

Ribot J. 1993. Market-state relations and environmental policy: Limits of state capacity in Senegal. In *The State and Social Power in Global Environmental Politics* (Lipschutz RD, Conca K, eds.). Columbia University Press, New York.

Sachs W. 1993. Global ecology and the shadow of development. In *Global Ecology: A New Arena of Political Conflict* (Sachs W, ed.). Zed Books, London.

Shell ER. 1997. Resurgence of a deadly disease. *Atlantic Monthly* 280 (2): 45–60.

Simon J. 1981. *The Ultimate Resource.* Princeton University Press, Princeton.

Simon J, Kahn H. 1984. *The Resourceful Earth.* Basil Blackwell, Oxford, U.K.

Suhrke A. 1991. *Pressure Points: Environmental Degradation, Migration, and Conflict.* Occasional Paper of the Project on Environmental Change and Acute Conflict. American Academy of Arts & Sciences, Cambridge, Mass.

Taylor D. 1997. Seeing the forests for more than the trees. *Environ Health Perspect* 105 (11): 1186–91.

Vitek CR, Wharton M. 1998. Diphtheria in the former Soviet Union: Reemergence of a pandemic disease. *Emerg Infect Dis* 4 (4): 539–50.

Wackernagel M, Rees W. 1996. *Our Ecological Footprint: Reducing Human Impact on the Earth.* New Society Publishers, Philadelphia.

Weiner M. 1992. Security, stability, and international migration. *Int Security* 17 (3): 91–126.

World Bank. 1991. *Urban Policy and Economic Development: An Agenda for the 1990s.* World Bank, Washington.

———. 1994. *Averting the Old Age Crisis: Policies to Protect the Old and Promote Growth.* Oxford University Press, New York.

World Commission on Environment and Development. 1987. *Our Common Future.* Oxford University Press, Oxford.

World Resources Institute. 1994. *World Resources, 1994–95.* Oxford University Press, New York.

———. 1996. *World Resources, 1996–97.* Oxford University Press, New York.

Zacher MW. 1992. The decaying pillars of the Westphalian Temple: Implications for international governance and order. In *Governance without Government: Order and Change in World Politics* (Rosenau JN, Czempiel E-O, eds.). Cambridge University Press, Cambridge.

Electronic References

Centers for Disease Control and Prevention. 2000. Preventing Emerging Infectious Diseases: A Strategy for the 21st Century. National Center for Infectious Diseases. http://www.cdc.gov/ncidod/emergplan/1toc.htm (Date Last Revised 2/1/2000).

National Science and Technology Council Committee on International Science, Engineering and Technology. 1997. Emerging Infectious Disease Task Force PDD/NSTC-7, Annual Report, December 19, 1997. http://www.whitehouse.gov/WH/EOP/OSTP/Security/html/eidann_rpt.html (Date Last Revised 12/19/1997).

Pew. 2000. Pew Center on Global Climate Change. http://www.pewclimate.org (Date Last Revised 2000).

United Nations Centre for Human Settlements. 1997. Human Settlements Basic Statistics, 1997. http://www.unchs.org/unon/unchs/unchs/english/stats/contents.htm (Date Last Revised 1997).

United Nations Environment Programme. 2000. Information Unit for Conventions. http://www.unep.ch/iuc/ (Date Last Revised 2000).

World Health Organization. 1999. WHO's Partners. http://www.who.int/ina/partners.html (Date Last Revised 1999).

World Resources Institute. 2000. World Resources Institute Home Page. http://www.wri.org (Date Last Revised 3/22/2000).

CHAPTER 7

The Changing Chemistry of Earth's Atmosphere

Steven A. Lloyd, Ph.D.

Since the dawn of time, our earth's atmosphere has been in a state of constant change. Its composition, temperature, and ability to cleanse itself chemically have co-evolved along with biological life on geological time scales. However, only since the advent of industrialization two centuries ago has our atmosphere experienced such remarkably *rapid* changes in composition.

Global environmental change can be defined as changes in the earth's ecosystem that are global in extent, arising from either anthropogenic or natural causes, including changes in local ecosystems caused by population pressure and consumption on local resources, which are becoming more widespread. Four of the most significant global change phenomena that scientists believe the earth is now experiencing are (1) stratospheric ozone depletion, (2) acid deposition, (3) urban air pollution, and (4) enhanced global warming. Each of these phenomena is the result of anthropogenically induced changes in the chemistry of our atmosphere. An understanding of the chemical bases of these phenomena will help us predict the potential future implications for the earth's fragile ecosystem.

Proven consequences of these atmospheric changes include (1) a seasonal thinning of the stratosphere's ozone layer, which protects the earth's surface from damaging solar ultraviolet radiation; (2) damage to our forests and aquatic life by acid deposition; and (3) respiratory and ocular distress induced by regional-scale urban smog events. Additionally, it is expected that the burning of fossil fuels and the release of trace gases related to increased industrialization and intensive agricultural practices may lead to enhancement of the earth's greenhouse effect, resulting in rapid warming of the planet.

Since 1987, it has been generally accepted among atmospheric scientists that increasing levels of manufactured chlorofluorocarbons (CFCs) are slowly releasing their chlorine in the earth's upper atmosphere, which catalytically destroys the ozone layer, especially over the polar regions (the so-called ozone hole). The consequences for human health include increased incidence of basal and squamous cell carcinomas, accelerated induction of cataracts, and suppression of the human immune response. Increased acid deposition (both acid rain and dry deposition) in industrialized countries over the last several decades has resulted in significant damage to forested areas and lakes, resulting in the premature deaths of forests (*Waldsterben* in Germany's Black Forest) and fish kills in overly acidified lakes and streams. Increasing urbanization, including the rise in automobile transport, during the last half-century has been achieved at the price tag of significant health risks associated with industrial and automotive exhausts. The most important consequence of urban air pollution for human health care is damage to the respiratory tract.

Although the scientific jury is still out as to the magnitude and time scale of enhanced global warming associated with increased atmospheric emissions of carbon dioxide and other industrial pollutants, most scientists agree that such warming is likely within the twenty-first century if present emission rates are not controlled. A wide spectrum of health risks can be anticipated, as well as socioeconomic risks arising from desertification, rising sea level, changing weather patterns, and other phenomena associated with a rapidly changing climate (see Chapter 13). A periodic meteorological phenomenon associated with tropical winds, sea surface temperature anomalies, and circulation patterns known as the El Niño/Southern Oscillation (ENSO) can also further complicate the consequences of enhanced global warming on a global scale (see Chapter 8).

These atmospheric phenomena, which have their source in chemical transformations induced by human activity, are the primary causes of global change in our environment. These processes merit further study to elucidate their future trends and associated health risks. It is surprising that each of these phenomena stems not from changes in the atmosphere's main constituents (nitrogen, oxygen, argon, and water vapor), but rather from rising levels of several of the minor constituents, or trace gases. These gases, such as nitrogen oxides (nitric oxide [NO] and nitrogen dioxide [NO_2]), sulfur dioxide (SO_2), and a variety of CFCs are typically present in the atmosphere at the 1–50 parts per billion level, while the most abundant trace gas, carbon dioxide (CO_2), is present at a level of only 370 parts per million. Although there is a natural component to many of the earth's trace gases, the observed rapid rise in their concentrations over the past two centuries is due to human intervention, or more specifically, to our current industrial and agricultural practices. For the first time in geological history, humanity has changed the earth's environment on a global scale, perhaps irreversibly.

Stratospheric Ozone Depletion

One of the most important of Earth's trace gases is ozone (O_3). Ozone is a more reactive allotrope of molecular oxygen (O_2), which is present in the natural atmosphere in trace amounts at the parts per million level. Ozone is important to the maintenance of life because it absorbs solar ultraviolet radiation, which would otherwise be absorbed by biological molecules termed *chromophores,* such as nucleic acids, thus damaging the genetic material and photosynthetic capacity of the earth's flora and fauna. About 90 percent of the ozone resides in the stratosphere, a region of the atmosphere roughly 10–50 kilometers above the surface of the earth. If the atmosphere were condensed down to sea-level pressure (1 atmosphere) and room temperature (25°C), the ozone would make a layer about 3 millimeters thick. This diffuse ozone layer, spread out principally in the lower stratosphere between 15 and 25 kilometers altitude, is all that separates life on earth from the sun's damaging ultraviolet (UV) rays. Our atmosphere is transparent to the least damaging of these rays, the UV-A region (wavelengths from 320 to 400 nanometers), since ozone does not absorb at these wavelengths. UV-C (<280 nanometers) is absorbed completely by molecular oxygen as well as by ozone, and hence none reaches the earth's surface. UV-B (280–320 nanometers), however, is only partially absorbed by the ozone layer; therefore, the ground-level flux of UV-B depends critically on the amount of ozone overhead. (See Table 3.2 for a classification of various portions of the electromagnetic spectrum.)

In the stratosphere, oxides of nitrogen, hydrogen, chlorine, and bromine act to limit the amount of ozone through catalytic cycles that remove a significant fraction of the ultraviolet-blocking compound. Anthropogenic increases in the levels of nitrogen, chlorine, and bromine oxides have greatly enhanced the ozone-destroying capacity of the stratosphere, and the halogen (chlorine and bromine) oxides lead to significant seasonal loss of ozone over Antarctica each austral spring, an event now known as the *ozone hole.*

One of the most pressing scientific questions in earth science today is: How is ozone changing in the stratosphere and troposphere around the globe? Observations show that, while the level of ozone is decreasing in the lower stratosphere, it is increasing in the upper troposphere. Rising levels of a host of trace gases within the last century have precipitated changes in the earth's global ozone balance at a rate never before observed. Policymakers around the globe are now addressing this issue from a regulatory perspective in an attempt to curb the current rate of change in both stratospheric and tropospheric ozone. Atmospheric scientists are now confronted with a political mandate both to explain the earth's ozone balance, which has both natural and anthropogenic components, and to formulate sound policy choices on the basis of accumulated scientific evidence.

The Development of the Stratospheric Ozone Issue:
Supersonic Transports

Concern was first raised in the 1960s that human activity could perturb the ozone layer to an extent that could endanger life on earth. Professor Hal Johnston of the University of California at Berkeley proposed that nitrogen oxides (NO_x, consisting of NO and NO_2) from the exhaust of proposed fleets of supersonic transports that fly in the lower stratosphere could pose a threat to the ozone layer globally (National Research Council [NRC] 1975). While this fleet of supersonic jets could not possibly produce nitrogen oxides at a level that would destroy significant amounts of ozone on a molecule-for-molecule basis, NO_x could potentially increase those naturally occurring cycles in the stratosphere that limit ozone abundance through the *catalytic* destruction of ozone. One such naturally occurring cycle involving NO_x is thought to be the major sink for ozone on a global basis:

$$
\begin{array}{lll}
& NO + O_3 & \rightarrow\ NO_2 + O_2 \quad (1) \\
\text{Mechanism I} & NO_2 + O & \rightarrow\ NO + O_2 \quad (2) \\
\hline
\text{net:} & O_3 + O & \rightarrow\ 2\,O_2 \\
\end{array}
$$

This cycle is most efficient in the middle stratosphere (around 40 kilometers altitude, above the peak of ozone concentration), where free oxygen atoms are sufficiently abundant to make this cycle a significant process for the loss of ozone. In the lower stratosphere, below the peak of the ozone layer and where the supersonic transports were designed to fly, free oxygen atoms are scarce, and the following cycle dominates:

$$
\begin{array}{lll}
& NO + O_3 & \rightarrow\ NO_2 + O_2 \quad (1) \\
\text{Mechanism II} & NO_2 + \text{sunlight} & \rightarrow\ NO + O \quad (3) \\
& O + O_2 + M & \rightarrow\ O_3 + M \quad (4) \\
\hline
\text{net:} & \text{null cycle} & \\
\end{array}
$$

In this cycle, where M is N_2 or O_2, NO_2 is photolyzed by near-ultraviolet and visible light to yield oxygen atoms, which then rapidly recombine to form ozone. The overall result is a null cycle, in which ozone is regenerated as fast as it is destroyed. Crucial concerns were the altitude at which the nitrogen oxides would be deposited into the atmosphere and how rapidly they would be redistributed within the troposphere and stratosphere. Therefore, this simple depiction of ozone photochemistry would indicate that NO_x deposited at higher altitudes would clearly represent a threat to the ozone layer, whereas NO_x deposited at lower altitudes (where the supersonic transports fly) might not have a substantial effect. Since only a few supersonic transports (the British-French Concorde and its Soviet analogue, the Tu-144) were ever manufactured, the issue of the effect of stratospheric transport on ozone became moot.

More recently, the issue of the effect of second- and third-generation supersonic transports on the earth's stratospheric ozone layer has been revisited. A series of aircraft studies, which have been funded by the U.S. National Aeronautics and Space Administration (NASA), use a modified high-altitude U-2 spy plane called the ER-2 to fly into the ozone layer (18–20 kilometers altitude) and make measurements of the chemical composition of the lower stratosphere and upper troposphere. These studies have shown that the catalytic cycles of ozone loss involving oxides of nitrogen, hydrogen, chlorine, and bromine are interconnected in a complex way. A more complete model of the ozone photochemistry of the stratosphere and troposphere now includes hundreds of coupled chemical reactions. The release of combustion products from high-speed aircraft can create both positive and negative feedbacks on the various catalytic cycles that control the abundance of ozone in the atmosphere, making evaluation of the *net* effect of proposed fleets of these aircraft, flying at a variety of altitudes over a wide range of atmospheric conditions of temperature and solar illumination, a considerably more difficult question.

NASA has commissioned extensive research into the potential effects of both subsonic and supersonic aircraft. In particular, the Atmospheric Effects of Aviation Project has evaluated the potential impact of aircraft that travel in both the troposphere and the stratosphere. The results of this research affect whether or not the U.S. government and aircraft industry will decide to fund the further development of such high-speed aircraft (Penner et al. 1999).

Concern has also been raised about the contribution of other anthropogenic sources of nitrogen oxides, including the use of nitrogen-based fertilizers and the large-scale burning of forests and grasslands. Biomass burning also contributes to increased levels of tropospheric ozone.

Another Threat to the Ozone Layer: Chlorofluorocarbons
Since 1974, attention has focused primarily on the ability of CFCs, chiefly used in refrigerators, air conditioners, and Styrofoam insulation, to enter the stratosphere and rise above the ozone layer, where ultraviolet light can release their chlorine atoms (Rowland and Molina 1974). Most stratospheric chlorine comes from CFCs, rather than from naturally occurring sources (such as volcanoes). The chlorine-catalyzed destruction of ozone is a standard textbook example of chemical catalysis, known since the

1930s (Hinshelwood 1933). In 1974 Mario Molina and Sherwood Rowland postulated that a chlorine-based catalytic cycle analogous to NO_x-catalyzed ozone loss cycle depletes the stratospheric ozone layer. They won the 1995 Nobel Prize in chemistry for this work (Rowland and Molina 1974).

$$Cl + O_3 \rightarrow ClO + O_2 \qquad (5)$$
$$\text{Mechanism III} \quad ClO + O \rightarrow Cl + O_2 \qquad (6)$$
$$\text{net:} \quad O_3 + O \rightarrow 2\,O_2$$

Using computer models originally developed to estimate the damage to the ozone layer from NO_x emissions, scientists began to elaborate on this new theory of chlorine-catalyzed ozone depletion. It soon became clear that there were key gaps in our understanding of stratospheric chlorine chemistry, specifically in the rates of key reactions and the role of secondary chlorine reservoirs, such as $ClONO_2$ and $HOCl$. While scientists grappled with these questions in the late 1970s, politicians in the United States debated the wisdom of placing limits on the manufacture of CFCs, which are a highly profitable and generally quite useful class of industrial compounds. Despite opposition from the chemical manufacturing industry, the use of CFCs as aerosol propellants was outlawed in 1978 in the United States, followed shortly by bans in Canada and Scandinavia. The elimination of this "frivolous" use of CFCs proved to be a Pyrrhic victory, since convincing evidence of the depletion of the ozone layer by CFCs was not forthcoming; legislators were hesitant to curtail the main uses of CFCs as refrigerants and cleaning solvents without such proof.

Although model calculations predicted relatively modest ozone depletions in the middle stratosphere (on the order of a few percentage points) over a period of decades, this small change eluded observations made during the 1970s and 1980s (World Meteorological Organization [WMO] 1986). First, no long-term database on stratospheric ozone was then available to provide a global baseline, and second, the ozone layer is subject to natural fluctuations much larger than the sought-after phenomenon. The issue of chlorine-catalyzed ozone depletion faded from the newspaper headlines and seemed destined to be relegated to history as either a false alarm or, at best, a minor environmental curiosity with little effect on the global ecosystem.

The Smoking Gun: The Antarctic Ozone Hole

In 1985, Farman et al. announced their findings of seasonal, dramatic depletions of ozone in the lower stratosphere over Antarctica (Fig. 7.1a), which were subsequently confirmed by archived NASA satellite data (Stolarski et al. 1986). This announcement provided the first publicly disclosed

(a)

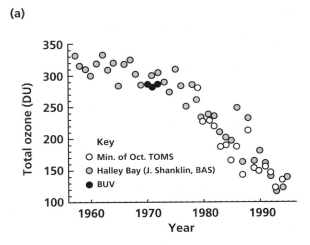

(b)

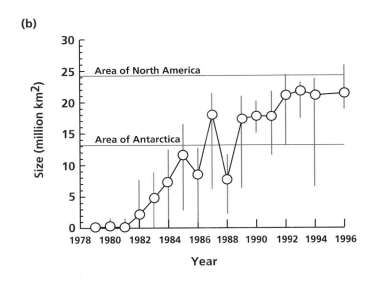

Figure 7.1 *a:* Total ozone observations over the Antarctic during the month of October in Dobson Units (DU) as measured by the British Antarctic Survey, 1957–95, and NASA satellites (Backscatter Ultraviolet [BUV], Nimbus 7 TOMS and Meteor 3 TOMS), 1970–95. The BUV mission is a predecessor of the SBUV (Solar BUV) and SSBUV (Shuttle Solar BUV) missions. Data were provided by the British Antarctic Survey (J. Shanklin) and the TOMS Ozone Data Processing Team, NASA Goddard Space Flight Center (P. Newman and R. McPeters). *Source:* Redrawn from *NASA/GSFC 2000a,* Figure 11.01. *b:* Annual maximum size of the Antarctic ozone hole in million square kilometers, 1978–96. Data were provided by the TOMS Ozone Data Processing Team, NASA Goddard Space Flight Center (P. Newman and R. McPeters). *Source:* Redrawn from *NASA/GSFC 2000a,* Figure 11.36.

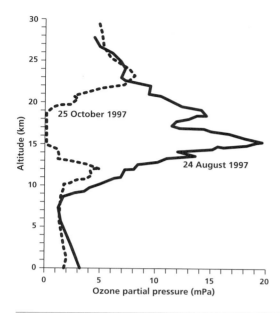

Figure 7.2 Ozone altitude profiles in August and October as measured by balloon-sondes at Mc-Murdo Station, Antarctica, 1997. Note that by October the ozone has been completely removed from altitude range 15–19 kilometers and mostly depleted just above and below these altitudes. *Source:* Data courtesy of Bruno Nardi, National Center for Atmospheric Research, Boulder, Colorado.

evidence for substantial perturbations to the ozone layer. The phenomenon, observed using Dobson spectrometers during routine ozone studies at Britain's Halley Bay Antarctic Station, was promptly dubbed the "ozone hole." This ozone hole is characterized by a substantial loss (greater than 50% since 1983) of the total column of ozone over an area roughly coterminous with the Antarctic continent or Antarctic circle. Losses are seasonal, appearing each austral spring in late September and persisting until sometime between late October and early December. Ozone loss is greatest at approximately 15–19 kilometers altitude (over 95% loss during some years), with substantial loss throughout the lower stratosphere (14–22 kilometers) (see Fig. 7.2 and Nardi et al. 1999).

The lower stratosphere also experiences dense polar stratospheric clouds (PSCs) throughout the polar winter and persisting into the early spring. These clouds are thought to provide sites upon which rapid heterogeneous reactions can convert the comparatively inert reservoirs of inorganic chlorine, HCl and $ClONO_2$, into the much more photochemically active species Cl_2 and HOCl, which can be rapidly photolyzed to chlorine atoms as soon as the sun appears over the horizon during the austral spring. The following two reactions, thought to be primarily responsible for this "reservoir exchange," occur on the surface of polar stratospheric clouds during the long polar night:

$$ClONO_2 + HCl \rightarrow Cl_2 + HNO_3 \tag{7}$$
$$ClONO_2 + H_2O \rightarrow HOCl + HNO_3 \tag{8}$$

Stratospheric Ozone Depletion Is Accelerated by Other Atmospheric Changes

Our quantitative understanding of stratospheric ozone depletion is now well established in terms of ozone loss catalyzed by halogens (chlorine and bromine). However, other processes that affect the conditions under which this chemical ozone loss occurs can have a significant influence on the amount of ozone destroyed. In particular, processes that affect aerosol loading and temperature in the stratosphere can substantially accelerate the rate of ozone loss. For example, the eruption of Mount Pinatubo in 1991 produced sulfate aerosols in the stratosphere that provide surfaces analogous to those of the icy polar stratospheric clouds (PSCs) over the South Pole. These surfaces facilitate chemical reactions that accelerate the destruction of ozone at latitudes too warm to produce PSCs. The increase in greenhouse warming gases also affects the chemistry of stratospheric ozone depletion. While greenhouse gases warm the earth's surface, they also act to cool the stratosphere. A cooler stratosphere would accelerate the rate of certain mechanisms of ozone loss (such as mechanism IV) and perhaps allow the spatial extent of the Antarctic ozone hole (which has reached its maximum at the margins of the Antarctic winter polar vortex; see Fig. 7.1b) to increase, leading to significant ozone loss at lower latitudes. A further concern is that colder temperatures in the Arctic would allow the formation of larger, water-encased type II PSC particles in the Northern Hemisphere, possibly leading to the regular seasonal appearance of an ozone hole over the North Pole as well (Shindell et al. 1998).

Without reactions such as 7 and 8 to convert relatively inert chlorine into a more reactive form, stratospheric ozone depletion via chlorine catalysis could not occur.

SO_2 emitted by the 1991 eruption of the Mount Pinatubo volcano in the Philippines seems to have generated a high concentration of stratospheric sulfate aerosols, which also served as surfaces upon which heterogeneous chemistry could take place, analogous to the PSCs (see Box 7.1). Satellite data (Gleason et al. 1993) and models indicate that global ozone depletions on the order of a few percentage points resulted from the interaction of high stratospheric chlorine loading (a recent phenomenon; CFCs have been manufactured only since the 1930s) with volcanically induced aerosol formation (a naturally occurring phenomenon).

The 1987 Antarctic Atmospheric Ozone Experiment (AAOE) sampled the polar ozone layer during the formation of the ozone hole, using NASA's ER-2 aircraft to carry a suite of instruments into the polar stratosphere. Simultaneous measurements were made of ozone and the reactive radical species ClO and BrO. Radicals are chemical species that contain an odd number of electrons and are thus especially reactive. These data demonstrated an association between high levels of halogenated species and low levels of ozone and gave strong support to the theory that chlorine- and bromine-catalyzed reactions are responsible for the more than 50 percent loss of ozone over the South Pole each austral spring (Anderson et al. 1989, 1991). Subsequent missions of the ER-2 to the Arctic stratosphere during the winters of 1988–89, 1991–92, and 1999–2000 confirmed that similar chemical preconditions for significant polar ozone loss also exist in the Northern Hemisphere.

After the discovery of the Antarctic ozone hole in 1985, it was quickly realized that conventional numerical models for stratospheric ozone depletion were incapable of explaining the comparatively sudden, large ozone losses experienced over several weeks each austral spring. The weak polar UV radiation field in the lower stratosphere during the early spring is incapable of providing sufficient quantities of free oxygen atoms to support mechanism III (above), which was originally proposed to explain stratospheric ozone losses at higher altitudes. Also, models that did not yet include heterogeneous chemical reactions on PSCs were incapable of explaining the large quantity of photochemically "active" chlorine species required to account for the sudden ozone loss.

In 1986 Molina and Molina (1987) proposed the following catalytic cycle as a mechanism for polar ozone depletion:

$$2 \ (Cl + O_3 \qquad \rightarrow \ ClO + O_2) \qquad (5)$$
$$ClO + ClO + M \quad \rightleftharpoons \ ClOOCl + M \qquad (9)$$
$$\text{Mechanism IV} \quad ClOOCl + \text{sunlight} \rightarrow \ Cl + ClOO \qquad (10)$$
$$ClOO + M \qquad \quad \rightarrow \ Cl + O_2 + M \qquad (11)$$

$$\text{net:} \quad 2 \, O_3 \qquad \qquad \rightarrow \ 3 \, O_2$$

This mechanism allows for the rapid depletion of ozone without necessitating the presence of free oxygen atoms, required in prior mechanisms proposed to account for the loss of stratospheric ozone. One feature that made it attractive as a candidate for explaining the sudden ozone losses is that the association reaction (reaction 9, where M is N_2 or O_2) proceeds *faster* at the much lower temperatures encountered in the polar lower stratosphere (whereas most other mechanisms would proceed too slowly under these conditions). In addition, the photolysis step (reaction 10) explains why the ozone loss begins only after the sun reappears over the horizon in August and September after the long Antarctic polar night.

Since the rates of reactions 5 and 11 are comparatively fast, we can say that reactions 9 and 10 are *both* rate limiting and form a photochemical steady state between the rates of dimer formation and destruction. The overall rate of the catalytic cycle, and hence of depletion of polar ozone, depends on the *photochemical balance* between ClO and its dimer. Within the chemically perturbed Antarctic polar vortex, the inorganic chlorine (Cl_x) budget is dominated by the reservoirs ClO and ClOOCl, rather than HCl and $ClONO_2$, as is the case elsewhere in the unperturbed stratosphere (Kawa et al. 1992).

Not only is the ozone hole devoid of most of the naturally occurring ozone in the lower stratosphere, it is also depleted of nitrogen oxides (Fahey et al. 1989). The "denitrification" of the Antarctic stratosphere is accomplished by the following sequence of reactions, where M is N_2 or O_2:

$$NO_2 + O_3 \rightarrow NO_3 + O_2 \tag{12}$$
$$NO_2 + NO_3 + M \rightleftharpoons N_2O_5 + M \tag{13}$$
$$N_2O_5 + H_2O \rightarrow 2\,HNO_3 \tag{14}$$

N_2O_5 is the anhydride of nitric acid; as soon as it comes in contact with water, either in the gas phase or more rapidly on the surface of PSC particles, it will form its corresponding acid. Much of the nitric acid forms a solid trihydrate, which forms the basis of type I PSC particles, and also forms the core of PSC type II particles, which are essentially type I particles enveloped in a thick casing of water ice when the temperature drops a few degrees cooler. This process effectively denitrifies the polar lower stratosphere; in the Antarctic, sedimentation of PSC type II particles removes nitrates from the stratosphere, while in the Arctic the erratic nature of the polar vortex allows incursions of nitrified air from farther south into the polar regions. Low levels of NO_2 are required for the rapid removal of ozone observed over the Antarctic, since NO_2 rapidly reacts with ClO to form $ClONO_2$; this reaction removes ClO from the catalytic cycles and stops the ozone loss. In the absence of interfering reactions with nitrogen oxides, which are in abnormally low concentration in the polar vortex, the catalytic cycle operates virtually in isolation.

Another halogen-catalyzed mechanism proposed by McElroy et al. (1986) posits that bromine radicals from halons (used primarily in fire extinguishers) react synergistically with chlorine radicals to deplete the ozone layer rapidly:

$$Cl + O_3 \rightarrow ClO + O_2 \tag{5}$$
Mechanism V $$Br + O_3 \rightarrow BrO + O_2 \tag{15}$$
$$BrO + ClO \rightarrow Br + Cl + O_2 \tag{16}$$
$$\text{net:}\quad 2\,O_3 \rightarrow 3\,O_2$$

In this mechanism, reaction 16 is rate limiting. Reaction 16 is actually a composite, since the reaction between ClO and BrO proceeds via three reaction pathways,

$$BrO + ClO \rightarrow Br + OClO \tag{16a}$$
$$BrO + ClO \rightarrow Br + ClOO \rightarrow Br + Cl + O_2 \tag{16b}$$
$$BrO + ClO \rightarrow BrCl + O_2 \rightarrow Br + Cl + O_2 \tag{16c}$$

only two of which (16b and 16c) contribute to the overall depletion mechanism. Since ClOO rapidly dissociates thermally (reaction 11) and BrCl photolyzes quite rapidly even in the polar twilight, the net products of channels 16b and 16c are the same, and the rate of this cycle simply becomes the sum of the rates of reactions 16b and 16c.

A third mechanism was proposed by Solomon et al. (1986) to account for the observed loss of polar ozone, involving the synergistic reaction of chlorine radicals and oxides of hydrogen:

$$
\begin{array}{lll}
 & Cl + O_3 & \rightarrow ClO + O_2 \quad (5) \\
 & OH + O_3 & \rightarrow HO_2 + O_2 \quad (17) \\
\text{Mechanism VI} & ClO + HO_2 & \rightarrow HOCl + O_2 \quad (18) \\
 & HOCl + \text{sunlight} & \rightarrow OH + Cl \quad (19) \\
\hline
\text{net:} & 2\,O_3 & \rightarrow 3\,O_2
\end{array}
$$

Simultaneous measurements of ozone, ClO, and BrO, along with other relevant parameters such as temperature and pressure, allow one to compare the calculated rates of ozone loss from the first two of these mechanisms with the observed rates of loss. Such measurements made during the 1987 AAOE indicate that the observed rates of ozone loss agree to within experimental error with those calculated from the simultaneously observed concentrations of radicals and the experimentally derived rate constants for the rate-determining steps of mechanisms IV and V. Mechanism IV accounts for about three-quarters of the loss of polar ozone, and mechanism V accounts for most of the balance. Although measurements of the oxides of hydrogen were not made during this mission, reasonable estimates of stratospheric HO_2 abundances lead one to conclude that mechanism VI is unlikely to account for more than about 5 percent of the observed ozone loss.

Perhaps the most important feature of each of these photochemical cycles is that they are *catalytic*. Radical intermediates are both consumed and regenerated within each cycle, such that each cycle can go round and round many times before a competing reaction interferes with the cycle (such as chlorine atoms reacting with methane, CH_4, to form HCl, a photochemically inactive species in the lower stratosphere). For example, the reaction rates within the Antarctic polar vortex are such that each chlorine atom can remove approximately 100,000 ozone molecules before forming an inactive HCl molecule. This means that the polar catalytic cycles provide an amplification factor of 10^5, so that a very small amount of chlorine (a few parts per billion) can remove much larger amounts of ozone (several parts per million).

The Global Extent of Stratospheric Ozone Depletion

Further concern was raised by the finding of the Ozone Trends Panel that ozone is also thinning at midlatitudes, as measured by the NASA Total Ozone Mapping Spectrometer (TOMS) satellite instrument (WMO 1990). The panel concluded in 1988 that, not only is the ozone layer thinning over heavily populated midlatitudes on the order of 1–6 percent per decade, but also the rate at which it is thinning is increasing. The fact that the loss is greatest at higher latitudes in the lower stratosphere during the winter leads one to speculate whether there might be a causal connection between losses of polar ozone and seasonal depletion of midlatitude ozone.

During the late 1980s, the ozone-destroying capacity of the ozone hole seemed to be rising slowly, following the steady increase in the release of CFCs and, consequently, the total chlorine budget of the atmosphere. During the mid-1990s, the spatial extent of the ozone hole seems to have reached a maximum limit, constrained by the size of the Antarctic polar vortex, roughly the size of the Antarctic continent (see Fig. 7.1*b*). However, ozone losses are now appearing at both higher and lower altitudes within the ozone hole, such that by late October all of the ozone between 14 and 22 kilometers has been photochemically removed. When the polar vortex breaks apart in the late austral spring (October through early December), this ozone-poor air mixes with the surrounding ozone-rich air and effectively dilutes the ozone levels over most of the Southern Hemisphere. Heavily populated regions of the Southern Hemisphere, in particular southeastern Australia, have reported significant losses of local ozone throughout the month of December. These seasonal losses can be expected to continue throughout much of the twenty-first century.

Ozone Loss in the Arctic

All of the chemical, meteorological, and dynamical preconditions that contribute to the formation of an ozone hole over the South Pole are also present at the North Pole, allowing for the possibility of seasonal reductions in ozone and consequent increases in the UV flux in the Arctic. Significant populations live in the Arctic and sub-Arctic regions of the Northern Hemisphere (while the subpolar region of the Southern Hemisphere is comparatively sparsely populated), and these populations are more genetically predisposed to skin cancers (especially the light-skinned Scandinavians and Celts) than are their neighbors to the south.

In general, the temperatures are not usually as cold over the Arctic Ocean as over the Antarctic continent, and the polar vortex is not usually as strong. These meteorological conditions lead one to expect that the loss of ozone would not be as severe over the North Pole as in the south, since extremely cold lower stratospheric temperatures and an intact polar vortex are preconditions for the chemical processing that defines an "ozone hole" event. However, in March 1997, two TOMS instruments on the U.S. Earth Probe and Japanese Advanced Earth Observing System (ADEOS) satellites observed significant losses of ozone over the North Pole during an unusually cold late winter and early spring. Color images of these data (polar projections centered about the North Pole) are publicly available at the TOMS website (*NASA/Goddard Space Flight Center [GSFC] 2000b*). These observations have been confirmed by the European Space Agency's Global Ozone Monitoring Experiment (GOME) instrument, which also monitors ozone levels globally.

In the springs of 1997 and 2000, the temperatures over the North Pole were sufficiently low and the dynamical strength of the polar vortex was sufficiently high that "ozone hole" events, similar to the one first publicly identified by Farman et al. in 1985 over the South Pole, were observed in the Northern Hemisphere. The extremely cold temperatures and dynamical stability of the vortex establish the meteorological preconditions under which photochemical depletion of ozone takes place. Although the sizes of these ozone holes at the North Pole were not as large as those over the South Pole, a recent cooling trend in the lower stratosphere would indicate that similar ozone losses could be expected over the next few decades (see Box 7.1).

Modeling the Effects on Human Health
of Stratospheric Ozone Depletion

The principal concern about stratospheric ozone depletion is that ozone loss will lead to enhanced levels of UV-B radiation, which may damage biological systems, many of which are very sensitive to UV-B flux. One of the marvels of nature is that DNA absorbs strongly in the UV-B region, with its absorption spectrum paralleling that of ozone; it is as if ozone provides a perfect umbrella to shield biological matter, wavelength for wavelength, from the sun's damaging UV-B. Figure 7.3 shows the absorption spectra for DNA (Setlow 1974), as well as the solar flux at the earth's surface for a "normal" atmosphere and one in which the ozone layer has been depleted by 30 percent (Anderson and Lloyd 1990). As the ozone layer is depleted, considerably more UV-B reaches the surface. It is the *overlap* of the ground-level flux and the relevant "action spectrum" (the wavelength-dependent measure of biological effects) that is important in predicting the consequences of enhanced UV-B.

Potential biological consequences of stratospheric ozone depletion include an increased incidence of nonmelanoma skin cancers, hastening of

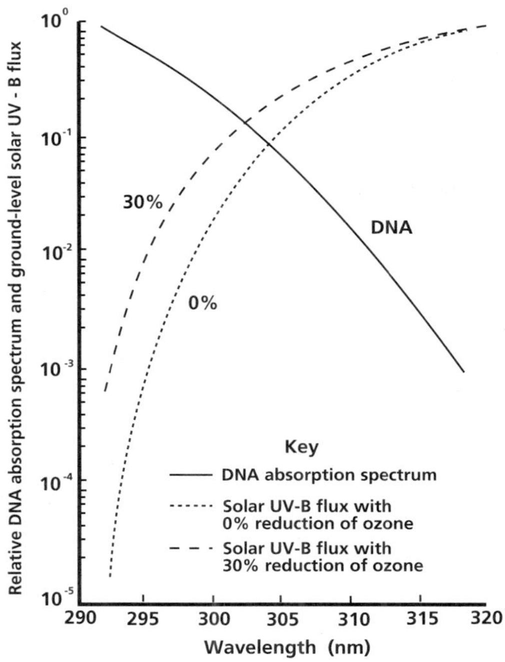

Figure 7.3 Relative DNA absorption spectrum (Setlow 1974) and ground-level solar UV-B flux (arbitrary units; Lloyd et al. 1994). Note that when the ozone abundance is reduced by 30 percent, the ground-level solar UV-B flux increases dramatically at shorter wavelengths. *Source:* Redrawn with permission from Lloyd 1993a, Figure 1. © The Lancet Ltd.

cataract formation, disruption of the aquatic food chain, and crop damage (Emmett 1986). In addition to attempting to predict the future course of global ozone depletion, one would also like to know what the potential consequences of such ozone depletion might be and how we can attempt to prepare for such consequences to both human health and the global environment. In recent years great strides have been made in understanding the link between UV-B exposure and nonmelanoma skin cancers (one of the few cases in which a direct biochemical mechanism for carcinogenesis is known), but most of the other risks to human health are relatively poorly understood. Of particular interest is suppression of the human immune response by excessive UV-B exposure and the resulting increased susceptibility to bacterial and viral infections.

In assessment of the human health effects of stratospheric ozone loss, numerical models have been useful in predicting changes in ambient UV insolation; however, other factors such as individual susceptibility, genetic predisposition, and personal behavior defy simple mathematical analysis. A primary biological assumption is that UV-B light can damage the DNA molecule via pyrimidine dimer formation or other covalent modification of the DNA helix (Francis et al. 1988). The hydrogen bonding of the DNA bases is shown in Figure 7.4. Pyrimidine bases cytosine (C) and thymine (T) pair up with guanine (G) and adenine (A), forming three or two hy-

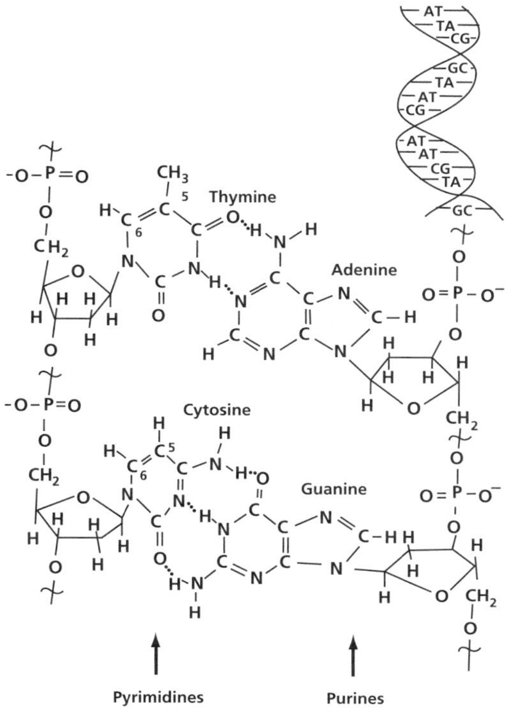

Figure 7.4 DNA base pairing of pyrimidine and purine bases. *Source:* Reproduced with permission from Lloyd 1993b, Figure 3.2.

drogen bonds (respectively) and the well-known spiral staircase double helix. Each base is like a step along the spiral staircase, positioned approximately 30° apart and slightly inclined at a 15° angle. If two pyrimidines happen to be adjacent, ultraviolet light can activate the double bond located at the 5–6 position on either base unit, resulting in cross-linking between the two bases. This covalent modification of the DNA strand, either a C—C, C—T, or T—T linkage, appears in a variety of forms; the best understood are the dithymine dimers.

A second assumption is that photochemical modification of DNA can result in deleterious biological consequences, such as nonmelanoma skin cancers (Scotto 1986; Longstreth et al. 1998) and damage to crops (Teramura 1986; Caldwell et al. 1998) and aquatic life (Thomson 1986; Häder et al. 1998). Setlow (1974) first reported that the absorption spectrum of DNA parallels the action spectrum for skin erythema in the UV-B region. Peak et al. (1984) more recently reported that the action spectra for lethality and mutagenesis of *Escherichia coli* exposed to UV light correspond to the DNA absorption spectrum, leading to the conclusion that these effects proceed via photochemical alteration of the DNA molecule.

A third assumption is that the biological effects of stratospheric ozone depletion scale linearly with changes in exposure to ground-level biologically effective dosages of UV-B light. Several studies have shown that the incidence rates for both nonmelanoma and melanoma skin cancers correlate inversely with latitude. Both Rundel (1983) and Scotto et al. (1982) took this a step further and compared data on the incidence of basal and squamous cell skin cancers in the United States to establish the linearity of the UV-B dose-response relationship.

Using the DNA absorption spectrum as the action spectrum for certain phenomena (such as nonmelanoma skin cancers), one finds that, for small ozone depletions (less than 5%), each 1 percent decrease in stratospheric ozone leads to about a 2 percent increase in biologically accumulated dosage (BAD). However, at larger ozone losses, the overlap integral between the action spectrum and the enhanced solar flux begins to increase exponentially; thus, a 10 percent ozone reduction results in a 24 percent increase in BAD, a 30 percent loss will result in a doubling, and a 50 percent loss yields a quadrupling. In the light of the significant polar ozone losses observed over the past decade (the Antarctic ozone hole), this nonlinearity of the biological consequences of ozone depletion should not be overlooked.

There are conflicting data in the literature connecting the loss of stratospheric ozone with increased ground-level UV-B radiation. Observations of ground-level UV-B flux using a network of Robertson-Berger meters failed to confirm the rise in ground-level UV-B as predicted by models using the TOMS satellite data (Scotto et al. 1988). Increased tropospheric ozone and aerosols (photochemical smog), cloud variability, questions about instrument calibration, and the lack of a well-established baseline are all factors implicated in accounting for this apparent discrepancy. More recently, a Canadian network of carefully calibrated Brewer spectroradiometers showed a correlation between diminishing abundances of atmospheric ozone and increases in ground-level UV-B fluxes (Kerr and McElroy 1993).

The Effects of UV-B Radiation on Human Skin

Of all the health-related consequences of enhanced UV insolation, nonmelanoma (basal and squamous cell) skin cancers provide the most firmly established link between UV exposure and biological consequences. Epidemiological data indicate that the risk of nonmelanoma cancers is correlated with *cumulative* lifetime UV exposure (Scotto et al. 1982). These carcinomas are thought to be initiated by UV-induced pyrimidine dimers or other covalent modification of the DNA strand (Saito et al. 1983). Since DNA appears to be the absorbing chromophore responsible for carcinogenesis, its absorption spectrum can be used as a model action spectrum (see also de Gruyl and van der Leun 1992).

A numerical model was developed to correlate epidemiological data of

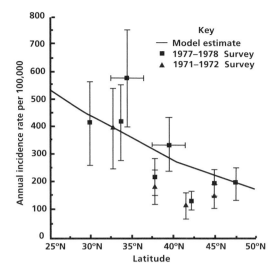

Figure 7.5 Latitudinal dependence of the estimated annual incidence rate for nonmelanoma skin cancers, per 100,000 population. See discussion in text. *Source:* Redrawn with permission from Lloyd 1993b, Figure 3.5.

nonmelanoma skin cancer incidence with annual BAD of UV-B (see Lloyd et al. 1994, Fig. 9), which is a strong function of latitude (Scotto and Fraumeni 1982). Comparison of the modeled BAD of UV-B light at 30–50 degrees north latitude with epidemiological data from the National Cancer Institute (NCI)'s study of nonmelanoma skin cancer incidence rates from June 1977 to May 1978 confirms the linearity of the dose-response relationship (Lloyd et al. 1994). Figure 7.5 illustrates the latitudinal dependence of the model-based estimates of the annual incidence rate per 100,000 population for nonmelanoma skin cancers. Also plotted are the annual age-adjusted incidence rates for nonmelanoma skin cancer (both basal and squamous cell) for the white population of North America (30–50 degrees north latitude) from the 1971–72 Third National Cancer Survey and the NCI's 1977–78 survey of nonrecurring skin cancers, with the incidence rate for males plotted as the upper "error bar" limit for each location and the incidence rate for females as the lower limit (Scotto et al. 1982). The earlier dataset indicates slightly lower incidence rates, which is consistent with the steady rise in nonmelanoma incidence rates during the six-year interval between the two studies.

At the equator the sun is directly overhead at midday, but at higher latitudes the midday sun is lower in the sky and passes through a much longer optical path of ozone, resulting in lower ground-level flux of UV-B. A subtle conclusion of this modeling is that a small loss of ozone at the equator provides as much *additional* biologically effective UV-B as a much larger loss of ozone at higher latitudes (Fig. 7.6). For example, a 10 percent loss of ozone at the equator will produce the same absolute annual increase in

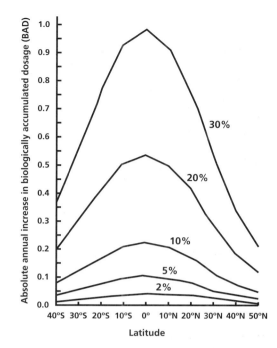

Figure 7.6 Latitudinal dependence of the absolute annual increase in the biologically accumulated dosage (BAD) of ground-level UV-B. *Source:* Redrawn from Lloyd et al. 1994, Figure 10.

the BAD as a 30 percent loss at 50 degrees north latitude. NASA has reported ozone losses of 2–3 percent in the tropics during 1991–92, possibly as a consequence of the eruption of Mount Pinatubo (Gleason et al. 1993). A 2 percent loss of ozone in the tropics is commensurate with a 6 percent ozone loss at 50 degrees north latitude; therefore, we can say that most of the Northern Hemisphere and tropics experienced a roughly equivalent absolute increase in the BAD of UV-B during this period. For temperate latitudes, each 10 percent increase in the nonmelanoma incidence rate among whites corresponds to a shift in latitude of 1.8 degrees equatorward (approximately 400 kilometers).

Although there is also a latitudinal gradient in the incidence of malignant melanoma (Scotto and Fraumeni 1982), this cancer seems to be well correlated not with total lifetime exposure, but rather with one or more severe overexposures (Scotto et al. 1982). Melanoma is of particular concern because of its current rapid rise in incidence. Other UV-related skin disorders (Emmett 1986) include precancerous actinic keratosis (which can develop into squamous cell carcinomas), actinic elastosis and photoaging of the skin, and simple solar erythema (sunburn).

Common-sense adaptive behaviors are to be recommended, especially to those genetically predisposed to a higher risk (fair-skinned individuals who burn easily rather than tan and those with green or blue eye color), such as wearing UV-B-blocking sunglasses, wearing additional protection from the sun (e.g., sunscreen or additional clothing or hats) when anticipating lengthy exposures to sunlight, and limiting time of exposure to the sun, especially during the summer months and at midday, when the sun's UV rays are the strongest. Parents should be encouraged to enforce these adaptive behaviors on their children. Patients taking photosensitizing medications (including tetracycline antibiotics, tricyclic antidepressants, barbiturates, estrogen, griseofulvin, isotretinoin, 8-methoxypsoralen, oral contraceptives, phenothiazines, sulfonamides, sulfonylureas, and thiazide diuretics) can have an enhanced susceptibility to sunburn and blistering and potentially also to nonmelanoma skin cancers and should be advised to exercise caution in their exposure to sunlight (Emmett 1986).

The Effects of UV-B Radiation on the Human Eye

Another significant concern of increased UV exposure is damage to various portions of the eye. Although the case for UV-induced damage to the eye is not as well established as is the case for skin cancers, a wealth of recent studies has provided substantial support in favor of a causal link.

The cornea absorbs almost all radiation below 290 nanometers and a portion of the remaining UV-B and is virtually transparent to UV-A (Boettner and Wolter 1962). Corneal photokeratitis ("snow blindness") has long been associated with exposure to UV-B (Rosenthal et al. 1985). More recently, pterygium and climate droplet keratopathy have been associated with UV exposure (Taylor et al. 1989).

The lens strongly absorbs UV-B, effectively shielding the retina from UV-B, while passing much of the UV-A (Boettner and Wolter 1962). While it has long been assumed that cataract formation is linked to cumulative exposure to sunlight (Zigman 1983), only recently has this link been more firmly established, both on biochemical and epidemiological levels. The mechanism of cataract formation is thought to originate in the UV-induced photochemical alteration of free and protein-bound tryptophan in the lens (van Heyningen 1976). Kynurenine photooxidation products of tryptophan may be responsible for lens protein aggregation via intermolecular cross-linking (Grover and Zigman 1972), as well as excessive pigmentation in brunescent nuclear cataracts (Taylor et al. 1988). Eyesight is impaired by significant scattering (by protein aggregation) or absorption (by pigmentation) of light. Recent epidemiological studies have found that high cumulative levels of UV-B exposure significantly increase the risk of cortical (Taylor et al. 1988) and posterior subcapsular cataracts (Bochow et al. 1989), as well as more severe (Hiller et al. 1986) or brunescent (Zigman et al. 1979) nuclear cataracts.

Only individuals whose lens has been surgically removed seem to be at any risk from UV-B damage to the retina, since the lens effectively filters out all UV-B. Age-related macular degeneration is not associated with cumulative exposure to UV (West et al. 1989). However, patients who received a polymethylmethacrylate intraocular lens (which absorbs all UV-B) with an additional UV-A-absorbing chromophore bonded to the implanted lens had a significantly decreased incidence of cystoid macular edema (Kraff et al. 1985).

Intraocular melanomas do not exhibit a strong latitude-dependent gradient, as do both melanoma and nonmelanoma skin cancers (Scotto et al. 1976). Nevertheless, UV exposure is an important risk factor for melanoma of the uveal tract (iris, ciliary body, and choroid) (Tucker et al. 1985).

Chlorofluorocarbon Regulation: An International Policy Success Story

A series of international treaties has resulted in near-global agreement to ban the further manufacture and use of CFCs. Under the auspices of the U.N. Environmental Programme, several international agreements to restrict the use of substances implicated in stratospheric ozone depletion have been reached (Benedick 1991). The 1985 Vienna Convention on the Protection of the Ozone Layer led directly to the 1987 Montreal Protocol on Substances That Deplete the Ozone Layer. Subsequent revisions in London in 1990 and in Copenhagen in 1992 called for the manufacture of all fully halogenated CFCs to cease by the end of the decade, with the gradual phaseout early in the twenty-first century of hydrogenated CFCs (HCFCs), which are less potentially damaging to the ozone layer than are CFCs and which are to be used as temporary replacements for them. Developing countries were given an extra decade in which to implement these changes, which consist primarily of changing the cooling mechanism for refrigerators and air conditioners to adjust to non-CFC coolants. The success of what has come to be called the "Montreal process" in regulating the manufacture of CFCs has led many to suggest that this process be considered a model for future international cooperation in developing and implementing an environmental consensus (see Chapter 6).

The Future of the Earth's Ozone Layer

Although the chlorine burden of the atmosphere is expected to begin to decrease after the turn of the millennium, the long atmospheric lifetime of CFCs (up to two centuries) means that the gradual recovery of the earth's ozone layer will span several generations before the atmospheric chlorine level approaches pre-1970s levels (WMO 1991). Assuming global compliance with the treaties, the stratospheric ozone layer will return to its "natural" state by the middle of the twenty-first century. Since many of the health consequences of enhanced UV-B are related to *cumulative* exposure over several decades, this means that the medical community must prepare to deal with these issues well into the twenty-first century.

Acid Deposition

Another aspect of global environmental change based on the changing chemistry of our atmosphere is the issue of acid deposition (Elsom 1992). This process is often commonly referred to by the term *acid rain*, which actually accounts for only part of the problem. In addition to the increasing acidity of rainwater downwind of urbanized areas (acid rain), particulate matter with a high acidity can deposit on plants and other surfaces in a process called *dry deposition*. Both the wet and dry processes together make up the issue of increased acid deposition.

"Normal" rainwater is always slightly acidic (about pH 5.6, where 7.0 is neutral and *lower* pH indicates *higher* acidity) because of the naturally occurring CO_2 in the atmosphere. Carbon dioxide is the anhydride of carbonic acid (H_2CO_3), which is formed when CO_2 is dissolved in liquid rain droplets. This carbonic acid naturally dissociates to an extent determined by the pK_a (acid dissociation constant) of carbonic acid, releasing acidic hydronium ions (H_3O^+) into the rainwater in the following sequence of reactions:

$$CO_2(g) + H_2O(l) \rightleftharpoons CO_2(aq) + H_2O(l) \quad\quad (20)$$
$$CO_2(aq) + H_2O(l) \rightleftharpoons H_2CO_3(aq) \quad\quad (21)$$
$$H_2CO_3(aq) + H_2O(l) \rightleftharpoons HCO_3(aq) + H_3O^+(aq) \quad\quad (22)$$

Each of the above reactions is an equilibrium reaction, and the abbreviations (g), (l), and (aq) refer to the phases—gas, liquid, and aqueous, respectively. One could, in principle, force the equilibria to the right and increase the acidity of rainwater by increasing the CO_2 content of the atmosphere. However, observations of increased rainwater acidity in industrialized areas (pH of less than 5.0) indicate that industrial CO_2 emissions cannot be the culprit here, since even a doubling of the CO_2 partial pressure would not decrease the pH of the rainwater below 5.0.

Analysis of acidic rainwater shows that the increased acidity is due to oxides of both nitrogen and sulfur produced in industrial combustion processes. High temperatures inside internal combustion engines (including automobile engines) have sufficient energy to break the strong nitrogen-nitrogen triple bond in molecular nitrogen gas (N_2), forming oxides of nitrogen (collectively known as NO_x, including NO, NO_2, and NO_3) by their reaction with molecular oxygen. Other nitrogen-containing impurities in fossil fuels can also lead to NO_x formation during combustion. Naturally occurring sulfur impurities, commonly found in coal and oil, lead to the production of foul-smelling sulfur dioxide (SO_2) during combustion. Depending on the source of the coal, sulfur content varies from about 0.5 percent to 5 percent by mass. The sulfur content of oil is similarly variable; Middle Eastern oil is low in sulfur content, whereas oil from Venezuela is relatively high. Refined gasoline is comparatively clean, containing only a few hundredths of a percent sulfur. Many metals exist in nature in combination with sulfur. The refining of these mineral ores by smelting in air at high temperatures also releases significant amounts of gaseous SO_2 into the air. Altogether, human activity releases over 200 million tons of SO_2 to the atmosphere globally each year, or over 20 times the "natural" background emission rate of SO_2 from volcanoes and biological activity.

Subsequent photochemical oxidation of NO_x and SO_2 leads to the formation of nitrogen- and sulfur-containing oxoacids, most notably nitric acid (HNO_3), sulfurous acid (H_2SO_3), and sulfuric acid (H_2SO_4). These acids are considerably stronger than carbonic acid and can lead to the observed rainwater acidities below pH 5.0.

Figure 7.7 illustrates the 1999 pH values of precipitation over the continental United States, the eastern half of which has been acutely affected by acid rain since about the 1930s. The area east of the Mississippi River is characterized by rainwater pH levels below 5.0. In addition, the rainfall in the highly industrialized Ohio River valley and most of Pennsylvania and upstate New York can reach a pH of less than 4.3. For comparison, this is about the same pH as beer, wine, or tomato juice. Exceptionally low pH values, as low as 2.0 (more acidic than lemon juice), have been occasionally measured during rainstorms; since the pH scale is logarithmic (pH = $-\log[H^+]$), this means that the acid concentration is over a thousand times what would be expected from an unpolluted atmosphere!

One of the primary concerns with acid deposition is that the effects are not always felt where the problem originates. What was originally thought to be a local, urban industrial problem has turned out to be a regional air pollution problem. High smokestacks, which are sometimes legally mandated to prevent factory exhausts from becoming a local nuisance, contribute to pollution problems downwind (see also Chapter 13). Weather systems often carry smokestack plumes and clouds hundreds of kilometers and across borders before depositing the acidified rainwater. This can result in much political argument between those who bear the brunt of the acid deposition and those who rely upon the industrial output that generates the pollution (Elsom 1992). For example, coal- and oil-burning factories in the heavily industrialized U.S. Midwest are largely responsible for the acid deposition in the U.S. Northeast, not more local sources. Almost 80 percent of the acidic sulfur compounds (mostly H_2SO_4, which is the largest contributor to acid deposition) in the atmosphere over Sweden is thought to originate outside its borders, primarily from industrial sources in Britain, the Ruhr Valley, and Eastern Europe.

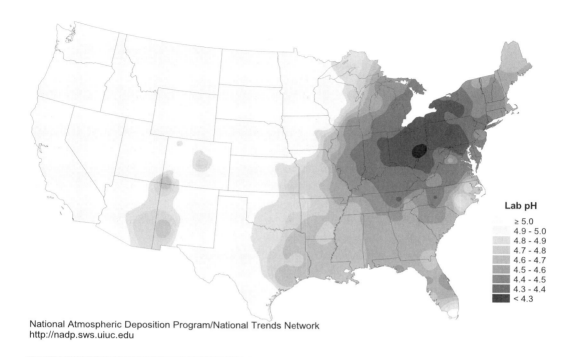

Lab pH

≥ 5.0
4.9 - 5.0
4.8 - 4.9
4.7 - 4.8
4.6 - 4.7
4.5 - 4.6
4.4 - 4.5
4.3 - 4.4
< 4.3

National Atmospheric Deposition Program/National Trends Network
http://nadp.sws.uiuc.edu

Figure 7.7 Hydrogen ion concentration as pH from measurements made at the Central Analytical Laboratory, 1999. The original map was produced by the National Atmospheric Deposition Program (NADP) (NRSP-3)/National Trends Network (2000). NADP Program Office, Illinois State Water Survey, 2204 Griffith Dr., Champaign, Illinois 61820. *Source:* Reproduced from *NADP 2000* converted to gray scale by Bob Larson, NADP.

The Consequences of Acid Precipitation

Two of the most prominent ecological consequences of increased acid precipitation are damage to forest ecosystems and freshwater fish habitats. Forests may be damaged not only directly by the combined effect of increased acidity of the ground water and damage to the leaves and needles by wet and dry deposition, but also indirectly by the leaching of soil nutrients, reduced availability of nitrogen, and decreased soil respiration. The effects of acid rain on forest ecosystems are especially evident in the northeastern United States, southern Scandinavia, and central Europe. Freshwater fish have been adversely affected by the declining pH of their ecosystems, both directly because of the toxicity of the excess acidity and indirectly by the loss of other organisms crucial to their food chains, such as pH-sensitive algae. A particularly good example of acid rain–induced ecological disaster is the collection of several hundred small lakes in the Adirondack Mountains of New York. In the 1930s, almost all of the lakes were populated with freshwater fish and only a handful had a pH of less than 5.0. By the mid-1970s, half of the lakes had a pH below 5.0, and 90 percent of these no longer supported fish populations.

The environmental consequences of acid deposition are sometimes not felt immediately but may be delayed years or decades, depending on the natural buffering capacity (ability to neutralize the acid rain) of the soils or lakes. For example, small lakes and streams in Vermont and New Hampshire underlain by granite rockbeds have a poor natural buffering capacity as compared to those underlain by limestone ($CaCO_3$, a naturally occurring base that can neutralize acid precipitation) in the soil and lake beds and are thus particularly sensitive to ecological damage by acid precipitation. Once the buffering capacity of the water or soil is exhausted, the pH begins to plummet. This can result in sometimes sudden ecological consequences, such as large-scale fish kills, as the lake is literally titrated to a lower pH by the acid deposition, in analogy to the standard freshman chemistry experiment: as the acid is added to the buffer solution drop by drop, no perceptible change occurs until a single drop prompts the color change and sudden drop in pH. Many of the smaller New England lakes and ponds are now acidified to the point that they can no longer support robust ecosystems.

A similarly sudden regional-scale ecological consequence of acid rain was the onset of *Waldsterben* (literally "forest death") in the Black Forest of southwestern Germany in the early 1980s. Even more dramatic are the devastated forests of Eastern Europe (most notably Czechoslovakia and Poland), which suffered from lax industrial pollution controls under their former Communist regimes. Complete forest death is rare in the United States; a few examples are visible at higher altitudes in the White Mountains of New Hampshire, where the buffering capacity of the soils has been exhausted.

Acid precipitation causes damage not only to the biosphere, but also to buildings and monuments made of limestone and marble, resulting in hundreds of millions of dollars in damage annually. Our cultural legacy from prior civilizations in the form of architectural structures, monuments, gravestones, and statues is being literally dissolved before our eyes by the reaction of the calcium carbonate stone with the sulfuric acid contained in acid rain:

$$CaCO_3(s) + H_2SO_4(aq) \rightarrow CaSO_4(s) + H_2O(l) + CO_2(g) \quad (23)$$

Note that the abbreviation (s) refers to the solid phase, which did not appear in reactions 20–22.

Atmospheric Aerosols
Our atmosphere has a natural background level of airborne particles, known as aerosol particles, which can contain either solids or liquids. In

general, larger aerosols can be formed by physically injecting nonvolatile particles into the gas phase, such as in dust storms or the action of the ocean's waves injecting sea salt aerosols into the marine boundary layer. Smaller submicron (less than a millionth of a meter) sized aerosol particles can be formed by combustion processes directly and grow by the absorption of water and other gaseous constituents from the air. In the troposphere, sulfate-based aerosol particles serve as cloud condensation nuclei and are rapidly incorporated into water droplets. In the plumes of industrial emissions, ammonium (NH_4^+) cations combine with sulfate (SO_4^{2-}) and nitrate (NO_3^-) anions to form additional aerosol particles, greatly enhancing the aerosol level above background by many orders of magnitude. Acidic sulfate aerosols contribute to the dry deposition of sulfuric acid onto plants and other surfaces and constitute a respiratory health hazard in themselves. Because their size is comparable to the wavelength of visible light, these sulfate aerosol particles are efficient scatterers of light, leading to decreased visibility and the photochemical haze associated with urban pollution events.

Acid precipitation is connected not only with cloud precipitation; similar processes leading to the formation of nitrogen- and sulfur-based oxoacids can occur in fogs. During the rapid condensation and evaporation that occurs in fogs, acidic sulfate aerosol particles can be absorbed into fog droplets. Because of the smaller liquid water content in fogs (up to an order of magnitude less than in cloud droplets), fogs can lead to exceedingly concentrated acid deposition, which threatens humans and plants. Particularly severe sulfurous smogs ("smog" being a combination of smoke and fog) occurred in London in 1952 and 1962 and in New York in 1953, 1963, and 1966, resulting in excess mortality. Most notably, the dense "pea-souper" fog in London during four days in December 1952 led to an additional four thousand deaths, primarily from bronchitis and other respiratory impairments (Elsom 1992). Increased mortality from heart disease has also been associated with these severe sulfurous fog events. These fogs or smogs were characterized by high levels of gaseous SO_2, as well as excess levels of suspended particles (smoke or soot) resulting from the uncontrolled burning of high-sulfur-content coal for domestic heating. Since the 1960s, pollution control measures and the use of alternative fuel sources have resulted in a dramatic improvement in the levels of these pollutants in most urban areas.

Urban Smog Formation / Tropospheric Ozone

In the last section we looked at one form of air pollution that has been called *London smog* or *classical smog*. A London smog is typified by high sulfur content in the air (especially SO_2) leading to the secondary pollutants H_2SO_4, sulfates, and aerosols. Conditions for this type of smog are generally cool temperatures (near the freezing point of water), high relative humidity, and a ground thermal inversion layer that keeps the pollutants close to the earth's surface.

By the middle of the twentieth century, it was becoming increasingly clear that a second type of smog, often called a *Los Angeles smog* or *photochemical smog,* was dominating the summer skyline of many of our more crowded urban areas. This type of smog is typified by high levels of oxidizing compounds (including ozone, NO_x, nitric acid, and peroxyacetyl nitrate) and a host of organic compounds. Typical conditions include generally warm or hot temperatures (above 20°C), low humidity, and a subsidence thermal inversion layer. This type of smog is particularly evident in the brick red–brown haze (NO_2) often seen in summertime thermal inversion layers trapped by geographic features, such as barrier mountains (surrounding Los Angeles on the north and east) or river valleys (such as the Potomac River valley in Washington, D.C.). In addition, contiguous urban sprawl like that found over the entire Boston-Washington corridor means that local photochemical smog events can easily expand to become regional issues (see also Chapter 13).

Tropospheric Pollutants

Whereas London smog is dominated by SO_2, Los Angeles smog is dominated by NO_x and organic compounds. America's love affair with the automobile is one of the primary causes of photochemical smog in the United States. Not only do automobile engines contribute a large fraction of the urban NO_x, but fumes from gasoline filling stations augment the natural levels of organic compounds in the air. In the United States, many gasoline pumps are now required by law to be fitted with emission control devices to minimize the release of gasoline vapors. One indication of the automobile's dominant role in urban air pollution is the fact that cities with no substantial urban industrial base but with considerable auto traffic (such as Washington, D.C.) also exhibit seasonal photochemical smog events. The combustion efficiency of the automobile internal combustion engine is a prime concern in curbing urban air pollution; only a small fraction of auto engines that are not functioning optimally produce the majority of pollutants. This is why regular auto emissions tests have become mandatory across the United States in recent years.

Urban pollutants can be categorized as either primary or secondary pollutants. *Primary pollutants* are those gases and particulates that are directly released to the atmosphere. These include NO_x (primarily in the form of nitric oxide), organic combustion products of gasoline (carbon monoxide, carbon dioxide, partially oxidized combustion products, and a small fraction of uncombusted hydrocarbons, usually categorized collectively as volatile organic compounds [VOCs] or nonmethane hydrocarbons [NMHC]), SO_2, and small airborne particles of soot and smoke. *Secondary pollutants* are produced in the air or on surfaces (e.g., aerosolized soot particles or rain droplets) by the reaction of primary pollutants over time in the presence of sunlight. Often secondary pollutants (such as ozone, nitric acid, and sulfuric acid) are considerably more difficult to control than are primary pollutants. In addition, a national ambient air quality standard (NAAQS) legislated in the United States establishes six *criteria pollutants* to be monitored and regulated for the health of the nation (O_3, NO_2, SO_2, CO, lead, and total suspended particulates; the compounds called NMHC or VOCs are often listed with the criteria pollutants because efforts to control smog often aim to reduce the emissions of NMHC/VOCs).

It has become apparent in recent years that small airborne aerosol particles (less than 2.5 micrometers in diameter), which can reach the respiratory tract, may represent as much (or more) of a threat to human health than do gaseous pollutants. Larger particles, such as marine aerosol particles and dust particles, are largely trapped by mucus in the tracheobronchial region and thus do not represent a significant threat to the respiratory tract. In addition, the smaller soot and smoke particles generated by fossil fuel combustion are more likely to contain toxic organic species than the coarser particles and can become lodged in the bronchioles of the lungs (Finlayson-Pitts and Pitts 2000). Literally thousands of organic compounds arising from combustion processes have been identified in urban air samples, including a full range of polycyclic aromatic hydrocarbons (PAH), many of which have been implicated as possibly carcinogenic or mutagenic agents. Some of these have nitrogen- and sulfur-containing substituents, which contribute to the noxious odor of polluted urban air.

Tropospheric Ozone Formation

Unlike stratospheric ozone, which protects life on Earth from excessive solar ultraviolet radiation, tropospheric ozone is a pollutant. While industrial chemicals (CFCs) act catalytically to destroy ozone in the stratosphere, other industrial and agricultural by-products (especially the oxides of nitrogen) act as precursors to the photochemical formation of ozone in the troposphere. Ozone causes a host of human health issues, including respiratory distress and eye irritation associated with urban smog events (see Chapter 13). Because of the complex interaction between the atmosphere's chemical composition and local weather patterns and their resultant effects on the biosphere, the study of urban- and regional-scale air pollution has become highly interdisciplinary, involving atmospheric chemists, meteorologists, industrialists, policymakers, and public health physicians.

The formation of tropospheric ozone is directly associated with high levels of NO_x, since the only known mechanism for the formation of ozone in the troposphere is the photolysis of NO_2 by sunlight, followed by the rapid reaction of free oxygen atoms with molecular oxygen (O_2) to form ozone, as seen in the following reaction couplet, where $M = N_2$ or O_2.

$$NO_2 + sunlight \rightarrow NO + O \qquad (3)$$
$$O + O_2 + M \rightarrow O_3 + M \qquad (4)$$

Since sunlight is needed to form ozone, the Los Angeles type of smog occurs primarily during the summer when sunlight is more plentiful, whereas the London type of smog is not dependent on sunlight and is more likely to occur during the cooler, moister months. Another important pollutant, peroxyacetyl nitrate (PAN), serves as a reactive reservoir for both NO_x and nonmethane hydrocarbons. PAN is a powerful lachrymator and is responsible for much of the ocular discomfort experienced in polluted urban environments.

While the connection between NO_x and the formation of tropospheric ozone is clearly direct, as shown in the preceding paragraph, the connection between the release of nonmethane hydrocarbons and ozone formation is considerably more complicated. The reaction couplet above, involving the photolysis of NO_2 in the production of ozone, is actually a small subset of a much larger reaction scheme, in which hydrocarbons combine with oxygen to yield CO_2 and water vapor, as in a flame burning natural gas. In a nutshell, the earth's atmosphere can be thought of as a slow-burning combustion (oxidation) process mediated by free radicals. All of the carbon in the hydrocarbon eventually ends up as CO_2, and all of the hydrogen ends up as water vapor.

The most abundant hydrocarbon in the troposphere is methane, CH_4, a compound that is rapidly increasing in concentration because of intensive agricultural and industrial activity. The primary oxidizing agent in the troposphere is the hydroxyl radical, OH. The oxidation of a hydrocarbon such as methane begins with OH removing a hydrogen atom from the methane, forming water. Subsequent reactions with molecular oxygen and the free radical species nitric oxide, NO, result in the formation of the key oxidation intermediate formaldehyde, CH_2O. Photolysis of formaldehyde with sunlight goes by two pathways, one of which produces two OH radicals; this step makes the entire process *autocatalytic*, since two OH radicals are formed for each OH radical consumed (at the beginning of the cycle). Subsequent reactions oxidize the formaldehyde further to CO_2.

The OH radical acts as a catalytic chain carrier, which is both consumed and produced within the cycle, much as ClO is the radical chain carrier in the catalytic destruction of polar stratospheric ozone. However, ozone is *produced* as a side product in this cycle, which dominates the urban troposphere. Hydrocarbons are important in the production of urban ozone because they are the fuel that keeps this slow combustion cycle going around. Nitrogen oxides are also key intermediates and are needed in this cycle. Thus, the combination of NMHC and NO_x in the presence of sunlight is the recipe for tropospheric ozone (NRC 1991).

Computer Modeling of Regional Air Pollution

One of the most important areas of air pollution research in the second half of the twentieth century was the development of quantitative models to predict the formation and transport of secondary pollutants (such as ozone) from their precursors. One of the primary computer models is the Empirical Kinetic Modeling Approach (EKMA). In this approach, the concentration of ozone is calculated as a function of both ambient NO_x and NMHC levels in the atmosphere. The results of this model for one set of initial conditions are shown in Figure 7.8 (Finlayson-Pitts and Pitts 1986; NRC 1991). Lines of constant ozone fraction (called *isopleths*, expressed in parts per million) are plotted as a function of NO_x and NMHC (y- and x-axes, respectively, both in units of parts per million). Note that the scales of the two axes are different. The line through the center of the plot (including point A) represents an 8:1 ratio of NMHC to NO_x. As one goes to a higher ratio (15:1), this line moves down the plot, pivoting from the origin, and a lower ratio (4:1) moves up the plot.

The interpretation of EKMA plots is central to policy decisions concerning the control of urban air pollution, particularly secondary pollutants such as ozone. One of the central issues is: What should be controlled—NO_x, NMHC, or both? Let's take a few examples to demonstrate the utility of these model systems (see Finlayson-Pitts and Pitts 1986). If we were to start at point A in Figure 7.8 with NMHC at 0.9 parts per million of carbon and NO_x at 0.12 parts per million and we were to lower the NMHC dramatically by 72 percent down to 0.25 parts per million of carbon at point B (by implementing legislated control measures to limit the release of NMHC to the atmosphere), then the tropospheric ozone content would drop from 0.28 parts per million to 0.12 parts per million, a substantial 57 percent reduction. One would say that this control strategy was a success. However, if we started with a much higher level of hydrocarbons (such as that found in the Great Smoky Mountains, where fir trees contribute to a higher than average level of hydrocarbon content in the air) with NMHC at 2.0 parts per million of carbon and much lower NO_x (0.03 parts per million) and were again to drop NMHC by 72 percent, the corresponding drop in ambient ozone would be inconsequential. Clearly, limiting NMHC in a low NO_x environment would not be an effective pollution control strategy. Similarly, if we start with low levels of NMHC (0.4 parts per million of carbon) and high levels of NO_x (0.24 parts per mil-

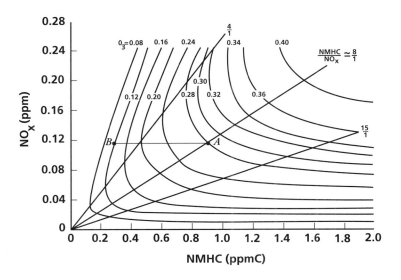

Figure 7.8 Ozone isopleths used in the Empirical Kinetic Modeling Approach (EKMA). Ozone isopleths are identified by ozone fraction, with the lowest level at 0.08 parts per million and the highest level at 0.40 parts per million. In the region toward the *upper left corner,* the formation of ozone is limited by the availability of non-methane hydrocarbons (NMHCs). In the region toward the *lower right corner,* the formation of ozone is limited by the availability of nitrogen oxides (NO_x). *Source:* Reproduced with permission from Finlayson-Pitts and Pitts 1986, Figure 10.5, from original data in Dodge 1977.

lion) and decrease only the NO_x, then the ozone levels would actually *increase.*

In the real world, different cities will have different combinations of NMHC and NO_x as industrial output, traffic patterns, weather conditions, and solar illumination change throughout the year. Clearly, the shape of the ozone isopleths indicates that different pollution control strategies (attempts to limit emissions of NMHC or NO_x) must be used for different locations and seasons. A single, national strategy might be effective at some locations but might likewise *increase* the ambient ozone levels in others. Although the EKMA approach has been exceedingly useful in developing policy, ever more sophisticated models are being developed in response to the limitations of this approach (NRC 1991; see also *North American Research Strategy for Tropospheric Ozone 2000*).

Enhanced Global Warming

The terms *global change, enhanced global warming,* and the *greenhouse effect* are often used casually to refer to global *climate* change, although global change also includes many other processes beyond those tradition-ally associated with climate, including stratospheric ozone depletion, acid precipitation, and regional photochemical smog formation, discussed above. In reality, Earth's climate is *always* in a state of change over the ge-ological time scale; recently, concerns have been raised that human activ-ity may be changing both the *rate* of climate change and the *magnitude* of temperature changes beyond that seen in the historical record.

Earth's climate is determined by astronomical parameters (solar heat-ing and the variability of the Earth-Sun distance) and a complex interac-tion of the earth's atmosphere, hydrosphere (the oceans), and biosphere. Figure 7.9 shows the distribution of wavelengths of light that Earth receives from the sun (primarily in the visible, peaking around 500 nm) and the distribution of wavelengths emitted by the earth (in the infrared, emitted as heat). These distribution curves can be approximated by the distribu-tion of black body radiation emitted by two spheres, one at about 6,000 K (the temperature of the outer surface of the sun) and another at an aver-age of 255 K (an average temperature for the earth). Our atmosphere is largely transparent to the visible and near-ultraviolet light of the sun. The only significant absorbers of near-ultraviolet sunlight in the atmosphere are ozone (O_3) and molecular oxygen (O_2); this absorption of ultraviolet sunlight by O_3 and O_2 is responsible for the warming of the stratosphere and upper atmosphere. From about 320 nanometers in the ultraviolet to about 750 nanometers in the near-infrared, the atmosphere is essentially transparent. A fraction of this sunlight is reflected back to space; the rest is absorbed by the earth's solid and liquid surfaces, converted to heat, and ra-diated back through the atmosphere to space. Naturally occurring "green-house gases," primarily water vapor and carbon dioxide, retain a portion of this heat radiated by the earth's surface and warm the atmosphere (see also Fig. 8.1*a*).

The existence of a greenhouse effect is not a matter of scientific dis-pute. Without the intervention of these naturally occurring greenhouse gases, Earth's surface would be about 33°C colder than at present, which would place the entire planet below the freezing point of water, and life as we know it could not exist on Earth (Houghton et al. 1990). What *is* a mat-ter of current scientific investigation is whether human activity has *added* to the natural burden of greenhouse gases so as to cause the earth to warm rapidly, perhaps beyond the envelope of natural climate variability, an ef-fect properly known as *enhanced global warming* or *enhanced greenhouse warming*.

The term *greenhouse effect* is somewhat of a misnomer because the pri-mary reason a greenhouse stays warm is that the glass roof prevents the evaporation/condensation cycle from leaving the greenhouse. The term is, however, appropriate in that the glass roof is transparent to visible sunlight

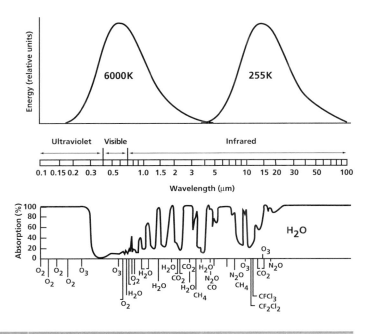

Figure 7.9 A comparison between the electromagnetic spectrum for black bodies at 6,000 K (the temperature of the outer surface of the sun) and 255 K (an average temperature for the earth) and the absorption spectrum of gases in the earth's atmosphere. Note that the atmosphere is practically transparent to black body radiation emitted at temperatures typical of the sun. *Source:* Redrawn with permission from Brimblecombe 1996, Figure 1.2. Reprinted with permission from Cambridge University Press.

and yet absorbs and reradiates infrared radiation (heat), much as greenhouse gases do higher in the atmosphere.

Climate Control and Climate Change

The earth is in thermodynamic equilibrium with its surroundings. If we believe that the earth's average global temperature is roughly constant over time, that means that the amount of energy received from the sun (as visible and ultraviolet radiation) must equal the amount of energy that the earth radiates back to space (as infrared radiation). If the earth absorbed more radiation than it emitted, the temperature would rise; conversely, if the earth emitted more energy than it received, it would cool down. The atmosphere/hydrosphere/biosphere provides both positive and negative feedback mechanisms to act as a thermostat that keeps the earth's climate within a carefully balanced range of temperatures.

Many factors may contribute to changing climate. The astronomical parameters of Earth's orbit around the sun allow for annual changes in the climate (the four seasons), as well as much slower changes over geological time scales, which are primarily responsible for the 100,000-year cycles of ice ages and interglacial warming periods. We are currently at a relative maximum of global temperatures within the context of the past few million years of geological history; in fact, recorded history (the Holocene period) begins about 12,000 years ago, with the retreat of the last ice age. The 11-year sunspot cycle does not seem to be correlated to currently measured climate variability; however, it has been proposed that the *lack* of sunspots is an indicator of changes in the solar output that may have been responsible for short-term climate change (on the order of a century or two) that wiped out the Viking colonies in Greenland and North America in the 1300s and froze the Thames River in the early 1700s.

Changes in the composition of the earth's atmosphere, specifically the level of carbon dioxide, have been positively correlated with excursions in the earth's climate over a few hundred thousand years (Houghton et al. 1990). Carbon dioxide concentrations measured from air bubbles trapped in polar ice cores indicate that colder climates correspond to lower CO_2 levels and warmer climates correspond to higher levels. The question still to be answered is one of causality (i.e., the chicken or the egg): Did the higher levels of CO_2 *cause* the climate to warm, or are these levels simply the *result* of a warmer climate?

Another factor that can contribute to climate change is the variability of the earth's *albedo,* or its effective reflectivity to sunlight. Any changes in the global extent of snow, ice, or cloud cover (which are all highly reflective compared to oceans or vegetation-covered land) could induce a change in the fraction of sunlight reflected back to space and, thus, a change in the global radiation balance. Stratospheric sulfate aerosols resulting from volcanic emissions of SO_2, which can both reflect the sun's visible light back to space and absorb the earth's emitted infrared radiation like the greenhouse gases, can also affect the climate for short periods (on the order of a few years at most). Perhaps the most dramatic example in recorded history was the 1815 eruption of the volcano Tamboora in Indonesia. The following year, 1816, was known as "the year without a summer" because of the dramatic cooling observed in North America and Eurasia. The cooling resulting from the 1991 eruption of Mount Pinatubo in the Philippines represents a complicating factor in trying to identify a clear signal of enhanced global warming in the recent temperature record.

The gist of the theory of enhanced global warming is that the addition of significant amounts of gases that absorb infrared radiation to the atmosphere will lead to a rise in the earth's average global temperature. This doesn't mean that every place on the globe will experience higher temperatures. As the *distribution* of solar heating changes, some places will be warmer, others cooler; *on the average,* however, the temperature will rise in response to rising levels of greenhouse gases. In general, this change will be more pronounced at the poles, as the changing climate funnels solar heating more efficiently away from the equator (where most of the warming occurs) to the polar regions. Changes in the climate may also induce changes in cloud cover globally and the extent of ice and snow cover in the polar regions; this, in turn, would change the earth's effective albedo, which may provide feedback to the enhanced global warming scenario.

Greenhouse Gases

The most important *naturally occurring* greenhouse gases are water vapor and carbon dioxide. Figure 7.9 shows the absorption features of these two gases along with a spectrum of the atmosphere's total absorption of infrared radiation. Note that many of the absorption features of both water vapor and carbon dioxide are *saturated,* meaning that all of the earth's emitted radiation at these wavelengths is absorbed by the atmosphere. This has the consequence that the emission of additional amounts of these gases into the atmosphere will only change the total atmospheric absorption by a minimal amount at these wavelengths, since the light is already maximally absorbed. Only at those wavelengths where all of the infrared radiation is *not* already absorbed (some of the smaller absorption features) will changing concentrations of these gases have an effect on global mean temperatures.

However, the *rate* of change of atmospheric carbon dioxide levels is currently well beyond the gradual change observed over geological time scales, and the absolute *amount* of CO_2 in the atmosphere is higher than at any time in recent geological history. This rapid rise is shown in Figure 7.10, as measured at the Mauna Loa Observatory in a continuous record dating back to 1958. Preindustrial (eighteenth century) tropospheric levels of CO_2 were about 280 parts per million and have risen by almost 30 percent to a current level of about 370 parts per million. Note that an annual cycle associated with the seasonal growth of vegetation in the Northern Hemisphere (as plants photosynthetically absorb CO_2 and release O_2) is superimposed on the steady growth curve associated with the ever-increasing combustion of fossil fuels. The correlation with fossil fuel burning becomes clear when one realizes that the slight dip in the early 1970s corresponds to the American fuel crisis.

Several gases associated with increasing industrialization and intensive agricultural practices absorb infrared radiation in the *atmospheric windows* where neither H_2O nor CO_2 absorbs significantly. The increasing concentrations of methane (CH_4) and nitrous oxide (N_2O) in the troposphere are shown in Figure 7.11, along with that of chlorofluorocarbon-11, another powerful greenhouse gas also responsible for the chlorine-catalyzed destruction of stratospheric ozone.

The net effect on the earth's radiation balance depends not only on the

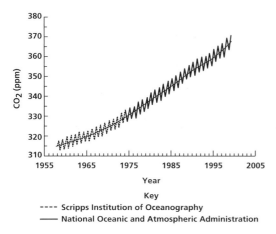

Figure 7.10 Observations of atmospheric carbon dioxide from the Mauna Loa Observatory, 1958–98. The original figure was produced by the National Oceanic and Atmospheric Administration (NOAA)'s Climate Monitoring and Diagnostics Laboratory, Carbon Cycle Group, Boulder, Colorado. *Source:* Redrawn from *NOAA 1999*.

Key
---- Scripps Institution of Oceanography
——— National Oceanic and Atmospheric Administration

concentration of greenhouse gases, but also on their relative effectiveness at absorbing infrared (IR) radiation. Since CH_4, N_2O, CFC-11, and several other anthropogenically produced gases absorb IR radiation in the atmospheric windows that are *not* saturated by H_2O and CO_2, their contribution to the total IR absorptivity of the atmosphere is large compared to that of H_2O and CO_2 on a molecule-for-molecule basis.

The concept of global warming potential (GWP) was developed to allow a quantitative comparison of the relative ability of different molecules to trap heat in the atmosphere. The GWP of a greenhouse gas is defined as the ratio of global warming induced by the emission of one kilogram of that gas into the atmosphere relative to the emission of one kilogram of CO_2 over some period of time (typically 100 years). Many carbon-based molecules (methane, CFCs, HCFCs, and perfluorinated compounds [PFCs]), as well as nitrous oxide (N_2O) and sulfur hexafluoride (SF_6), have significantly higher GWPs than does CO_2; methane is about 21 times more effective than CO_2, and CFC-11 is about 12,000 times more effective. Therefore, because these molecules are more effective atmospheric absorbers, the addition of a small amount of these gases (on the parts per billion level) can have as much effect as the addition of much larger amounts (on the parts per million level) of CO_2.

Because the CFCs are relatively nonreactive and not water soluble (meaning that they are unlikely to dissolve in rainwater and be washed out of the atmosphere), they persist in the atmosphere for many decades. This long atmospheric lifetime contributes to their comparatively large GWPs. Inventories of industrial and agricultural emissions of greenhouse gases are often presented in units of million metric tons of carbon-equivalent (MMTCE), which take into account both the total mass of each gas emitted and its GWP (*Environmental Protection Agency 2000*). As the bottom

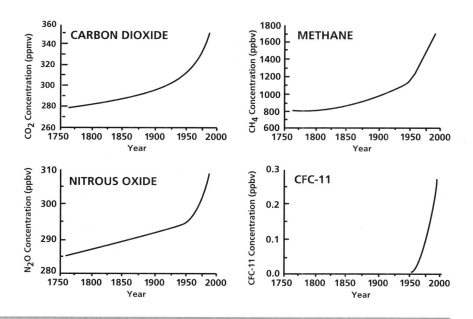

Figure 7.11 Increasing atmospheric concentrations of greenhouse gases. *CFC-11,* chlorofluorocarbon-11. *Source:* Reproduced with permission of the Intergovernmental Panel on Climate Change from Houghton et al. 1990, Figure 3.

line, model calculations show that the decadal *increase* in radiative forcing (warming due to the presence of gases that absorb and emit solar and terrestrial radiation) during the 1980s is attributable to CO_2 (56%), CH_4 (11%), N_2O (6%), and the CFCs (24%), with the remainder attributed to increases in stratospheric water vapor (Houghton et al. 1990).

Issues of Current Controversy

There is little doubt in the scientific community that the addition of significant amounts of greenhouse warming gases to the atmosphere will eventually lead to a warming of the surface of the planet. What remains controversial is the magnitude of the warming (Will it be enough to produce undesired ecological consequences?) and the time scale over which the effect will be induced (Is enhanced global warming already here? Will we see climatic changes within our lifetime?).

The environmental and societal consequences of a potentially sudden change in Earth's climate are discussed more fully in Chapters 11 to 14. Global circulation models (GCMs) of the earth's atmosphere predict an increase in the average surface temperature in the range of 1–3.5°C for a doubling of the preindustrial CO_2 concentration over the twenty-first century (Houghton et al. 1996). Global consequences of a warming of this magnitude might include melting of the polar ice caps, a rise in the sea level, the flooding of low-lying coastal areas and island communities, increased desertification of the interiors of the continents of Africa, Asia, and North America, changing patterns of precipitation, changes in the variability of weather and the frequency of extremes, and a geographic shift of ecosystems. As with any aspect of global change, a key factor is timing— will the earth warm slowly enough to allow humans and other elements of the various ecosystems to adapt gradually to these changes?

A current debate surrounds the question of whether the effects of enhanced global warming are already apparent. There are several issues in this debate, perhaps the most important of which is the natural variability of the climate. If the climate normally experiences temperature fluctuations of more than a degree (the cumulative enhanced greenhouse warming up to the present) from year to year, then the identification of a signal smaller than a degree will require a sufficiently long baseline of data to make such a small trend statistically significant. Many of the older databases of long-term temperature trends have uncertainties large enough to bring into question the significance of any observed trend. Furthermore, the use of a few select locations to identify a global trend is not sufficient proof. In addition to natural variations of the *global* climate over geological time scales, *localized* climate variations with a periodicity on the order of decades are not uncommon. Satellite measurements of the entire earth's temperature variability are a significant advance over compilations of observations from selected weather observation sites but have not yet accumulated enough data to indicate an unambiguous trend. It is clear that the decade of the 1990s is the warmest in the past millennium (Mann et al. 1999). However, to determine whether enhanced global warming or a natural climate cycle is responsible for this warming, we probably need another decade or more until enough data have been accumulated to resolve this issue convincingly.

Feedback Mechanisms

If the only factor in enhanced global warming were the concentration and atmospheric lifetime of the various greenhouse gases, the calculation of the temperature effects of additional inputs of these gases would be relatively straightforward. However, further complicating factors include the various feedback mechanisms in the connected atmosphere/hydrosphere/biosphere system, which are currently poorly parameterized in global circulation models that attempt to predict the earth's future climate. These feedback mechanisms can provide both positive feedback (reinforcing the phenomenon and amplifying the initial warming) or negative feedback (moderating the rise in temperature).

Water vapor exhibits both positive and negative feedback. As the atmosphere warms, the relative humidity (water-bearing capacity of the atmosphere) will increase. Since water is an important greenhouse gas, it can further amplify the warming. However, as the humidity rises, so might the global distribution of cloud cover. These additional clouds can both trap additional IR radiation (positive feedback) and reflect more of the incoming solar radiation back to space (negative feedback). Since the magnitude (and perhaps even the sign) of this *cloud radiative feedback* is unknown, the net effect on the earth's future climate remains uncertain. Another positive feedback mechanism involves the change in the earth's albedo (reflectivity) as the highly reflective polar ice caps melt. As the polar regions warm, an unknown amount of organic matter may be volatilized, releasing methane to warm the atmosphere further. Additional feedback mechanisms involve the change in rates of photosynthesis under conditions of enhanced levels of CO_2 (the CO_2 *fertilization effect*), changes in carbon fixation due to changes in soil moisture, the vast amounts of CO_2 released during biomass burning events in the tropics, the decreased sink for atmospheric CO_2 from deforestation and biomass burning, changes in the distribution of vegetation, changes in ocean circulation due to changing atmospheric temperature distributions, and changes in marine biogeochemical cycling.

The earth's atmosphere is oxidizing and acidic, while the oceans are reducing and basic. The global carbon cycle can be cast as a global oxidation/reduction cycle. The hydrosphere/biosphere emits reduced carbon compounds (hydrocarbons), which are photochemically oxidized in the atmosphere to CO_2, much as natural gas is burned in a flame. Oxidized carbon in the form of CO_2 is absorbed by the larger carbon reservoir of the oceans, where it is reduced by a combination of biological, chemical, physical, and geological processes back to reduced organic matter.

The oceans are by far the largest reservoir of the global carbon cycle (see Chapter 8). As a rule of thumb, the oceans absorb about half of the CO_2 that enters the atmosphere, through a series of equilibrium reactions (see equations 20–22 under "Acid Deposition," above). Eventually the carbon is stored in mineral form as $CaCO_3$, and some of it is incorporated into biological matter as shells and skeletal matter. As the oceans warm, the equilibrium in reaction 20 shifts to the left, leaving more CO_2 in the atmosphere to contribute to enhanced global warming. It is estimated that the future atmospheric CO_2 increase may be amplified by about 5 percent by this effect.

Unfortunately, the uncertainties associated with the various feedback mechanisms have cast a pall over the public's impression of all aspects of atmospheric chemistry. Stratospheric ozone depletion, acid precipitation, and urban smog are scientific certainties. The existence of the greenhouse effect is also a scientific certainty (without which the earth would be little more than a barren, frozen rock), even if the details of the magnitude and timing of *enhanced* global warming are yet to be elucidated.

Conclusion

This chapter has shown how anthropogenically induced changes in a variety of atmospheric trace gases, such as ozone, nitrogen oxides (nitric oxide and nitrogen dioxide), volatile organic compounds, carbon dioxide, nitric and sulfuric acids, chlorofluorocarbons, and halons, can induce dramatic changes in the earth's global ecosystem. For the first time in human history, our species has produced substantive changes in the earth's atmospheric composition, perhaps irreversibly. These changes in composition have resulted in rapid changes in the livability of our planet, including stratospheric ozone depletion, acid precipitation, urban smog, and enhanced global warming. The consequences of these changes for the biosphere in general and for human health in particular are only beginning to be understood. Even if humanity were to place rational limits on how we alter our only atmosphere, we would still have to deal with the serious consequences of the present pollution of our planet for many generations to come.

SUGGESTED STUDY PROJECTS

Suggested study projects provide a set of options for individual or team projects that will enhance interactivity and communication among course participants (see Appendix A). The Resource Center (see Appendix B) and references in all of the chapters provide starting points for inquiries. The process of finding and evaluating sources of information should be based on the principles of information literacy applied to the Internet environment (see Appendix A).

The objective of this chapter's study projects is to create enhanced understanding about atmospheric changes and their biological and societal effects.

PROJECT 1: Effects of Industrial Production and Land Use Practices on Atmospheric Change

The objective of this project is to establish the effects of industrial production and land use practices on four atmospheric changes that have regional and global consequences—stratospheric ozone depletion, acid deposition, the production of tropospheric ozone, and increased concentrations of greenhouse gases.

Task 1. Select two or three regions and identify industrial production and land use practices that affect one of the four atmospheric changes.

Task 2. Compare your findings across the selected regions and present them in one of the following options:

a. written paper
b. Power Point presentation (lecture)

c. visually oriented presentation (charts, statistics, photographs, or
    videos)

PROJECT 2: Biological Consequences of Stratospheric Ozone Depletion

The objective of this project is to assess the biological consequences of
stratospheric ozone depletion and increased exposure to UV-B radiation,
considering direct effects on human tissues as well as effects on other biological organisms.

Task 1. Identify the effects of UV-B radiation on eyes, skin, and the
immune system in humans.

Task 2. Identify the effects of UV-B radiation on other biological organisms.

Task 3. Identify the uncertainties of projecting the biological effect of
stratospheric ozone depletion.

Task 4. Summarize your findings.

PROJECT 3: Societal Consequences of Atmospheric Change

The objective of this project is to establish the societal consequences of four
atmospheric changes that have regional and global effects—stratospheric
ozone depletion, acid deposition, the production of tropospheric ozone,
and increased concentrations of greenhouse gases.

Task 1. Select two or three regions and identify the possible societal
consequences of one of the four atmospheric changes.

Task 2. Compare your findings across the selected regions.

Task 3. Identify the uncertainties of assessing the societal consequences described in task 1.

Task 4. Summarize your findings.

References

Anderson DE, Lloyd SA. 1990. Polar twilight UV-visible radiation field: Effects due to
    multiple scattering, ozone depletion, clouds and surface albedo. *J Geophys Res*
    95:7429–34.

Anderson JG, Brune WH, Lloyd SA, Toohey DW, Sander SP, Starr WL, Loewenstein M,
    Podolske JR. 1989. Kinetics of O_3 destruction by ClO and BrO within the Antarctic
    vortex: An analysis based on in situ ER-2 data. *J Geophys Res* 94:11480–520.

Anderson JG, Toohey DW, Brune WH. 1991. Free radicals within the Antarctic vortex:
    The role of CFCs in Antarctic ozone loss. *Science* 251:39–46.

Benedick RE. 1991. *Ozone Diplomacy.* Harvard University Press, Cambridge.

Bochow TW, West SK, Azar A, Muñoz B, Sommer A, Taylor HR. 1989. Ultraviolet light
    exposure and risk of posterior subcapsular cataracts. *Arch Ophthalmol* 107:369–72.

Boettner EA, Wolter JR. 1962. Transmission of the ocular media. *Invest Ophthalmol*
    1:776–83.

Brimblecombe P. 1996. *Air Composition and Chemistry,* 2d ed. Cambridge University
    Press, Cambridge.

Caldwell MM, Björn LO, Bornman JF, Flint SD, Kulandaivelu G, Teramura AH, Tevini M. 1998. Effects of increased solar ultraviolet radiation on terrestrial ecosystems. *Photochem Photobiol* 46:40–52.

de Gruyl FR, van der Leun JC. 1992. Action spectra for carcinogenesis. In *Biological Responses to Ultraviolet A Radiation* (Urbach F, ed.). Valdenmar, Overland Park, Kans., pp. 91–97.

Dodge MC. 1977. Combined use of modeling techniques and smog chamber data to derive ozone-precursor relationships. In *Proceedings of the International Conference on Photochemical Oxidant Pollution and Its Control*, Vol. 3 (Dimitriades B, ed.). EPA-600/3–77–001b. U.S. Environmental Protection Agency, Environmental Sciences Research Laboratory, Research Triangle Park, N.C., pp. 881–89.

Elsom DM. 1992. *Atmospheric Pollution: A Global Problem*, 2d ed. Blackwell, Oxford, U.K.

Emmett EA. 1986. Health effects of ultraviolet radiation. In *Effects of Changes in Stratospheric Ozone and Global Climate*, Vol. 1 (Titus JG, ed.). Environmental Protection Agency, Washington, pp. 129–45.

Fahey DW, Murphy DM, Kelly KK, Ko MKW, Proffitt MH, Eubank CS, Ferry GV, Loewenstein M, Chan KR. 1989. Measurements of nitric oxide and total reactive nitrogen in the Antarctic stratosphere: Observations and chemical implications. *J Geophys Res* 94:16665–81.

Farman JC, Gardiner BG, Shanklin JD. 1985. Large losses of total ozone in Antarctica reveal seasonal ClO_x/NO_x interaction. *Nature* 315:207–10.

Finlayson-Pitts BJ, Pitts JN. 1986. *Atmospheric Chemistry: Fundamentals and Experimental Techniques*. J. Wiley & Sons, New York.

———. 2000. *Chemistry of the Upper and Lower Atmosphere: Theory, Experiments, and Applications*. Academic Press, San Diego.

Francis AA, Carrier WL, Regan JD. 1988. The effect of temperature and wavelength on production and photolysis of a UV-induced photosensitive DNA lesion which is not repaired in xeroderma pigmentosum variant cells. *Photochem Photobiol* 48:67–71.

Gleason JF, Bhartia PK, Herman JR, McPeters R, Newman P, Stolarski RS, Flynn L, Labow G, Larko D, Seftor C, Wellemeyer C, Komhyr WD, Miller AJ, Planet W. 1993. Record low global ozone in 1992. *Science* 260:523–26.

Grover D, Zigman S. 1972. Coloration of human lenses by near-UV photooxidized tryptophan. *Exp Eye Res* 13:70–76.

Häder D-P, Kumar HD, Smith RC, Worrest RC. 1998. Effects on aquatic ecosystems—environmental effects of ozone depletion 1998 assessment. *Photochem Photobiol* 46:53–68.

Hiller R, Sperduto RD, Ederer F. 1986. Epidemiologic associations with nuclear, cortical, and posterior subcapsular cataracts. *Am J Epidemiol* 124:916–25.

Hinshelwood CN. 1933. *The Kinetics of Chemical Change in Gaseous Systems*. Clarendon Press, Oxford, U.K., pp. 280–81.

Houghton JT, Jenkins GJ, Ephraums JJ, eds. 1990. *Climate Change: The IPCC Scientific Assessment*. Cambridge University Press, Cambridge.

Houghton JT, Meira Filho JG, Callender BA, Harris N, Kattenberg A, Maskell A, eds. 1996. *Climate Change, 1995: The Science of Climate Change. Contribution of Working Group I to the Second Assessment Report of the Intergovernmental Panel on Climate Change*. Cambridge University Press, Cambridge.

Kawa SR, Fahey DW, Heidt LE, Pollock WH, Solomon S, Anderson DE, Loewenstein M, Proffitt MH, Margitan JJ, Chan KR. 1992. Photochemical partitioning of the reactive nitrogen and chlorine reservoirs in the high-latitude stratosphere. *J Geophys Res* 97:7905–23.

Kerr JB, McElroy CT. 1993. Evidence for large upward trends of ultraviolet radiation linked to ozone depletion. *Science* 262:1032–34.

Kraff MC, Sanders DR, Jampol LM, Lieberman HL. 1985. Effect of an ultraviolet-filtering intraocular lens on cystoid macular edema. *Ophthalmology* 92:366–69.

Lloyd SA. 1993a. Health and climate change: Stratospheric ozone depletion. *Lancet* 342:1156–58.

———. 1993b. Issues in Stratospheric Ozone Depletion. Ph.D. diss., Harvard University, Cambridge.

Lloyd SA, Im ES, Anderson DE. 1994. Modeling the latitude-dependent increase in non-melanoma skin cancer incidence as a consequence of stratospheric ozone depletion. In *Stratospheric Ozone Depletion/UV-B Radiation in the Biosphere (3–540–57810–2)* (Biggs RH, Joyner MEB, eds.). NATO Advanced Science Institute Series I: Global Environmental Change, Vol. 18. Springer Verlag, New York, pp. 329–37.

Longstreth J, de Gruijl FR, Kripke ML, Abseck S, Arnold F, Slaper HI, Velders G, Takizawa Y, van der Leun JC. 1998. Health risks—environmental effects of ozone depletion 1998 assessment. *Photochem Photobiol* 46:20–39.

Mann ME, Bradley RS, Hughes MK. 1999. Northern hemisphere temperature during the past millennium: Inferences, uncertainties and limitations. *Geophys Res Lett* 26:759–62.

McElroy MB, Salawitch RJ, Wofsy SC. 1986. Reductions of Antarctic ozone due to synergistic interactions of chlorine and bromine. *Nature* 321:759–62.

Molina LT, Molina MJ. 1987. Production of Cl_2O_2 from the self-reaction of the ClO radical. *J Phys Chem* 91:433–36.

Nardi B, Bellon W, Oolman L, Deshler T. 1999. Spring 1996 and 1997 ozonesonde measurements over McMurdo Station, Antarctica. *Geophys Res Lett* 26:723–26.

National Research Council. 1975. *Environmental Impact of Stratospheric Flight: Biological and Climatic Effects of Aircraft Emissions in the Stratosphere.* National Academy of Sciences, Washington.

———. 1991. *Rethinking the Ozone Problem in Urban and Regional Air Pollution.* National Academy Press, Washington.

Peak MJ, Peak JG, Moehring MP, Webb RB. 1984. Ultraviolet action spectra for DNA dimer induction, lethality, and mutagenesis in *Escherichia coli* with emphasis on the UVB region. *Photochem Photobiol* 40:613–20.

Penner JE, Lister DH, Griggs DJ, Dokken DJ, McFarland M, eds. 1999. *Aviation and the Global Atmosphere: A Special Report of IPCC Working Groups I and III.* Cambridge University Press, Cambridge.

Rosenthal FS, Safran M, Taylor HR. 1985. The ocular dose of ultraviolet radiation from sunlight exposure. *Photochem Photobiol* 42:163–71.

Rowland FS, Molina MJ. 1974. Stratospheric sink for chlorofluoromethanes: Chlorine atom-catalysed destruction of ozone. *Nature* 249:810–12.

Rundel RD. 1983. Promotional effects of ultraviolet radiation on human basal and squamous cell carcinoma. *Photochem Photobiol* 38:569–75.

Saito I, Sugiyama H, Matsura T. 1983. Photochemical reactions of nucleic acids and their constituents of photobiologic relevance. *Photochem Photobiol* 38:735–43.

Scotto J. 1986. In *Effects of Changes in Stratospheric Ozone and Global Climate,* Vol. 2 (Titus JG, ed.). Environmental Protection Agency, Washington, pp. 33–61.

Scotto J, Cotton G, Urbach F, Berger D, Fears T. 1988. Biologically effective ultraviolet radiation: Surface measurements in the United States, 1974 to 1985. *Science* 239:762–64.

Scotto J, Fears TR, Fraumeni JF. 1982. In *Cancer Epidemiology and Prevention* (Schottenfeld D, Fraumeni JF, eds.). W. B. Saunders Co., Philadelphia, pp. 254–76.

Scotto J, Fraumeni JF. 1982. Skin (other than melanoma). In *Cancer Epidemiology and Prevention* (Schottenfeld D, Fraumeni JF, eds.). W. B. Saunders Co., Philadelphia, pp. 996–1011.

Scotto J, Fraumeni JF, Lee JAH. 1976. Melanomas of the eye and other noncutaneous sites: Epidemiological aspects. *J Natl Cancer Inst* 56:489–91.

Setlow RB. 1974. The wavelengths in sunlight effective in producing skin cancer: A theoretical analysis. *Proc Natl Acad Sci USA* 71:3363–66.

Shindell DT, Rind D, Lonergan P. 1998. Increased polar stratospheric ozone losses and delayed eventual recovery owing to increasing greenhouse-gas concentrations. *Nature* 392:589–92.

Solomon SR, Garcia RR, Rowland FS, Wuebbles DJ. 1986. On the depletion of Antarctic ozone. *Nature* 321:755–57.

Stolarski RS, Krueger AJ, Schoeberl MR, McPeters RD, Newman PA, Alpert JC. 1986. Nimbus 7 satellite measurements of springtime Antarctic ozone decrease. *Nature* 322:808–11.

Taylor HR, West SK, Rosenthal FS, Muñoz B, Newland HS, Abbey H, Emmett EA. 1988. Effect of ultraviolet radiation on cataract formation. *N Engl J Med* 319:1429–33.

Taylor HR, West SK, Rosenthal FS, Muñoz B, Newland HS, Emmett EA. 1989. Corneal changes associated with chronic UV irradiation. *Arch Ophthalmol* 107:1481–84.

Teramura AJ. 1986. The potential consequences of ozone depletion upon agriculture. In *Effects of Changes in Stratospheric Ozone and Global Climate* (Titus JG, ed.). Environmental Protection Agency, Washington, Vol. 2, pp. 255–62.

Thomson BE. 1986. Is the impact of UV-B radiation on marine zooplankton of any significance? In *Effects of Changes in Stratospheric Ozone and Global Climate* (Titus JG, ed.). Environmental Protection Agency, Washington, Vol. 2, pp. 203–9.

Tucker MA, Shields JA, Hartge P, Augsburger J, Hoover RN, Fraumeni JF. 1985. Sunlight exposure as a risk factor for intraocular malignant melanoma. *N Engl J Med* 313:789–805.

van Heyningen R. 1976. What happens to the human lens in cataract. *Sci Am* 233:70–81.

West SK, Rosenthal FS, Bressler NM, Bressler SB, Muñoz B, Fine SL, Taylor HR. 1989. Exposure to sunlight and other risk factors for age-related macular degeneration. *Arch Ophthalmol* 107:875–79.

World Meteorological Organization. 1986. *Global Ozone Research and Monitoring Project, Report No. 16: Atmospheric Ozone 1985*, Vol. 1. NASA, Washington.

———. 1990. *Global Ozone Research and Monitoring Project, Report No. 18: Report of the International Ozone Trends Panel 1988*. NASA, Washington.

———. 1991. *Global Ozone Research and Monitoring Project, Report No. 25: Scientific Assessment of Ozone Depletion: 1991*, Vol. 1. NASA, Washington.

Zigman S. 1983. The role of sunlight in human cataract formation. *Surv Ophthalmol* 27:317–26.

Zigman S, Datiles M, Torczynski E. 1979. Sunlight and human cataracts. *Invest Ophthalmol* 18:462–67.

Electronic References

Environmental Protection Agency. 2000. U.S. Emissions 1998. Inventory of U.S. Greenhouse Gas Emissions and Sinks, 1990–1996 (March 1998). EPA 236-R-98–006. Publications-GHG Emissions. http://www.epa.gov/globalwarming/publications/emissions/us1998/index.html (Date Last Revised 7/12/2000).

National Aeronautics and Space Administration/Goddard Space Flight Center. 2000a.

Stratospheric Ozone: An Electronic Textbook. Chapter 11: The Antarctic Ozone Hole. http://see.gsfc.nasa.gov/edu/SEES/strat/class/S_class.htm (Date Last Revised 5/9/2000).

———. 2000b. Total Ozone Mapping Spectrometer Home Page. http://toms.gsfc.nasa.gov (Date Last Revised 8/3/2000 [revised daily]).

National Atmospheric Deposition Program/National Trends Network. 2000. Hydrogen Ion Concentration as pH from Measurements Made at the Central Analytical Laboratory, 1999. [original version in color] http://nadp.sws.uiuc.edu/isopleths/maps1999/phlab.gif (Date Last Revised 7/26/2000).

National Oceanic and Atmospheric Administration. 1999. Mauna Loa Monthly Mean Carbon Dioxide. http://www.cmdl.noaa.gov/ccg/figures/co2mm_mlo.gif (Date Last Revised 11/17/1999).

North American Research Strategy for Tropospheric Ozone. 2000. Welcome to NARSTO. http://www.cgenv.com/Narsto/ (Date Last Revised 7/12/2000).

CHAPTER 8

An Earth Science Perspective on Global Change

George W. Fisher, Ph.D.

To understand the processes of global change and their consequences for human health, we need to see Earth as a system of physical, chemical, and biological processes, radically interconnected on every scale. This way of looking at Earth is relatively new. A half-century ago, many thought of Earth as having limitless resources of land, energy, water, lumber, and ore. Earth scientists understood Earth in terms of processes operating within the traditional subdivisions of atmosphere, biosphere, hydrosphere, and solid earth. But the Apollo missions of the late 1960s made those views untenable. We began to see Earth as small, isolated, and vulnerable. We began to appreciate the importance of processes operating across the traditional disciplinary boundaries and to see Earth as a system in which virtually all processes are directly or indirectly connected to one another (e.g., Mackenzie 1998). We began to understand that changes in one part of the system will eventually affect virtually every other part.

To portray those connections, we begin by focusing on the chemical, physical, and biological cycles that carry energy, water (H_2O), and carbon dioxide (CO_2) through the earth system, then look at the links between individual cycles, and finally outline processes within the terrestrial biosphere that relate to human health. We will represent each of the three major cycles as a system of sources, sinks, and reservoirs connected to one another by fluxes. The energy cycle is open to large energy gains (incoming solar radiation) and losses (outgoing infrared radiation). The geochemical cycles are essentially closed, except for very minor losses of hydrogen (H_2) to space and occasional small additions of meteoritic material. Consequently, the fluxes in each geochemical cycle must sum to zero, a condition that helps us to estimate the values of fluxes that cannot be measured directly.

The Energy Cycle

In the simplest terms, Earth's energy cycle involves a delicate balance between incoming solar radiation and outgoing infrared radiation. Solar radiation warms Earth's surface, which then emits energy just sufficient to balance the incoming radiation. A complication arises because the incoming radiation is mostly visible light, which passes readily through the atmosphere, while the outgoing energy is primarily infrared radiation, which is absorbed by the atmosphere (see Fig. 7.9). As it absorbs this energy, the atmosphere itself begins to warm and emit infrared radiation. Some is directed outward and lost to space, but more than half is directed downward and further warms the earth (Fig. 8.1*a*). The effect of this secondary warming is substantial. Without an atmosphere, Earth's surface temperature would average only 255 K ($-18°C$), but with an atmosphere, it averages about 288 K ($15°C$), a difference of $33°C$. Although the mechanism behind this supplemental warming is different from that which warms a greenhouse, it has come to be called the *greenhouse effect*. Roughly 90 percent of this "greenhouse" warming is due to energy absorption by water vapor in the atmosphere; about 10 percent is due to absorption by CO_2, methane (CH_4), and the other gases associated with the greenhouse effect (see Fig. 7.9).

The greenhouse effect is often portrayed in negative terms, but, without it, Earth would be too cold for liquid water to exist at the surface and could not sustain life as we know it. Concern centers on the fact that humans have sharply increased the levels of CO_2 and CH_4 in the atmosphere since the beginning of the industrial revolution and continue to do so. The balance of evidence now suggests that these increases are affecting global climate (Houghton et al. 1996), and there is widespread concern because the consequences of this anthropogenic component of the greenhouse effect are poorly understood. We are in effect tinkering with our life support system without fully understanding the consequences.

The simple model of Figure 8.1*a* yields an average global temperature because it implicitly assumes that solar radiation warms all parts of Earth uniformly. That is, of course, an oversimplification. In equatorial regions, the sun is nearly overhead, and so the ground surface receives the full intensity of incoming solar radiation per unit area. In polar regions, the sun is near the horizon, and so the incoming bundle of solar radiation is strongly inclined to the ground surface, and the amount of energy received per unit area is correspondingly reduced (Fig. 8.1*b*). The ice and snow of the near-polar regions also reflect sunlight, further reducing the solar energy absorbed at high latitudes. Because the outgoing infrared radiation varies less strongly with latitude, there is a net energy loss from the poles and a net energy gain at the equator (Fig. 8.1*b*).

Those differences are made up by poleward transport of energy by at-

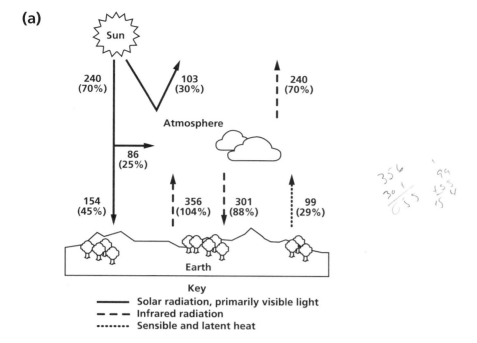

(a)

Key
—— Solar radiation, primarily visible light
– – – Infrared radiation
········ Sensible and latent heat

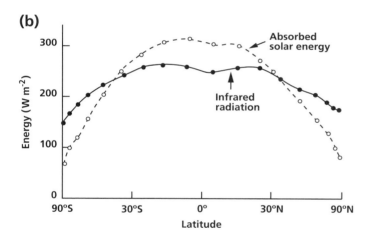

(b)

Figure 8.1 *a:* Earth's energy cycle. Solar radiation provides 343 Wm⁻² to Earth, of which 103 (30%) is reflected by the atmosphere, 86 (25%) is absorbed by the atmosphere, and 154 (45%) is absorbed by Earth. In the steady state, Earth emits infrared radiation equal to 356 Wm⁻², nearly all of which is absorbed by the atmosphere. Evaporation of water transfers another 99 Wm⁻² of sensible and latent heat to the atmosphere. The atmosphere then returns infrared radiation equal to 301 Wm⁻² to Earth and loses the balance (240 Wm⁻²) to space. *Source:* Data from Schneider 1993. *b:* Latitudinal variation in solar radiation absorbed and infrared radiation emitted by Earth. *Source:* Redrawn with permission from Vonder Haar and Suomi 1971, Figure 1.

235

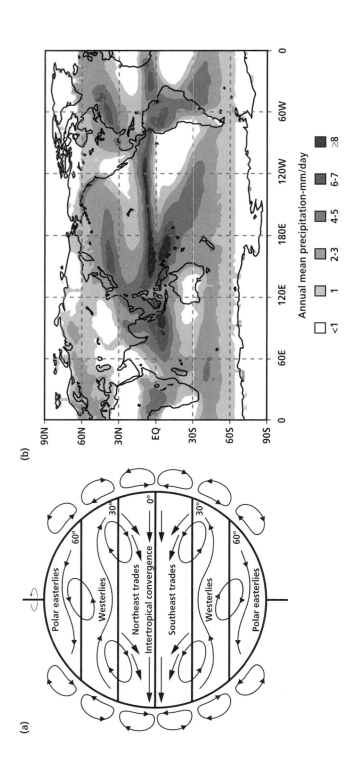

Figure 8.2 *a*: Atmospheric circulation on a rotating Earth, showing the atmospheric convection cells in cross-section and the major surface wind systems. *Source:* Redrawn from Schneider and Londer 1984, Figure 4.2. Reprinted with permission from Sierra Club Books. *b*: Average annual global precipitation (millimeters per day) for the period 1987 through 1998. Note that precipitation is concentrated near the equator, where Hadley cell upwelling is most intense, while deserts are most extensive near 30 degrees north and south and at the poles, where Hadley cell downwelling is most pronounced. The technique for producing these data is described in Huffman et al. (1997). The original figure was produced by the Global Precipitation Climatology Project (GPCP), which is directed by Dr. Arnold Gruber of the National Environmental Satellite Data and Information Service, National Oceanic and Atmospheric Administration (NOAA). The GPCP is a component of the Global Energy and Water Cycle Experiment of the World Climate Research Program, World Meteorological Organization. *Source:* Redrawn from *NOAA 1999.*

mospheric circulation and, to a lesser degree, by ocean circulation. The transfer is driven mostly by upward convection of air near the equator and downward convection near the poles, compensated by pole-to-equator net air flow at the land surface and by equator-to-pole net air flow near the top of the troposphere (lower atmosphere). The pole-equator distance is too long relative to the thickness of the troposphere for this circulation to form a single convection cell, however, and the motion is expressed as a series of three smaller cells in each hemisphere (Fig. 8.2*a*).

Earth's rotation and the resulting Coriolis effect modify this simple pattern in important ways. As the air at the bottom of the downward convection zones at 30 degrees north and south latitude in Figure 8.2*a* begins to move toward the equator, the air and the land beneath it are both moving east at a velocity of 1,446 kilometers per hour because of Earth's rotation (the circumference of the small circle at latitude 30 degrees is 34,704 kilometers, and a point on that circle moves through that distance every 24 hours). As the air moves toward the equator, momentum tends to keep its velocity the same, but the eastward velocity of the ground beneath increases as Earth's circumference increases. By the time the air mass reaches the equator, the ground beneath it is moving eastward 224 kilometers per hour faster than the air, so the air is moving *west* relative to the ground beneath it. Consequently, as air masses move toward the equator, their trajectories curve systematically westward, changing the simple roll-shaped convection cells into westward-moving spirals, which produce both the easterly winds at the equator and the trade winds flanking the equator (Fig. 8.2*a*). Between 30 and 60 degrees latitude, the winds move west to east because convection is moving the surface air from south to north.

The Water Cycle

Earth is the only planet in the solar system with surface pressure and temperature conditions in the range in which liquid water is stable, and consequently it is the only planet capable of supporting life as we know it. The processes that control the movement of water through the earth system do much to determine the distribution and diversity of life. The global geochemical cycle of H_2O involves net evaporation from the oceans, transport through the atmosphere, and net precipitation over the continents, from which the groundwater and river systems carry water back to the oceans (see Figs. 8.3 and 9.1).

We can assess the vulnerability of steady-state reservoirs to a change in fluxes by calculating the average residence time of a molecule in a reservoir, obtained by dividing the net input flux into the content of the reservoir. A single-reservoir system will undergo 95 percent of the change to a new steady state in a period equal to three residence times (Mackenzie 1998, 159). In multiple-reservoir systems, the relationship is more com-

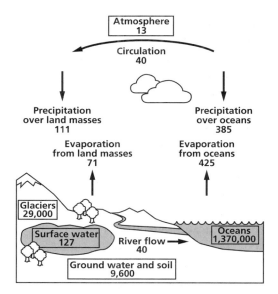

Figure 8.3 The global H_2O cycle, showing annual fluxes (in 10^{18} grams of H_2O per year) and average reservoir contents (in 10^{18} grams of H_2O). *Source:* Data from Mackenzie 1998, 89–91.

plex, but the residence time still provides a rough index of reservoir vulnerability. In the case of the H_2O cycle, the ocean as a whole has a residence time of approximately 34,000 years, when calculated using the river flux as input, and is very stable to changes in the terrestrial system. Consequently, the oceans tend to dominate the H_2O system as a whole, and ocean properties and dynamics have a profound effect on global climate. In contrast, the surface water system has a residence time of only about three years and so is vulnerable to changes in the amount, distribution, and regularity of rainfall on time scales of years to decades.

The pattern of global precipitation is controlled in large part by the atmospheric circulation discussed under "The Energy Cycle" (above). As the humid tropical air rises in the equatorial regions (Fig. 8.2*a*), it expands and cools below the dew point, forcing the water vapor it carries to condense as rain in a nearly continuous belt of heavy rainfall near the equator (Fig. 8.2*b*). The water-depleted air above these rain belts then moves north or south to the down-going limbs of the convection cells at about 30 degrees latitude (Fig. 8.2*a*). As it descends, this air, which has already lost most of its water, is subjected to higher pressure and forced to contract, warming it well above the dew point, so that very little rain falls at these latitudes (Fig. 8.2*b*). This global pattern of rainfall is locally modified by the pattern of land masses (which tend to warm more quickly than oceanic areas and so promote more vigorous upwelling and rainfall in the Amazon, the Congo, and Indonesia) and by the orographic effect (air rises on the upwind side of mountain ranges, concentrating rainfall there, and descends on the downwind side, reducing rainfall there).

These spatial and temporal patterns control the amount of rainfall available for agricultural, domestic, and industrial use. But in areas where rainfall is inadequate to meet current needs, the groundwater system is often used to supply a substantial proportion of water needs. Use of ground water at a rate comparable to that at which water is resupplied to the groundwater reservoir can be sustained, but in areas like the Middle East and the southern Great Plains of the United States, groundwater withdrawals considerably exceed resupply rates and cannot be continued indefinitely. Water shortages in parts of the developing world are rapidly approaching crisis proportions (see Frederiksen 1996 and Chapter 9).

The ocean can be visualized as a two-layer system involving a thin, relatively warm upper layer and a lower, much colder zone, separated by a thin region of variable temperature, known as the *thermocline*. Because the warmer surface water is generally less dense than the colder deep water, this stratification tends to be stable, and exchange between the two subsystems is slow and limited to areas in which the surface water becomes abnormally cold or salty or to areas where the surface layer is thinned by persistent wind systems, as in El Niño, discussed below.

Currents within the upper layer of the ocean are driven by the surface wind system of Figure 8.2a. The equatorial and trade winds drive the equatorial surface waters westward while the winds at latitudes just above 30 degrees blow the surface waters eastward, combining to form the huge circular current systems, called *gyres,* that dominate the shallow ocean (Fig. 8.4). The patterns of ocean currents in Figure 8.4 and winds in Figure 8.2a represent long-term averages, which are responsible for much of the world's long-term climate. Seasonal changes in these wind and current patterns produce seasonal variations in weather and precipitation. Longer-term rhythmic fluctuations in ocean and atmosphere produce multiyear climate fluctuations.

El Niño is the best understood of these longer-term fluctuations and serves as a useful example. It originates in closely coupled fluctuations of wind and sea surface temperature (SST) in the equatorial Pacific but has worldwide consequences for weather and rainfall. Much of the time, large SST gradients across the Pacific are associated with vigorous atmospheric convection, producing intense rainfall over the rain forests of the Amazon and Indonesia and strong east-to-west surface winds over much of the Pacific (Fig. 8.5a). This configuration involves a strong positive feedback: the strong SST gradient tends to strengthen the surface winds and intensify convection, while the winds tend to keep the warm surface waters in the western Pacific, maintaining the temperature gradient. Neither the SST gradient nor the winds should be viewed as the sole cause of the pattern; the two simply reinforce each other and stabilize the configuration.

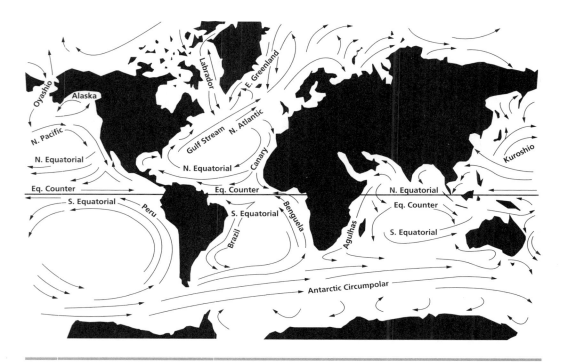

Figure 8.4 Major oceanic current systems, showing circular gyres driven by the surface wind systems of Figure 8.2*a*. *Source:* Redrawn from Knauss 1997, Figure 1.9. Reprinted with permission from Prentice-Hall, Inc.

Periodically, however, the configuration becomes destabilized and the warm surface water moves east across the Pacific almost to Peru, in the phenomenon known generally as El Niño or, more technically, as a warm episode. The resulting weaker SST gradients produce weaker surface winds and convection, and so rainfall tends to be more widely dispersed. Some locations, such as northern Peru, may experience unusually heavy rains while other locations, such as Indonesia, may experience severe droughts; temperature may also be unusually warm or cool in some areas (Glantz 1996, 93). As with the large SST gradient, the configuration involves a positive feedback: the weak SST gradient produces less vigorous winds, which are unable to blow the warm surface waters to the west and so allow the weak SST gradient to persist. Eventually, the conditions of El Niño reverse and the waters of the eastern Pacific become cooler; in its most acute phase, this pattern has come to be called La Niña or, more technically, a cold episode.

Because both configurations involve an element of positive feedback, the problem is not how either configuration could exist on its own but rather how either configuration could break down and change to the other. Transitions occur at intervals varying from two to seven years (Fig. 8.5*b*),

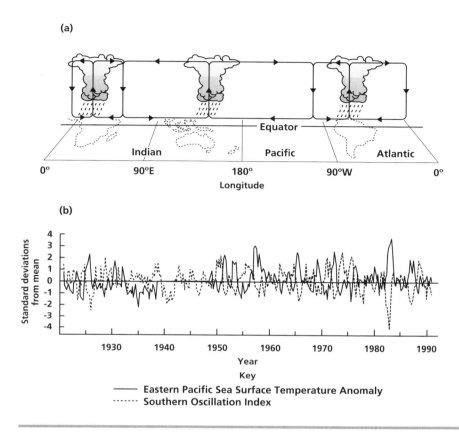

Figure 8.5 *a:* Equatorial atmospheric circulation in the Pacific and Indian Oceans. Normally the circulation-driven winds are strong, blowing the warm surface water into the western Pacific and concentrating rainfall there. During El Niño, circulation is less intense and winds are weaker, allowing warm water to move farther east, and rainfall is more widespread. *Source:* Redrawn with permission from Webster 1983, Figure 9.4. *b:* Southern Oscillation Index (SOI, the difference between sea level pressure in Tahiti and Darwin, Australia) and the sea surface temperature anomaly (SSTA) in the eastern Pacific (an index of El Niño). The indices are shown normalized in terms of standard deviations from the mean. The two indices tend to be out of phase and to represent the periodicity of El Niño. *Source:* Redrawn from Cane 1993, Figure 18.1*b*. Reprinted with permission from Cambridge University Press and the University Corporation for Atmospheric Research/National Center for Atmospheric Research.

and the patterns tend to become most pronounced near the first of the calendar year. The reasons behind the transition are complex and involve coupling between slow east-west wave systems in the equatorial Pacific, the movement of the warm surface water, and changes in the vigor of atmospheric circulation (Cane 1993). Though much remains to be learned about the details of this process, the pattern of change from El Niño to La Niña and the reverse is now so well known that computer predictions of SST and rainfall distribution can be made a year or more in advance, as was done with some success during the 1997–98 El Niño and the 1998–99 La Niña. The public health community is paying more attention to the possible use of forecasts of El Niño and La Niña to reduce the adverse effects of climate variability on human health (see Kovats et al. 1999 and Chapters 2, 11, 12, 13, and 14).

The Carbon Cycle

Like the H_2O cycle, the carbon cycle is essentially closed; the only systematic loss is due to burial of a small amount of carbon in ocean sediments, offset on geological time scales by additions of volcanic CO_2 to the atmosphere. The major reservoirs and fluxes in the short-term carbon cycle are summarized in Figure 8.6. As in the H_2O cycle, the deep ocean is the major reservoir and the systems that are of direct importance to humans are much smaller and more vulnerable. Unlike the H_2O cycle, however, the amount of carbon carried by the river systems is so small that the terrestrial and marine portions of the cycle are connected almost entirely through the atmosphere.

The key to the carbon cycle is the photosynthetic reaction, in which plants assimilate CO_2 and H_2O and capture solar energy, which converts them to organic matter (approximately CH_2O) and oxygen (O_2):

$$CO_2 + H_2O \quad \rightarrow \quad CH_2O + O_2$$

Allowing for losses due to respiration, the terrestrial biosphere converts about 55 petagrams of carbon to organic matter each year, a quantity termed net primary productivity (NPP). Approximately the same amount of carbon falls to the ground as litter and becomes part of the soil system. Roughly the same amount of carbon in the soil is then oxidized by organic decay processes in the reverse of the photosynthetic reaction,

$$CH_2O + O_2 \quad \rightarrow \quad CO_2 + H_2O$$

Terrestrial decomposition returns approximately 52 petagrams of carbon to the atmosphere each year. A similar loop operates in the marine biosphere, with the important difference that the entire oceanic biological carbon cycle takes place within the water. The substrate for oceanic production is dissolved CO_2 in the water; gas exchange between the air and water is entirely abiological. Moreover, the carbon pool in the marine biosphere is very small because of rapid turnover by microscopic plants, which accomplish most marine production.

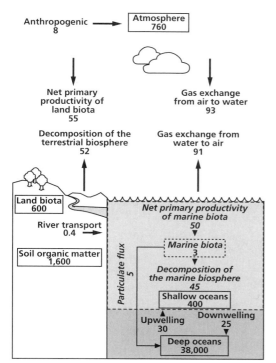

Figure 8.6 The global carbon cycle, showing annual fluxes (in petagrams of carbon per year) and reservoir contents (in petagrams of carbon). Some sources present data in terms of metric gigatons instead of petagrams; one metric gigaton and one petagram are both equivalent in weight to 10^{15} grams. *Source:* Data from Mackenzie 1998, 167 and 169; estimates (rounded) of net primary productivity in Field et al. 1998, 238; and data courtesy of Pieter Tans, Carbon Cycle Group, National Oceanic and Atmospheric Administration, Boulder, Colorado.

For two millennia before the industrial revolution, atmospheric CO_2 levels held steady at about 280 ± 5 parts per million (Indermuhle et al. 1999), and it is generally assumed that exchanges with the atmosphere were similarly steady. But the burning of fossil fuel and deforestation are now adding about 8 petagrams of carbon to the atmosphere each year, an amount equal to about 6 percent of the fluxes from the marine and terrestrial systems combined. Part of this addition is accounted for by the increase in CO_2 observed in the atmosphere (see Fig. 7.10). But the increase in atmospheric CO_2 during the 1980s accounts for slightly less than half of the carbon emitted. It is generally thought that about half of the "missing carbon" is dissolved in the ocean, and, although the details of the process are still uncertain, it is now suspected that the remainder is being sequestered in the forest systems of the northern hemisphere (Mackenzie 1998, 375–78). Understanding the fate of this anthropogenic carbon is one of the major problems in assessing the risk of global warming; until we understand how anthropogenic carbon emissions affect the global carbon cycle, we cannot really be sure what their overall effect will be.

The NPP of the biosphere depends strongly upon temperature (Fig. 8.7*a*) and rainfall (Fig. 8.7*b*), and so the global distribution of ecosystems (Fig. 8.8) is determined by the global patterns of climate and rainfall (see

(a)

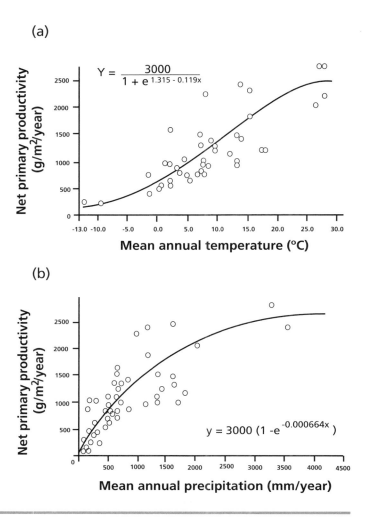

(b)

Figure 8.7 The dependence of net primary productivity upon temperature (a) and rainfall (b). *Source:* Redrawn from Lieth 1975, Figure 12.3a and Figure 12.4a. Reprinted with permission from Springer-Verlag GmbH & Co. KG.

Fig. 8.2b) produced by the pattern of atmospheric circulation (see Fig. 8.2a). The major tropical rain forests correspond to the zone of intense equatorial rainfall associated with upward convection, the world's major deserts are located in the dry zone produced by downward convection near 30 degrees north and south, the belt of temperate and boreal forests corresponds to the secondary rain belt near 60 degrees north and south, and the polar regions are in effect cold deserts because of downward convection at the poles. Many local variations on this theme reflecting the orographic effect and differences in the degree of warming of continent and ocean waters can be detected in a critical comparison of Figures 8.8 and

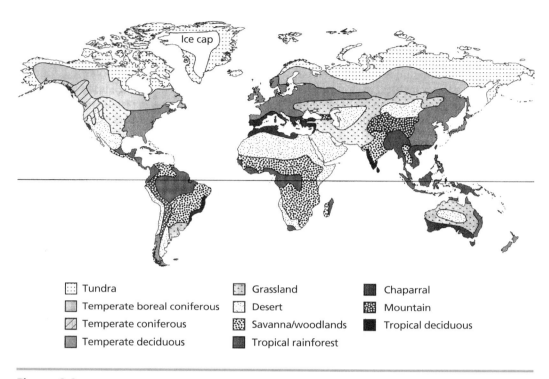

Figure 8.8   Global distribution of the major biomes. Note the close correlation between biome distribution and the global pattern of rainfall in Figure 8.2*b*. *Source:* Redrawn from Mackenzie 1998, Figure 4.8. Reprinted with permission from Prentice-Hall, Inc., Upper Saddle River, N.J.

8.2*b*. And because the freshwater system of the H_2O cycle is vulnerable to annual to decadal changes in rainfall, terrestrial ecosystems are similarly vulnerable. The effects of these fluctuations tend to be especially pronounced where rainfall is already low. For example, the boundary between the southern Sahara desert and the Sahel region of Africa shifts its position in response to changes in rainfall and has been observed to move as much as 100 kilometers in one year (Tucker et al. 1991).

This discussion emphasizes the fact that, although the portrayals of the energy cycle (see Fig. 8.1*a*), the H_2O cycle (see Fig. 8.3), and the carbon cycle (see Fig. 8.6) seem to imply that the cycles can be treated independently, they are in fact linked in very complex ways. The NPP in the biosphere depends critically upon the movement of water and energy, the cycling of water depends upon the flow of energy, and the atmospheric circulation, which redistributes energy, depends upon local patterns of precipitation and evaporation, which in turn depend upon local variations in NPP. So, although it is convenient to step through the overall system by examining

Table 8.1   Anthropogenic and Photosynthetic Production of Carbon, Nitrogen, and Phosphorus in Relation to the Size of Their Reservoirs in the Terrestrial Biosphere

	Million Metric Tons of the Element		
	Carbon	Nitrogen	Phosphorus
Annual global anthropogenic production	8,000[a]	91.6[b]	40.1[b]
Annual net primary productivity by terrestrial biosphere	56,400[c]	580[d]	320[d]
Size of reservoir in terrestrial biosphere	600,000[e]	10,000[e]	3,000[e]

Source: Data courtesy of Pieter Tans, Carbon Cycle Group, National Oceanic and Atmospheric Administration, Boulder, Colo.; World Resources Institute 1998, 338; Field et al. 1998, 238; Mackenzie 1998, Tables 5.1 and 5.2.

[a]Data courtesy of Pieter Tans, Carbon Cycle Group, National Oceanic and Atmospheric Administration, Boulder, Colo., fossil fuel combustion and biomass burning.

[b]World Resources Institute 1998, 338, with data for the year 1995.

[c]Field et al. 1998, 238.

[d]Mackenzie 1998, Table 5.1.

[e]Mackenzie 1998, Table 5.2.

each individual cycle in isolation, as we have done here, we must bear in mind that these three cycles are interconnected at all scales of time and space. And, of course, NPP and the carbon cycle depend also on the geochemical cycles of nitrogen, phosphorus, and other nutrients needed for plant growth. Table 8.1 compares the fluxes of carbon, nitrogen, and phosphorus involved in NPP with the amount produced anthropogenically in 1995, showing that human activity has become a global geochemical process (Mackenzie 1998, 419–38).

Ecosystem Dynamics

The biomes of Figure 8.8 are regional ecosystems, each dominated by characteristic plant assemblages such as rain forest, savannah, grassland, or desert. Each is delicately adjusted to a particular range of environmental conditions determined by the movement of energy, water, and carbon through the earth system, and each supports a fauna attuned to the local plant assemblage. Current biome boundaries mark places where environmental conditions gradually shift from values to which one biome is best attuned to values to which another is more suited. Biome boundaries are dynamic, and even small changes in environmental conditions can cause them to shift, as in the case of the Sahara, discussed under "The Carbon Cycle" (above). The geological record, especially that of the past 100,000 years, shows a constantly shifting pattern of biomes (Mackenzie 1998, 353–63).

Each biome is made up of a grid of ecosystems composed of specific species assemblages. Ecosystems function by cycling energy and nutrients so that each species obtains the energy and nutrients that it needs to live (Mackenzie 1998, 127–34). Each ecosystem can thus be visualized as a complex mechanism for cycling energy and nutrients to the benefit of each species. Every species has a role to play in the cycle and contributes in some way to the operation of the overall system.

The starting point for the energy and nutrient cycle in each terrestrial ecosystem is its assemblage of plants, which capture solar energy by photosynthesis and extract nutrients from the soil; green plants are therefore referred to as the *primary producers* of an ecosystem. Organisms that feed on other organisms are known as *consumers*. Primary consumers are *herbivores*, which feed directly on plants; secondary consumers, *carnivores*, feed on the primary consumers. Both primary producers and consumers die and produce waste materials, of course. Organisms like bacteria, fungi, crabs, and vultures, which feed on the refuse of an ecosystem, are known as *detritivores* and play a crucial role by returning energy and nutrients to the soil system (the bacteria and fungi) or by converting detritus to a form (crab meat, for example) that can be utilized by consumers. Without detritivores, vital nutrients would be gradually buried, dimming and eventually extinguishing the life system.

Each ecosystem and each species in an ecosystem is responsive to changes in environmental conditions controlled by the global energy and nutrient cycles, and each plays a role in determining environmental conditions by affecting the amount of runoff, the solar energy absorbed, and so on. Even apparently static ecosystems involve an extraordinarily rich variety of interactions among the component species (Stiling 1992, 202–93). The feeding patterns (*predation*) that sustain an ecosystem by benefiting one organism at the expense of another may involve *parasitism* or *cannibalism* in addition to the herbivory and carnivory already mentioned. But not all interactions result in harm. *Mutualism* (interactions like those between plants and pollinators) benefit both species, while *commensalism* (interactions like the dispersal of plant seeds as burrs transported by passing animals) benefits one without affecting the other. Since Darwin's seminal work, *competition* (which may be between species or within a species and which may involve direct competition between individuals or indirect competition by resource consumption) has been seen as a major element of this dynamism. The forms of competition and the ways of representing them are legion, but one useful scheme is to contrast species that reproduce rapidly but ultimately compete poorly (e.g., weeds that quickly colonize a burned forest patch but are eventually excluded by slower-growing species) with species that compete more effectively but reproduce and approach resource limits more slowly (e.g., most primates). These two competitive strategies are known respectively as *r*-selection and *K*-selection (Stiling 1992, 225–27).

This scheme is too simplistic to provide a comprehensive analysis of competition, and few species can be unambiguously identified with either of these strategies. Nevertheless, the distinction is helpful in identifying two strongly contrasted styles of competition and the characteristics of species that tend to adopt those styles. Species that opt for *r*-selection strategies tend to be small, to reproduce only once but have many offspring, to develop rapidly, and to have a short life span. They tend to move into unoccupied niches, expand rapidly, and then die out almost completely until other unoccupied niches present themselves. They use resources rapidly but inefficiently and tend to prosper in rapidly changing environments. Species that rely on *K*-selection strategies tend to the opposite in most respects. They are typically large, reproduce often but produce few offspring each time, develop slowly, and have long life spans. They tend to be skilled competitors, to exploit subtle competitive advantages, and to expand their territory slowly. They use resources efficiently and tend to prosper in stable environments.

Oversimplified though it is, this model provides a useful metaphor for the interaction between humans, who tend to behave like *K*-strategists, and infectious agents, many of which behave like *r*-strategists. A warning implicit in this metaphor is that increasing rates of deforestation, climate change, urbanization, and population movement are all likely to enhance opportunities for infectious agents at the expense of humans. The metaphor also suggests that humans may succeed or fail *as a species* to a degree that government policies have thus far failed to acknowledge. Infectious agents, such as the virus causing acquired immune deficiency syndrome (AIDS), that emerge or evolve in response to changes in one part of the world can easily affect another. These implications show how important it is that we understand disease in an ecological context (see Chapter 10) and that we regard public health as a global issue (see Chapter 6).

Conclusion: An Ecological Perspective on Health

In a sense, the themes of this book are that humans live in a complex, dynamic ecosystem, that our health depends on that system, and that our ability to manage health depends upon our understanding of how that system works. Each of the chapters in the book explores individual components of our relationship to that system. The theme of this chapter is that those components must always be considered against the background of the global system as a whole. The question of how individual diseases and health measures will interact with each other and with the system as a whole must always be asked and must be asked against an awareness that the system itself is changing, in ways that are not yet fully understood, in response to both natural fluctuations and human activities.

SUGGESTED STUDY PROJECTS

Suggested study projects provide a set of options for individual or team projects that will enhance interactivity and communication among course participants (see Appendix A). The Resource Center (see Appendix B) and references in all of the chapters provide starting points for inquiries. The process of finding and evaluating sources of information should be based on the principles of information literacy applied to the Internet environment (see Appendix A).

The objective of this chapter is to develop a better understanding of an ecological perspective on health. The study projects are designed to achieve this objective.

PROJECT 1: Interrelation of Energy, Water, and Carbon Cycles

The objective of this project is to develop an understanding of how changes in any part of the energy, water, and carbon cycles can affect all of the cycles.

Task 1. Describe the interrelationship of the energy, water, and carbon cycles.

Task 2. Describe how those cycles create the global climate system.

Task 3. Describe how the global climate system creates biomes.

PROJECT 2: Effects of Natural Variability in Climate

The objective of this study project is to develop an understanding of the influence of natural variability in climate on environment and health.

Task 1. Trace three examples of environmental changes (disasters) connected with El Niño or La Niña.

Task 2. Describe the health effects caused by your examples in task 1.

Task 3. El Niño/La Niña forecasting:

a.   Describe an application of El Niño/La Niña forecasting.
b.   Find the current El Niño/La Niña forecast and its projection for the future.

PROJECT 3: Ecological Perspective on Health

The objective of this study project is to deepen an understanding of the concept of an ecological perspective on health.

Task 1. Explain what is meant in this chapter by an ecological perspective on health.

Task 2. How does material from other chapters fit into this perspective?

Task 3. Are there other aspects of an ecological perspective on health that you would add?

References

Cane MA. 1993. Tropical Pacific ENSO models: ENSO as a mode of the coupled system. In *Climate System Modeling* (Trenberth KE, ed.). Cambridge University Press, New York, pp. 583–616.

Field CB, Behrenfeld MJ, Randerson JT, Falkowski P. 1998. Primary production of the biosphere: Integrating terrestrial and oceanic components. *Science* 281 (5374): 237–40.

Frederiksen HD. 1996. Water crisis in developing world: Misconceptions about solutions. *J Water Resources Planning Management* 122 (2): 79–87.

Glantz MH. 1996. *Currents of Change: El Niño's Impact on Climate and Society.* Cambridge University Press, Cambridge.

Houghton JT, Meira Filho LG, Callander BA, Harris N, Kattenberg A, Maskell K. 1996. *Climate Change, 1995: The Science of Climate Change.* Cambridge University Press, New York.

Huffman GJ, Adler RF, Arkin P, Chang A, Ferraro R, Gruber A, Janowiak J, McNab A, Rudolf B, Schneider U. 1997. The Global Precipitation Climatology Project (GPCP) combined precipitation data set. *Bull Am Meteorol Soc* 78 (1): 5–20.

Indermuhle A, Stocker TF, Fisher H, Smith HJ, Wahlen M, Deck B, Mastoianni D, Tschumi J, Blunier T, Meyer R, Stauffer B. 1999. Holocene carbon-cycle dynamics based on CO_2 trapped in ice at Taylor Dome, Antarctica. *Nature* 398:121–26.

Knauss JA. 1997. *Introduction to Physical Oceanography,* 2d ed. Prentice Hall, Upper Saddle River, N.J.

Kovats RS, Bouma M, Haines A. 1999. *El Niño and Health.* WHO/SDE/PHE/99.4. World Health Organization, Geneva.

Lieth H. 1975. Modeling the primary productivity of the world. In *Primary Productivity of the Biosphere* (Lieth H, Whittaker RH, eds.). Springer Verlag, New York, Chap. 12.

Mackenzie FT. 1998. *Our Changing Planet: An Introduction to Earth System Science and Global Environmental Change,* 2d ed. Prentice Hall, Upper Saddle River, N.J.

Schneider SH. 1993. Introduction to climate modeling. In *Climate System Modeling* (Trenberth KE, ed.). Cambridge University Press, New York, pp. 3–26.

Schneider SH, Londer R. 1984. *The Coevolution of Climate and Life.* Sierra Club Books, San Francisco.

Stiling PD. 1992. *Introductory Ecology.* Prentice Hall, Englewood Cliffs, N.J.

Tucker CJ, Dregne HE, Newcomb WW. 1991. Expansion and contraction of the Sahara desert from 1980 to 1990. *Science* 253:299–301.

Vonder Haar TH, Suomi VE. 1971. Measurements of the earth's radiation budget from satellites during a five-year period. Part I: Extended time and space means. *J Atmospheric Sci* 28:305–14.

Webster P. 1983. The large-scale structure of the tropical atmosphere. In *Large Scale Dynamical Processes in the Atmosphere* (Hoskins B, Pierce R, eds.). Academic Press, San Diego, pp. 235–76.

World Resources Institute. 1998. *1998–99 World Resources: A Guide to the Global Environment.* Oxford University Press, New York.

Electronic References

National Oceanic and Atmospheric Administration. 1999. Annual Average GPCP Precipitation (mm/day), 1987–98. Global Precipitation Climatology Project. http://orbit-net.nesdis.noaa.gov/arad/gpcp/maps/grayyear.gif (Date Last Revised 8/4/1999).

CHAPTER 9

Water Resources Management

Nicolaas J. P. M. de Groot, M.Sc.

All life on Earth depends on the availability of water. Our planet is the only planet on which the existence of water in liquid phase is known. In its processes of precipitation and flow, water is an excellent solvent and transports nutrients that are essential to life. Part of the precipitated water returns directly to the atmosphere through evaporation. The remaining precipitation infiltrates into the soil or flows over it, is recycled by organisms, recharges aquifers, fills rivers and lakes, and enters the ocean to be returned to the atmosphere. In continuous movement under and above the surface of the earth, water serves as an indispensable resource in the ecosystems of this planet. Water can also cause destruction and death. Floods are one of the worst natural disasters and have claimed more victims and have caused more material loss than have earthquakes, eruptions of volcanoes, and other similar catastrophes.

The hydrological cycle, which has always been taken for granted, has been modified since early history by the construction of dams, wells, canals, and systems for irrigation and drainage to serve the interests of people. Nowadays, governments and private entities are spending increasing amounts of money to design, construct, and maintain these installations. These anthropogenic activities not only modify the hydrological cycle but also disturb the natural sources of water and, because of poor practices of land use and water management, both the quantity and quality of this "renewable" natural resource are decreasing. Overuse of ground water causes intrusion of salt water in coastal areas and alarming rates of decline in the level of ground water in many regions of the world. All around the world, the quality of water is decreasing because of ongoing contamination. This degradation of water increases the danger of diseases that are transmitted through water or are related to water and threatens not only people but all parts of the ecosystems that surround us. Shortage of water and abusive use of water create a serious threat to environmental protection and sustainable development. Public health and well-being, food security, industrial development, and the ecosystems on which they depend are all put in danger if human societies are not able to find better ways to manage existing water and improve land use (see also Chapter 14).

In spite of early warnings regarding lack of available water as a constraint to development, water issues still tend to be seen as mainly technical problems. Many water generalists complain about widespread water illiteracy. For example, in the discussion of rain-fed agriculture, water is taken more or less for granted and the focus is put on other agricultural factors (e.g., agrochemicals, plant breeding). In environmental politics, land and water issues are still seen as belonging to different worlds, taken care of by different professions with distinctly different educational and professional cultures. In the documentation that followed from the United Nations Conference on Environment and Development, for instance, water issues were largely neglected in the land chapters of Agenda 21— although land use generally both depends on water and affects water. In a report to the United Nations Commission on Sustainable Development some years later, detailing the follow-up activities related to the land use planning chapter 10 of Agenda 21, water appears only as a footnote (Food and Agriculture Organization [FAO] 1995).

One general challenge of governance is to cope successfully with environmental systems in the landscape while seeking to meet relevant societal needs. Important questions to be answered and tasks to be faced are how to implement a price for water and how to regard water as an economic good.

The Hydrological Cycle

The hydrological cycle on our planet is induced by energy supplied by our sun (see also Chapter 8). Figure 9.1 is a schematic drawing of the hydrological cycle, which includes the basic processes of precipitation, infiltration, surface and groundwater flow, evaporation, transpiration, and condensation. The hydrological cycle moves enormous quantities of water around the world. This movement is often relatively fast, as one drop of water stays an average of 14 days in a river and about 10 days in the atmosphere. But these units of time can change into centuries for glaciers and thousands of years for water that moves slowly through a deep aquifer.

Some important differences in how hydrological processes operate must be well understood. First, the genesis of rainfall in different zones depends on different processes, such as tropical cyclones, monsoons, trade winds, convection, and frontal movement. The intensities of rainfall may therefore be very different, and the variability of rainfall changes greatly

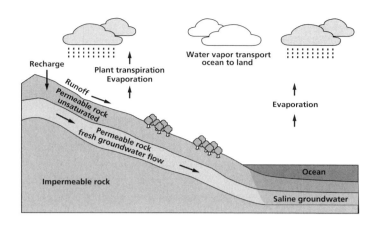

Figure 9.1 Basic processes of the hydrological cycle.

between zones. Patchiness or spatial variability is another aspect, and differences in the seasonal pattern are fundamental to the whole organization of society.

The infiltration resulting from rainfall will depend on the hydraulic conductivity of the soil, which varies among soil types and depends also on the structure and density of the soil and its pattern of macropores. The capacity to store water in soil is a function of soil type as well as the structure and organic content of soil. The type of aquifer may also be quite different in different locations. The depth of the water table is a vital factor for the accessibility of water in wells and can vary considerably between humid and arid conditions. Depending on the characteristics of the soil and the duration and intensity of rainfall, the depth of the water table may vary seasonally as well. The general pattern of the movement of ground water may differ between different geological settings. Outflow areas may have very different characteristics in different environments; they may be wetlands, oases, coastal springs, and so forth.

Runoff is generally referred to as one distinctive element in the water balance. There are, however, some basic differences between types of runoff formation. Runoff can occur as flood flow, which is rapidly transferred along a river during rainy periods, or a less variable base flow, which is produced by outflow of ground water and provides a more or less dependable source of water flowing into a river. The base flow may be increased by contributions from water stored in lakes and reservoirs. In arid zones, rivers have no base flow, as the level of ground water lies below the streambed. Flood flows in arid areas often do not reach the drainage terminal of a river as surface flow because the water infiltrates into the ground before it reaches its destination, such as an ocean. Understanding interregional differences in the process of runoff formation is fundamental as a basis for the development and management of water resources.

Finally, river flow may also be classified according to its genesis, depending on whether it comes from storage in solid phase, such as glaciers and snow packs; in liquid phase, such as lakes and ground water; or in atmospheric water vapor. The primary effect on river flow is that its stability increases with an increase in the turnover time of the source, where turnover time is defined by the United Nations Educational, Scientific and Cultural Organization (UNESCO) as the conventional residence time of water in a specific system (UNESCO 1971). In other words, river flow would vary more if it received water only from the atmosphere, where the residence time is short, and would be very stable if it received water only from a groundwater system with a long residence time (Table 9.1). In areas where water is stored in solid phase, flow is fairly regular from permanent snow or glaciers, but melting of the snow pack due to seasonal changes in temperature can cause floods in the spring and summer.

Where snow and ice are not factors, there is a marked contrast between the perennial regimes of rivers in the humid zone, which are fed by base flow from ground water, and the large fluctuations of rivers in the arid zone, which exhibit flash flows that may recharge ground water before reaching their drainage terminals. A particular case of great historical and social significance is that of rivers that run through the arid zone but have their sources in the humid zone, either in snow packs (e.g., Indus and Colorado Rivers) or in large lakes (e.g., Nile River).

These hydrological differences tend to complicate the transfer of hydrological understanding, methods, and theories between zones. Basic differences in terms of climate and geology between the different regions of the world produce different soil water regimes, differences in potential evaporation, different genesis and behavior of ground water, different regimes of river flow, and so on. Hydrological differences are part of fundamental ecohydrological differences, since differences in temperature, soil water, and ground water affect vegetation biomes, root development, and root depth (see also Chapter 8). Patterns of vegetation, in turn, are related to differences in population pressures and intensities of human activities, which contribute to major differences in the environmental problems encountered.

Determining How Much Water Is Available

Water is one of the most common chemical compounds on Earth, and it can occur in three different phases:

1. it can be found in liquid phase on the surface in rivers, lakes, and oceans or as groundwater reservoirs below the surface;

Table 9.1   Water Volumes and Residence Times of the Hydrological Cycle

Type of Storage	Total Cycle Volume 10^6 km³	%	Freshwater Volume Only (%)	Freshwater Volume without Icecaps and Glaciers (%)	Residence Times
Oceans and seas	1,370	94	—	—	~4,000 yr
Lakes and reservoirs	0.13	<0.01	0.14	0.21	~10 yr
Swamps and marshes	<0.01	<0.01	<0.01	<0.01	1–10 yr
River channels	<0.01	<0.01	<0.01	<0.01	~2 wk
Soil moisture	0.07	<0.01	0.07	0.11	2 wk–1 yr
Ground water	60	4	66.5	99.65	2 wk–50,000 yr
Icecaps and glaciers	30	2	33.3	—	10–1,000 yr
Atmospheric water	0.01	<0.01	0.01	0.02	~10 days
Biospheric water	<0.01	<0.01	<0.01	<0.01	~1 wk

Source: Reproduced from Chapman 1996, Table 1.1, with permission of Taylor & Francis, Inc.,/Routledge, Inc., http://www.routledge-ny.com.

Note: Although the general patterns are clear, different procedures for estimating cycle volumes and residence times may produce different results. For example, see Falkenmark and Chapman (1989), Vereniging voor Landinrichting (1992), and Mackenzie (1998). Some sources present data in terms of grams instead of cubic kilometers; one cubic kilometer of water is equivalent to 10^{15} grams of water.

2. it can be found as water vapor in the atmosphere; and
3. it can be found in its solid phase covering the polar regions and high mountains where the temperature is low enough to maintain that phase.

Furthermore, water forms an important part of the physical construction of all that lives on the earth, as it constitutes almost 80 percent of all living tissue.

The total amount of water on Earth is about $1,384,000 \times 10^{12}$ cubic meters. Although figures given in the literature do not coincide completely (World Meteorological Organization [WMO] and UNESCO 1997; Vereniging voor Landinrichting 1992; Chapman 1996; Mackenzie 1998), it can be said that only a small part of the earth's water is directly available as suitable drinking water. About 94–97 percent of all water on our planet is located in the oceans and seas (Chapman 1996; Mackenzie 1998). Only 0.26 percent of the sweet water is directly available as surface water in (possibly artificial) lakes and rivers or as shallow ground water; the remaining

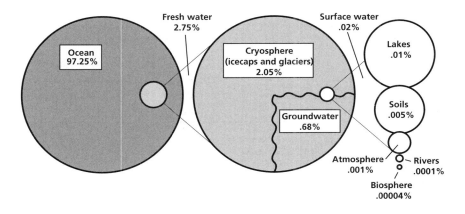

Figure 9.2 Distribution of the earth's water resources. *Source:* Data from Mackenzie 1998, Table 3.2.

sweet water, about 2.24 percent of all water, is confined in polar ice or glaciers and deep aquifers (Shiklomanov 1996). Table 9.1 and Figure 9.2 indicate some of the numbers that are available in the literature (Chapman 1996; Mackenzie 1998). Although these numbers must be used with caution, as the procedures for estimating the components of water resources do vary, they give a clear picture of how little water is readily available for human consumption. In addition, Table 9.1 indicates the residence time of water in different reservoirs. As with the estimates of the reservoirs themselves, estimates of residence times are a product of the definitions of the different reservoirs. For example, Falkenmark and Chapman (1989) estimate a turnover time of 28 years for the oceans and a turnover time of 1–4 weeks for the plant root zone, which differ from the values in Table 9.1.

Only from lakes, rivers, and relatively shallow aquifers is it possible to produce large quantities of sweet water. These most important components of the hydrological resources of the earth are continuously replenished by precipitation and melting water from glaciers and, in some areas, are supplemented by dew and mist. However, the quantity of sweet water is decreasing all around the world because of an increase of evaporation and transpiration caused by anthropogenic modifications of the environment. The construction of artificial lakes facilitates the generation of electricity and the storage of water for the purposes of drinking or irrigation. Although these artificial lakes recharge the aquifers, surface water runoff decreases at coastal areas; the resulting diminished recharge of ground water causes intrusion of salt water. The hydrological cycle varies from place to place and day to day; thus, these hydrological resources are constant neither in quantity nor in quality. Nevertheless, they are the most accessible sources of sweet water and are without any doubt very important to human society.

As stated before, the mismanagement of land use can seriously affect the hydrological cycle and thus the available amount of accessible water. Deforestation, expansion of paved regions, construction of canals and artificial lakes, communication between river basins, irrigation and drainage, and many other activities modify the hydrological balance. To evaluate the effects of those modifications and how water can be used in the industrial, domestic, and agricultural sectors, one needs detailed regional information on quantity as well as quality of water.

Using information made available by fieldwork, hydrologists have derived global estimates of the total discharges of the rivers of the world, which can be taken as the ultimate limit of water resources in the world. These discharges add up to a figure between 35,000 and 50,000 cubic kilometers per year, less than 1 percent of the total volume of the sweet water, and there exists a great variation between years and from region to region. The Amazon basin, which covers 5,870,000 square kilometers, or 4 percent of the total global surface not covered by water, has a discharge that equals about 16 percent of the total global river discharge. On the other hand, the arid zones that cover about 40 percent of the earth's land surface produce only 2 percent of the total river discharge (WMO and UNESCO 1997). Also, the discharge can fluctuate greatly throughout the year. Heavy rainfall or defrost of glaciers in one season can cause a river to swell, while drought and frost during another season can decrease the discharge of a large river to nothing more than a small waterway and create water shortages.

Other problems in the usage of surface water are great distances from the natural course of the water to the end-users and upstream activities that can decrease the quality of the water that reaches downstream users. Transport of water is a very costly procedure, and large investments are often necessary to assure a sufficient supply of water to satisfy demand. Even if transport is not a problem, industrial use, disposal of nontreated sewerage waters, and excessive use of pesticides and insecticides in agriculture can make the water less suitable for further usage.

Finally, the availability of water per person depends not only on the presence of water resources but also on the number of people who depend on the water supply from a particular source. For example, population density in Australia is much lower than in Asia and the availability of water per person is higher in Australia, despite a substantially lower amount of available resources (Fig. 9.3).

Because of these problems in distribution and quality, not all of the discharges can be effectively utilized; the total potential resource is only

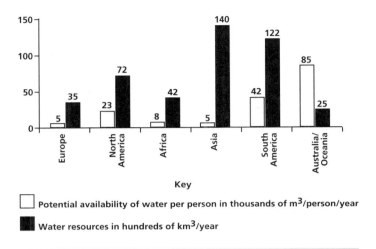

Figure 9.3 Availability of water resources per continent and per person. *Source:* Data from Shiklomanov 1996, as they appear in World Meteorological Organization, United Nations Educational Scientific and Cultural Organization (1997).

about 12,500 cubic kilometers per year and will continue to decrease if no proper action is taken to reduce contaminating activities (WMO and UNESCO 1997). Even now, although the actual global use of 4,000 cubic kilometers per year stands well below the total potential resource, the uneven distribution of resources means that many areas have to cope with shortages of water.

The Quality of Water

It is not an easy task to provide a definition of water quality. With the expansion of water use requirements and the ability to measure and interpret water characteristics, our understanding of water quality has evolved over the past century. Furthermore, the complexity of factors determining water quality and the large choice of variables used to describe the status of water bodies in quantitative terms only make the task more difficult. These variables include concentrations, specifications, physical partitions of inorganic or organic substances, and the composition of the aquatic biota present in a water body. Because of internal and external factors, the defined quality is constant neither in space nor in time. Nevertheless, the pollution of aquatic environments can be defined as indirect or direct introduction of substances or energy that may harm living resources or form hazards to human health. This pollution can also endanger aquatic activities, such as fisheries, or impair the quality of water with respect to its use in agricultural, industrial, and other economic activities. The quality of water can be described in several ways, through quantitative measurements of specified substances or by semiquantitative and qualitative descriptions based on biotic indices, visual aspect, odor, species inventories, and so forth.

With the initiation of the industrial revolution and increasing population numbers, the requirements for water in quantitative as well as qualitative aspect have increased (see also "Increasing Competition among Users of Fresh Water," below). Over time, water requirements have emerged for drinking water and personal hygiene, fisheries, agriculture, livestock, navigation for transport of goods, industrial production, cooling in power plants, hydropower generation, and recreational activities. Fortunately, agricultural irrigation and industrial cooling, which are the largest users, do not require very high quality water, at least not for more than one or two factors. On the other hand, water for drinking and sophisticated industrial applications requires very high standards of quality; fortunately, however, these uses require relatively low quantities in comparison with the larger users. In addition to these uses, water has been considered since ancient times the most suitable medium with which to clean, disperse, transport, and dispose of wastes, including domestic and industrial wastes, mine drainage waters, irrigation returns, and so forth.

Each use of water, including extraction of water and discharge of wastes, leads to specific and generally rather predictable effects on the quality of water. In addition to these intentional uses of water, several human activities have indirect and undesirable, if not devastating, effects on the aquatic environment. Examples are uncontrolled land use for urbanization or deforestation, accidental (or unauthorized) release of chemical substances, discharge of untreated wastes, or leaching of noxious liquids from solid waste deposits. Similarly, the uncontrolled and excessive use of fertilizers and pesticides has long-term effects on ground and surface water resources.

Structural interventions in the natural hydrological cycle through drainage basins and the overpumping of aquifers are usually undertaken with a beneficial objective in mind. In many development programs, however, the environmental degradation caused by the structural intervention has been so serious that reestablishing environmental circumstances would cost a lot more money than has been gained from the interventions. In some cases, the intervention in the natural cycle and its resulting influence on the quality and quantity of the water has caused complete destruction of the capacity for primary production in those regions. The most important anthropogenic effects on water quality, on a global scale, are summarized in Table 9.2. Meybeck et al. (1989) provide a fuller discussion of the sources and effects of pollutants.

Table 9.2 also gives an indication of the relative severity of deterioration found in four different types of water bodies. A similar table can be

Table 9.2 Major Freshwater Quality Issues at the Global Scale

| | Water Body | | | |
Issue	Rivers	Lakes	Reservoirs	Ground Waters
Pathogens	xxx	x[a]	x[a]	x
Suspended solids	xx	na	x	na
Decomposable organic matter[b]	xxx	x	xx	x
Eutrophication[c]	x	xx	xxx	na
Nitrate as a pollutant	x	0	0	xxx
Salinization	x	0	x	xxx
Trace elements	xx	xx	xx	xx[d]
Organic micropollutants	xxx	xx	xx	xxx[d]
Acidification	x	xx	xx	0
Modification of hydrological regimes[e]	xx	x		x

Source: Reproduced from Chapman 1996, Table 1.2, with permission of Taylor & Francis, Inc.,/Routledge, Inc., http://www.routledge-ny.com.

Note: This is an estimate for the global scale. At a regional scale these ranks may vary greatly according to the stage of economic development and land use. Radioactive and thermal wastes are not considered here. xxx, severe or global deterioration found; xx, important deterioration; x, occasional or regional deterioration; 0, rare deterioration; na, not applicable.

[a]Mostly in small and shallow water bodies.

[b]Other than resulting from aquatic primary production.

[c]Algae and macrophytes.

[d]From landfill, mine tailings.

[e]Water diversion, damming, overpumping, etc.

composed on local and regional as well as international scales to indicate the comparative consequences for different types of uses of water. As is indicated in Table 9.3, some uses of water are more sensitive to the presence of pollutants than are others. Water quality criteria, standards, and related legislation are used as the main administrative means to manage water quality to achieve user requirements. The most common requirement is for drinking water of suitable quality, and many countries base their own standards on the World Health Organization (WHO) guidelines for the quality of drinking water (WHO 1984). Requirements for quality of water should be treated on a case-by-case basis as natural water can be unsuitable for certain uses while heavily polluted water may well be suitable for uses in industrial or other applications.

Most of the worldwide degradation of water quality is a result of unbalanced activities that have disturbed the natural equilibrium. However, severe contamination can be the result of natural disasters. Flooding, earthquakes, and eruptions of volcanoes, with the mudflows that sometimes accompany them, can seriously disturb the natural equilibrium and change the concentrations of substances, biotic as well as abiotic. Some of

Table 9.3　Limits of Water Uses Due to Water Quality Degradation

Pollutant	Drinking Water	Aquatic Wildlife, Fisheries	Recreation	Irrigation	Industrial Uses	Power and Cooling	Transport
Pathogens	xx	0	xx	x	xx[a]	na	na
Suspended solids	xx	xx	xx	x	x	x[b]	xx[c]
Organic matter	xx	x	xx	+	xx[d]	x[e]	na
Algae	x[e,f]	x[g]	xx	+	xx[d]	x[e]	x[h]
Nitrate	xx	x	na	+	xx[a]	na	na
Salts[i]	xx	xx	na	xx	xx[j]	na	na
Trace elements	xx	xx	x	x	x	na	na
Organic micropollutants	xx	xx	x	x	?	na	na
Acidification	x	xx	x	?	x	x	na

Source: Reproduced from Chapman 1996, Table 1.3, with permission of Taylor & Francis, Inc./Routledge, Inc., http://www.routledge-ny.com.

Note: xx, marked impairment requiring major treatment or excluding the desired use; x, minor impairment; 0, no impairment; na, not applicable; +, degraded water quality perhaps beneficial for this specific use; ?, effects not yet fully realized.

[a]Food industries.

[b]Abrasion.

[c]Sediment settling in channels.

[d]Electronic industries.

[e]Filter clogging.

[f]Odor, taste.

[g]In fish ponds higher algal biomass can be accepted.

[h]Development of water hyacinth (*Eichhomia crassipes*).

[i]Also includes boron, fluoride, etc.

[j]Ca, Fe, Mn in textile industries, etc.

these natural disasters can be caused in part or aggravated by human activities. Mudflows can be more intense as a result of increasing erosion, which in turn can be the result of deforestation. Depending on the intensity and the spatial scale of a disaster, the restoration of damages to water quality may take many years. Eruptions of volcanoes and subsequent mudflows, for example, can have profound and lasting effects on downstream water quality.

How do pollutants reach aquatic environments? Unfortunately, contaminating substances can be released into the environment as gases, as liquids, dissolved in liquids, and in solid phase. Through the atmosphere and the soil, these contaminating compounds ultimately reach the aquatic environment. Pollution is the result of the release of contaminants by point sources and by diffuse sources. There is no clear demarcation between these two forms of pollution, as diffuse sources on large scales may be an assembly of numerous point sources. An important difference between a point and a diffuse source is that a point source in principle can be con-

Table 9.4   Anthropogenic Sources of Pollutants in the Aquatic Environment

Source	Bacteria	Nutrients	Trace Elements	Pesticides/ Herbicides	Industrial Organic Micropollutants	Oils and Greases
Atmosphere		x	xxxG	xxxG	xxxG	
Point sources						
Sewage	xxx	xxx	xxx	x	xxx	
Industrial effluents		x	xxxG		xxxG	xx
Diffuse sources						
Agriculture	xx	xxx	x	xxxG		
Dredging		x	xxx	xx	xxx	x
Navigation and harbors	x	x	xx		x	xxx
Mixed sources						
Urban runoff and waste disposal	xx	xx	xxx	xx	xx	xx
Industrial waste disposal sites		x	xxx	x	xxx	x

Source: Reproduced from Chapman 1996, Table 1.4, with permission of Taylor & Francis, Inc./Routledge, Inc., http://www.routledge-ny.com.

Note: x, low local significance; xx, moderate local/regional significance; xxx, high local/regional significance; G, globally significant.

trolled by either collecting or treating the contaminating substances. Some diffuse sources or, more specifically, diffuse sources of which all point sources can be identified, can also be controlled; unfortunately, this is not always a realistic possibility. The major point sources of pollution to fresh water originate from the collection and discharge of domestic wastewater, industrial wastes, or certain agricultural activities, such as animal husbandry. Most other agricultural activities, such as the spraying of pesticides or the application of fertilizers, are considered to be diffuse sources. The atmospheric fallout of pollutants also leads to diffuse pollution of the aquatic environment. Table 9.4 provides a comparative indication of the significance of the atmosphere, point sources, diffuse sources, and mixed sources in the pollution of aquatic environments.

Increasing Competition among Users of Fresh Water

In the first phase of the development of water resources, when the availability of the resource is far greater than the existing demand, small structures (e.g., single wells, slight modifications of river beds, small barrages) with only local effects are sufficient to meet the demands of society. A second phase of the development of water resources could be characterized by the evolution of the various branches of water management—flood control, drainage and irrigation, water supply and sanitation engineering, river training, navigation and utilization of hydropower, water quality control, and protection of the environment—when the construction of large systems is required to satisfy the increasing demands. The integration of the systems and their multipurpose utilization becomes necessary in what could be defined as a third phase, which is to eliminate undesirable interactions between the systems and to minimize investment and operational costs.

With the onset of the industrial revolution and with the increasing number of people inhabiting our planet, the range of requirements for water has increased together with the greater demands for higher-quality water. Water is a limited resource; hence, the production that depends on water has to deal more and more with constraints. According to statistics from the United Nations Development Programme (UNDP), the world population grew at a rate of 1.74 percent per year from 1970 to 1995. Between 1995 and 2025, the annual growth rate is projected to be 1.23 percent. The total population was estimated to be 5.6 billion in 1995, and, with the projected growth rate, the earth will be inhabited by close to 7.2 billion people by the year 2015; at the current growth rate, the population of 1995 would be expected to double to over 11 billion by the year 2046 (Table 9.5).

It is inevitable that this rapid increase in population will increase the pressure on natural resources. Between 1900 and 1995, the increase in water demand exceeded the rate of population growth. The demand has increased six to seven times, or roughly twice the rate of population growth over that same period, and it is not expected that this will change in the future. So, in contrast to the decreasing availability of water, the demand will continue to increase (WMO and UNESCO 1997).

Of the actual estimated yearly use of 4,000 cubic kilometers, more than 80 percent is directed to agriculture, mainly to irrigation. But significant quantities of water are used for industrial purposes, for the generation of electricity, and for other domestic purposes. The available information on the use of water is more scarce and in general less reliable than are data on the water resources themselves because of the lack of measurements in numerous countries around the world. Nevertheless, the largest users that can be consolidated as groups are agriculture, industry, and municipalities, and the water demand of these three groups increased during the twentieth century from about 600 cubic kilometers to about 4,000 cubic kilometers per year (Fig. 9.4).

Human Intervention in the Terrestrial Water Cycle

Although a general distinction may be made between the natural and modified forms of the water regime, it should be recognized that natural

Table 9.5 Population Trends for Years 1970, 1995, and 2015

Region	Estimated Population (millions)			Annual Population Growth Rate (%)		Population Doubling Date (at current growth rate),
	1970	1995	2015	1970–95	1995–2015	1995
All developing countries	2,616.06	4,394.061	5,892.131	2.1	1.48	2037
Least developed countries	285.661	542.486	873.726	2.6	2.41	2022
Industrial countries	1,043.536	1,233.064	1,294.742	0.67	0.24	2223
World	3,659.596	5,627.125	7,186.873	1.74	1.23	2046

Source: Data from *United Nations Development Programme 1999a.*

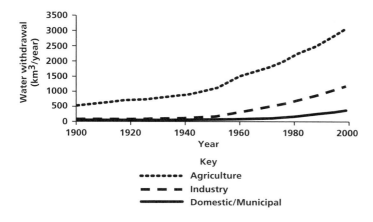

Figure 9.4 Global water withdrawal by sector, 1900–2000. *Source:* Data from Shiklomanov 1996, as they appear in World Meteorological Organization, United Nations Educational Scientific and Cultural Organization (1997).

conditions hardly exist on the continents because any utilization of the land surface by people influences the elements of the water balance. It is obvious that the size of such changes depends on the level of development in the region in question. Human activities in the landscape are driven by societal demands for life support as influenced by expectations in society, by population growth increasing the number of individuals expecting life support, and by growth aspirations within the economy, in particular the industrial sectors (see Fig. 9.4). In studying the influence of human activities on the quantification of hydrological variables, the most serious problems to be solved are those related to the modification of land use because the associated hydrological changes are not directly measurable and their development usually requires a long period of transition. Therefore, an analysis of the hydrological records may lead to erroneous conclusions if the data are not correctly related to such changing conditions. These aspects are especially important in developing countries where both water and land management are in the second phase of development, with rapid evolution of the various branches of water management and construction of large systems to satisfy increasing demands.

In developing a policy for properly coping with anthropogenic interventions, four main functions of water in the water cycle must be taken into account (Lundqvist and Falkenmark 1997):

1. *health function* as manifested in the fundamental importance of safe drinking water as a basic precondition for socioeconomic development;

2. *habitat function* of water bodies, hosting aquatic ecosystems, which are easily disturbed when the water bodies become polluted;
3. *carrier function* for dissolved and suspended material picked up and carried by water moving along its pathway through atmosphere, landscape, and water courses—this function plays a central role in the process of land degradation (leaching of nutrients, erosion, and sedimentation); and
4. *production function* in economic development: (a) biomass production, operated by a flow of "green" water entering through roots and leaving through foliage (in the absence of "green" water, photosynthesis stops altogether and vegetation wilts); and (b) societal production in households and industry, based on "blue" water, withdrawn while passing through the landscape and delivered to cities and industries through water supply systems.

Different functions of water are relevant for different sectors of society, each driven by its own political driving forces. The fact that the integrity of the water cycle links all of these sectors together is a tremendous challenge for interdisciplinary, interprofessional, and intersectoral communication. The final objective in developing a water policy and strategy must be to master the whole system.

Agricultural Activities and Forest Management

Agricultural Activities

Intensive agriculture requires the maintenance of optimum conditions in the system composed of soil, water, and plants. The soil should have a well-defined structure that can store a sufficient amount of water and at the same time have an efficient transport capacity to carry away excess water from the pores. The nutrients needed for the development of the plants should be present in the soil, but the latter must not contain materials toxic to the plants or elements decreasing the stability of the soil structure. Water, which passes through the transpiration pathway soil-roots-stems-leaves, is the main carrier of dissolved nutrients. Lack of water causes the wilting of vegetation, but serious damage may also be caused by waterlogging. Intensive agriculture also requires the selection of high-yielding varieties and the protection of plants against pests, insects, and weeds.

It is inevitable that agricultural production must be increased to ensure a food supply for the prospective world population. This objective can be achieved by increasing the productivity on land already cultivated and by expanding the area of arable land. Irrigation plays a paramount role in both activities. Worthington estimated in 1977 that about 13 percent of the arable lands of the world were irrigated, using about 1,400 cubic kilometers of water per year. In the developing countries, the yearly increase of

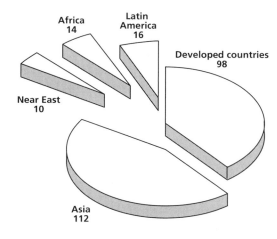

Africa
14

Latin
America
16

Developed countries
98

Near East
10

Asia
112

Figure 9.5 Irrigated area (in million hectares) by region of the world, 1990. *Source:* Data from *Food and Agriculture Organization 1996.*

productivity on irrigated land is about 2.9 percent, while it is only 0.7 percent in nonirrigated areas. According to the FAO (1995), a total area of 255,446,000 hectares was under irrigation in 1990 (Fig. 9.5). Irrigated land is more than twice as productive as rain-fed cropland. Only 16 percent of the world's croplands are irrigated, but those lands yield some 36 percent of the global harvest. Apart from this very positive primary effect, it is necessary to recognize, however, that the secondary effects of irrigation may be undesired, harmful, and irreversible.

In the developing countries, irrigation increases yields for most crops by 100–400 percent. Irrigation also allows farmers to reap the economic benefits of growing higher-value cash crops. Half or even two-thirds of future gains in crop production are expected to come from irrigated land. In the developing world, where about 20 percent of arable land is irrigated, the prevalence of irrigation varies widely within and among countries and crops. A World Bank/UNDP study estimates that irrigation could be extended over an additional 110 million hectares in developing countries, producing enough additional grain to feed 1.5–2 billion people (FAO 1996).

To determine the influence of irrigation on the hydrological regime, one should consider both the primary and the secondary effects involved in modifying the water regime. These changes can be divided into three groups according to the water horizons where they develop:

1. changes in interactions between atmosphere and land and water surfaces;
2. quantitative and qualitative modifications of surface runoff; and
3. changes in the regimes of soil water and ground water.

The flux of actual evapotranspiration is raised considerably by irrigation because its primary purpose is to maintain a relatively high water content in the root zone and prevent plants from being subjected to water stress. Hence, there is always water available for vaporization from the surface, and actual evapotranspiration approaches the potential value in each season of the year. Due to the higher evapotranspiration, the vapor content stored in the atmosphere increases. However, it was observed that, in and around a large area in Texas, no significant differences in the amount or the pattern of precipitation existed between areas under or without irrigation (Schickendanz and Ackermann 1977).

The effects of irrigation on surface runoff can be observed both over the complete river basin and in the rivers. The amount and the rate of surface runoff increase because of the higher moisture content in the upper soil layers. The undesired consequences of the higher runoff rates are the increased potential for erosion, which may cause the degradation of soils by carrying away organic materials, and the increase of sediment transport in rivers.

The flow of rivers is directly modified by irrigation. Part of this change is planned a priori when reservoirs are constructed to augment low river flows and to raise the amount of water available for irrigation in dry periods. However, there is generally a considerable decrease in river discharge due to the high water consumption of irrigated areas. Of all types of water use, the difference between water intake and effluent discharge is the highest with irrigation. The difference may reach 70–80 percent of the intake for irrigation, whereas the value is usually less than 10 percent for industrial and community water supplies.

Other important changes in rivers are caused by the possible increase of sediment load and decrease in the capacity of sediment transport due to the increase of discharge. The result of these effects is the rapid increase of siltation and the aggradation of river beds. Even more serious changes in water quality are caused by the high salt content of effluents from irrigated fields.

Forest Management

The type of vegetative cover influences considerably the development of the response of the catchment/river basin to precipitation. The largest differences occur perhaps between forests and cultivated lands, and hence any change due to forest management (deforestation, thinning, clear cutting, reforestation, introduction of exotic species, etc.) may have considerable effects on local hydrological phenomena and water resources.

The first difference is caused by the greater interception by forest than by grasslands or cultivated crops. The surface litter in a forest protects the soil against the splashing effects of raindrops, and the surface mulch of decomposing vegetation increases infiltration relative to surface runoff. Evapotranspiration is also increased, not only because of the direct reevaporation of intercepted water from the canopy, but also because of the higher consumptive use of trees and the greater amount of water available in the root zone, which is much deeper than that of crops and pastures. As a consequence of the influence of forest cover on the various elements of water balance, the general regime of surface runoff from forested catchments differs considerably from the catchment response of agricultural areas. In humid areas, the most striking differences are much less flood discharge and greater base flow.

In areas of low rainfall, the situation with regard to the effect of forest on base flow is quite different. In Australia, for example, the general result of replacing deep-rooted native forests and woodland with pastures or other shallow-rooted vegetation has been to cause an increase in groundwater recharge, often resulting in increased saline base flow in streams and salinization of lower land areas (Peck 1983).

In deforested areas, the loss of the protection for the soil provided by the forest means that the increasing intensity of surface runoff causes an extremely high increase of erosion. For example, in Brazil, 5–6 tons of total solid discharge per square kilometer was measured from very steep deforested areas during heavy rainfalls, while in the control area (where the forest remained intact) the value was 0–2 tons per square kilometer (Bordas and Canali 1980).

It is very difficult to extend observations made in small experimental catchments because the spatial scale strongly influences hydrological variables. At the outlet of a large basin, the discharge data provide integrated information on the behavior of the catchment. The structural variation of the surface (the proportions of forest, cultivated area, and grassland and the variability of slope conditions) eliminates the extreme effects of special local conditions. However, the two strongest influences of forest cover, which are the decrease of both flood volume and soil erosion, can be verified from the data of large rivers.

The data that have been analyzed indicate some aspects of the relation between forests and the local water balance. In general, the water regime of forested catchments is more favorable from the point of view of water resources development (decrease of flooding and erosion, increase of base flow in humid areas, only a slight decrease of average runoff) than that in cultivated or grassland areas. It is necessary, therefore, to predict all of the expected effects of forest management and particularly those of planned changes in land use, such as deforestation. On the other hand, reforestation can be used in some situations to improve environmental conditions and produce a more favorable water regime.

Urban Activities

For many centuries people have been living together in cities. Often, their choice of a certain location was based on the presence of water: for navigation and trade, for domestic and craft use, for food production (fish and crops), or even for defense of the inhabitants against enemies. The presence of urban areas affects the hydrology of local and adjacent regions in several ways, such as groundwater withdrawal for water supply, surface and groundwater pollution, increased precipitation downwind of cities, and increased flood flow due to a decrease of the area where water can infiltrate.

Various facets of the urban hydrological cycle are being studied in urban water balances and budgets. These studies must also include the influence of engineering on flows of water due to systems for water supply and wastewater disposal. This complicates the approach to hydrological processes, compared to the processes that are followed in rural hydrology.

Although urban hydrology, with numerous artificial features, may seem superficially to be simpler than that of natural areas, a closer comparison of the hydrological systems leads to an opposite conclusion. The urban hydrological system comprises all the basic elements of the natural system plus many features due to human activity, some of which serve to manipulate the urban water system for society's advantage.

Urbanization, which is characterized by a large concentration of people in a relatively small area, is now recognized as an inevitable historical process. In 1965, about one-third of the world's population lived in cities. In the 1998 U.N. Human Development Report, it is estimated that, in 1970, 36.8 percent and, in 1995, 45.4 percent of the total world population lived in urban areas; for 2015, the urban population is expected to be 54.6 percent of the world's total (Table 9.6). If the U.N. projections become true, by the year 2015 a little under 7.2 billion people will inhabit the earth (see Table 9.5), meaning that 3.9 billion people will be living in cities. This increase is especially due to the growth of middle large cities. Today, only about 15 percent of urban dwellers live in cities with more than 5 million people, while over 60 percent live in towns and cities with populations of 1–5 million. At the same time, the relative concentration of the poor in towns and cities is increasing. U.N. estimates for the year 2000 suggest that half the world's poorest people, or some 420 million, are living in urban settlements.

Situations in, especially, the major cities in the humid tropics are already unstable in many perspectives (e.g., health, pollution, water supply), and it is inevitable that the even higher population growth in these urban areas will increase the pressure on their natural resources. In the absence of proper solutions, this will encourage unsustainable development poli-

Table 9.6 Urbanization Trends for Years 1970, 1995, and 2015

Region	Urban Population (as % of total)			Urban Population Annual Growth Rate (%)		Population in Cities of More than 750,000 in 1995	
	1970	1995	2015	1970–95	1995–2015	% of total population	% of urban population
All developing countries	24.7	37.4	49.3	3.8	2.885	15.6	41.2
Least developed countries	12.7	22.9	34.9	5.05	4.6	9.5	41.4
Industrial countries	67.1	73.7	78.7	1.05	0.57	29.8	40
World	36.8	45.4	54.6	2.59	2.17	18.7	40.8

Source: Data from *United Nations Development Programme 1999b*.

cies and practices concerning water management in the developing nations in the region (Donoso and de Groot 1996).

High concentrations of population in urban areas have several implications for water resources. The urban population acts as a driving force that changes the landscape and hydrological cycle of the area, and it further requires certain services, such as water supply, flood protection and drainage, and water-based recreation. Because of the large populations of urban areas, all urban water resource projects have a potential to benefit a large population, but failures in planning or design will, on the other hand, lead to large damages.

Development in some cities of the world is very dynamic. Some areas are expected to double their population in periods as short as 20 years (see Table 6.4). This dynamic character leads to changing hydrological data and therefore makes it difficult to analyze information needed to conduct assessments of water resources.

Lindh (1983) stated that 80 percent of the world's available water resources was used to irrigate food crops, much of which is consumed in cities. In this sense, cities make use of the water resources of quite distant areas. However, they have a more direct effect on the hydrological conditions and water resources of adjacent areas.

Concentrations of people in urban areas lead to dramatic changes in land surface cover. In particular, the permeability of the surface is dramatically reduced because of the use of impervious cover and compaction of the top layer of the soil. Extensive impervious surfaces are typical of urban centers throughout the world. Such surfaces comprise roofs, roads and streets, sidewalks, and parking areas. The extent of the impervious area in an urban catchment is usually described by the *catchment imperviousness,* defined as the ratio of the impervious area to the total catchment area.

The imperviousness ratio in an urbanized area is characteristic for the amount of rain that has to be drained as surface water. The higher the imperviousness ratio, the less water will infiltrate into the soil and recharge the ground water. Because of the high rate of human interventions, urban areas are characterized by a special hydrological system that is more complex that that of natural areas. The additional engineered features that must be included in the analysis of urban hydrological processes are, among others, water import from other areas for water supply, withdrawal of ground or surface water, drainage, and wastewater systems.

Water quality in urban streams and rivers is strongly affected by urban areas. Such effects include degradation of water quality due to the discharges of wastewater and municipal effluents, combined sewer overflows, and storm water. The municipal effluents include treated or untreated discharges of domestic, commercial, and industrial wastewater. Their composition varies considerably, depending on the source and the level of the treatment. Generally, untreated or poorly treated effluents may be strong sources of pathogens, nutrients, solids, and some toxic contaminants. Where solid waste is not collected or is disposed of without proper protection against leaking, such wastes may find their way into water bodies and contribute to their pollution.

The Overexploitation of Groundwater Resources

Ground water has always been considered to be a readily available source of water for domestic, agricultural, and industrial use. In many parts of the world, extracted ground water has made a major contribution to the improvement of social and economic circumstances of human beings. Projects of various types and scales have been developed and managed in response to the growing demand for water by communities and industries. In spite of bringing many benefits, the increase in demand for ground water is leading to its overexploitation in many areas, resulting in a permanent depletion of the aquifer system and associated environmental consequences, such as subsidence of the land and deterioration of water quality. Moreover, with changes in land use and a vast increase in the quantities and types of industrial, agricultural, and domestic effluent entering the hydrological cycle, water quality is gradually declining because of surface and subsurface pollution.

The world is becoming more concerned with sustainable and environmentally sound development. The goal of environmentally sound and sustainable development of water resources is to develop and manage them in such a way that the resource base is maintained and enhanced over the long term. The development of ground water typically begins with a few pumping wells and initially, in many cases, the management of ground water is geared to facilitate its usage and development. As development progresses, with more drilled wells scattered over the basin, issues such as overexploitation, equitable sharing of water, and degradation of water quality become apparent in many basins. Sustainable development of ground water depends on the understanding of processes in the aquifer system, quantitative and qualitative monitoring of the resource, and interaction with land and surface water development. The following key principles reflect different aspects of concern in the evolution of sustainability in groundwater development:

1. long-term conservation of groundwater resources;
2. protection of groundwater quality from significant degradation; and
3. consideration of environmental impacts of groundwater development.

Long-term sustainable development of groundwater resources implies that the rate of extraction should be equal to or less than the rate of recharge. When the rate of extraction is higher than the rate of recharge, a continual lowering of water level is expected. Such a continual lowering of the water table steadily increases the pumping cost and then, at a certain level, it is no longer economical to pump water for many uses, such as agricultural production.

Also, the qualitative aspect of resource availability for sustainable use is of critical importance and is closely linked to the quantitative aspect. The quality of ground water in aquifers can be affected by natural and human activities, and the extent to which the quality is affected varies with the hydrogeological and climatic settings (Todd 1980; Fetter 1993). A deterioration of groundwater quality is detected only when there is a characteristic odor, color, or taste in consumed water or when the presence of a pollutant has an immediate effect on users of water. With the presence of various sources of contamination and with the complexity of the hydrogeological environment and transport processes, sometimes it is difficult to establish a simple cause-and-effect relationship. Moreover, the consequences of the deterioration of groundwater quality could be widespread and difficult to control.

Pressure from Climatic Change

Climatic change will probably alter regional precipitation and evaporation patterns (Intergovernmental Panel on Climate Change [IPCC] 1996). According to climate models, rising levels of greenhouse gases (see Chapter 7) are likely to raise the global average surface temperature by 1–3.5°C over the next 100 years (United Nations Environment Programme [UNEP] and WMO 1997). A more indirect result of climatic change, but not with a less dramatic effect on the availability of water, is the rising of the sea level, influencing negatively the availability of fresh water, especially in coastal regions.

Precipitation and Evapotranspiration

In general, higher temperatures should increase evaporation and therefore precipitation. Climate models indicate that a doubling of atmospheric concentrations of carbon dioxide (CO_2) would increase global precipitation by about 5 percent. Nevertheless, the results from different General Circulation Models on a CO_2-doubling scenario indicate that the influence of CO_2 doubling on precipitation is still very unclear.

Although changes in temperature and precipitation would clearly have profound effects on the water cycle, the current generation of climate models still cannot make regional forecasts. It is likely that precipitation would increase in some areas and decline in others. Even in areas where precipitation increases, higher evaporation rates may still lead to reduced runoff. In areas where climatic change causes reduced precipitation, freshwater storage reserves, primarily in the form of ground water, will steadily shrink. Areas where more precipitation is not matched by increased evaporation will experience floods and higher lake and river levels. Some projections indicate an increase in the frequency of extreme events, such as droughts and floods, which would undermine the reliability of many critical sources.

Warming would also tend to reduce the accumulation of winter snow in mountains and other cold regions. Water resources would become even more vulnerable than they are now. Diminished accumulation of snow in winter would reduce spring runoff, which can be vital in replenishing lakes and rivers; a 10 percent decline in precipitation and a 1–2°C rise in temperature could reduce runoff by 40–70 percent in drier basins (UNEP and WMO 1997).

Worsening droughts combined with overexploitation of water resources would cause salt to leach from the soil, thus raising the salinity of the unsaturated zone (the layer between the ground and the underlying water table). In coastal zones, a lowered water table would also draw salt water from the sea into the fresh ground water. At the same time, higher levels of CO_2 in the atmosphere are expected to improve the efficiency of photosynthesis in plants, which could in turn cause more rapid evapotranspiration.

Together, these various effects would have extremely negative consequences for river watersheds, lake levels, aquifers, and other sources of fresh water, leading to a reduced water supply that would place greater stress on people, agriculture, and the environment. Leaching and intrusions of salt water into freshwater stores would make ground water unfit for household and agricultural use. (This has already occurred in many parts of the world.) Reduced precipitation and increased evaporation would damage croplands, forests, marshes, and other ecosystems. Falling water levels would also require major adjustments by urban settlements, particularly those located on the shores of rivers and lakes. The costs of such adjustments may prove too high for many poorer nations. Conflicts over water resources are likely to worsen in fertile basins such as the Nile and Mekong and in regions with rapid population growth and increasing problems with drought. Because clean fresh water is so vital to health, some developing countries could face reduced health standards and worsening epidemics. Marginal areas such as the Sahel would be at particular risk.

Improved water management is needed to minimize the consequences of climatic change. Around the world, growing populations and rising living standards are increasing humanity's consumption of water. If climatic change does reduce precipitation or increase the frequency of drought in some areas, catastrophic situations are certain to arise. To respond to this threat, we need to upgrade our existing infrastructure for water storage. In particular, more facilities are needed for storing water during the spring to help alleviate problems during summer droughts. Other important steps would be to sensitize people to the problems of water wastage and to introduce policies or taxes that would cut waste and constrain demand. Finally, a survey of the vulnerability and resiliency of water basins is needed to support water use planning in the event that climatic change has its predicted effects.

The Rise in Sea Level

The global mean sea level may have already risen by around 15 centimeters during the past century. According to several studies, the sea has been rising at the rate of 1–2 millimeters per year over the past 100 years. Measuring past and current changes in sea level, however, is extremely difficult. There are many potential sources of error and systematic biases, such as the uneven geographic distribution of measuring sites and the effect of the land itself as it rises and subsides.

Climatic change is expected to cause a further rise of about 20 centimeters by the year 2030. Forecasts of a rising sea level are based on the results of climate models, which indicate that the earth's average surface temperature may increase by 1–3.5°C over the twenty-first century. This warming would cause the sea to rise in two ways: through thermal expansion of ocean water and through shrinking of ice caps and mountain glaciers. According to the IPCC, if no specific measures are taken to abate greenhouse gas emissions, these two factors are likely to cause the sea to rise by about 15–95 centimeters from current levels by the year 2100. This expected rate of change is significantly faster than that experienced over the last 100 years (IPCC 1996).

Forecasting a rise in sea level involves many uncertainties. While most scientists believe that anthropogenic emissions of greenhouse gases are changing the climate, they are less sure about the details, and particularly the speed, of this change. Global warming is the main potential consequence of greenhouse gas emissions, but other aspects of the climate besides temperature may also change. For example, some studies suggest that changes in precipitation will increase the accumulation of snow in Antarctica, which may help to moderate the net rise in sea level. Another complication is that a rise in sea level is not uniform all over the globe because of the effects of the earth's rotation, local coastline variations, changes in major ocean currents, regional land subsidence and emergence, and differences in tidal patterns and seawater density.

Higher sea levels threaten low-lying coastal areas and small islands. The figures for sea level rise given by the IPCC may seem modest. However, a one-meter rise would put millions of people and hundreds of thousands of square kilometers of land at risk (Table 9.7). The most vulnerable land is unprotected, densely populated, and economically productive

Table 9.7   Land and Population at Risk from a 1-Meter Rise in Sea Level

Country	Land at Risk		Population at Risk	
	sq km	%	millions	%
Argentina	>3,430–3,492	>0.1	—[a]	—[a]
Bangladesh	25,000	17.5	13	11
China	125,000[b]	1.3	72.0[b]	6.5
Egypt	4,200–5,250	12–15[c]	6.0	10.7
Malaysia	7,000	2.1	—[a]	—[a]
Nigeria	18,398–18,803	2.0	3.2	3.6
Senegal	6,042–6,073	3.1	0.1–0.2	1.4–2.3
Uruguay	94	<0.1	0.01	0.4
Venezuela	5,686–5,730	0.6	0.06	0.3
Total	194,852–196,498		94.4–94.5	

Source: Reproduced from Nicholls and Leatherman 1995, Table 5, with permission of Cambridge University Press.

[a]Not available.

[b]In China, land and population at risk include increased flooding due to sea-level rise.

[c]Percentage of arable land.

coastal regions of countries with poor financial and technological resources for responding to a rise in sea level. Some megacities may be particularly at risk (see Table 6.4). For example, in China, the entire city of
Shanghai could be inundated by a one-meter rise in sea level (Wang et al.
1995), and a similar rise combined with a severe storm surge would sub-
merge the city of Tianjin (Han et al. 1995). A rise in sea level would create
irreversible problems for low-lying island nations such as the Maldives and
the Pacific atolls. Elsewhere, tourist beaches, cultural and historical sites,
fishing centers, and other areas of special value are potentially at risk. The
costs of protecting this land from a rise in sea level and preventing constant erosion are enormous. Additional investments would also be needed
to adapt sewage systems and other coastal infrastructure. On the other
hand, some localities, such as shallow ports, would benefit from a higher
sea level.

Ground water in some coastal regions would become more saline. Rising seas would threaten the viability of freshwater aquifers and other
sources of fresh ground water. Communities might have to pump out less
water to prevent aquifers from being refilled with seawater. Coastal farming would face the triple threat of inundation, freshwater shortages, and
salt damage.

The flows of estuaries, coastal rivers, and low-lying irrigation systems
would be affected, and tidal wetlands and mangrove forests would face ero-

sion and increased salinity. Wetlands not only help to control floods, but also are critical to biodiversity and to the life cycles of many species. Flat river deltas, which are often agriculturally productive, would be at risk. Among the most vulnerable are the Amazon, Ganges, Indus, Mekong, Mississippi, Niger, Nile, Po, and Yangtze.

The damage caused by floods, storms, and tropical cyclones might worsen. Major harbor areas would experience more frequent flooding during extreme high tides and, in particular, during storm surges. Countries already prone to devastating floods, such as low-lying Bangladesh, would be the most affected. Warmer water and a resulting increase in humidity over the oceans might even encourage tropical cyclones, and changing wave patterns could produce more swells and tidal waves in certain regions.

A rise in sea level poses a threat especially to low-lying areas, which seem to be very vulnerable and sensitive to climatic change. On small islands in developing stages, climatic change may invoke serious disturbances in socioeconomic development. Summarizing, freshwater resources of coastal areas, and small islands in particular, are endangered by climatic change due to the following five mechanisms:

1. a rise in sea level endangers fresh groundwater sources through reduction in island area and intrusion of saline water;
2. a change in the amount of rainfall and the duration of its seasonal distribution may affect the water supplies of small islands that use rainwater-harvesting systems and groundwater recharge;
3. higher sea surface temperatures may trigger greater frequency and severity of storms caused by tropical cyclones, hurricanes, and typhoons, thus increasing the risk of flooding and storm damage, which would be exaggerated by higher sea levels;
4. demand for water (for public supply and agriculture) may grow as temperature rises with the increase of levels of greenhouse gases, adding to a growth in demand expected from a rise in population; and
5. water quality may deteriorate because of a decreased dilution of wastes discharged to surface water, ground water, and the marine environment.

Where these changes occur together, the effects of climatic change will be most severe. Scientists increasingly recognize the need for a study of the consequences of climatic change for the freshwater resources of small islands. Such a study should develop a method for identifying and estimating effects, apply the method, generalize results, outline a response strategy, and indicate the needs in systems for collecting data and monitoring outcomes.

Conclusion

The hydrological cycle has been taken for granted for too long. International conferences on the environment and sustainable development have not given to water management the attention that it deserves. However, during the last decade, there has been a growing understanding of the existing problems and the challenges we face to ensure that all socioeconomic activities will not be limited by either the quantity or quality of available water. More attention is being given to decreasing contamination from numerous sources and to mitigating or preventing possible effects of changes in land use on water resources. Water resources are more often taken into account in a multicriteria decision analysis (see Chapter 5). Nevertheless, we still have a long way to go.

With the ongoing growth of population in many countries, especially in areas that already have to cope with shortages of potable water, and with the continuing increase of competition among different uses of water, more attention should be given to minimizing the exploitation of the resource rather than inventing more ingenious technologies for deeper depletion of the available sources. The availability of water is influenced by many different anthropogenic and natural elements, so the problems we face can be solved only by including as many of these elements as possible. An interdisciplinary management should incorporate contingency plans to mitigate the effects of natural events that cannot be controlled, such as floods, eruptions of volcanoes, or larger-scale climatic and oceanographic phenomena, among which the El Niño/Southern Oscillation is the best understood (see Chapter 8).

Finally, we have to face the possible changing climate with all its negative and positive results. An increasing temperature will increase evapotranspiration and probably intensify the hydrological cycle. More rain could easily become too much rain, but areas that always had to face droughts might receive some more precipitation, enabling them to increase their agricultural or industrial production. We must study more intensively all of these effects on water as a natural resource, never losing sight of the final objective of the research, which should be the use of scientific knowledge in the process of multidisciplinary decision making.

SUGGESTED STUDY PROJECTS

Suggested study projects provide a set of options for individual or team projects that will enhance interactivity and communication among course participants (see Appendix A). The Resource Center (see Appendix B) and references in all of the chapters provide starting points for inquiries. The process of finding and evaluating sources of information should be based on the principles of information literacy applied to the Internet environment (see Appendix A).

The combined objective of the following study projects is a deepened understanding of the hydrological cycle, its interrelationship with climatic changes, and its effects on sustainability.

PROJECT 1: Anthropogenic Effects on the Hydrological Cycle

The objective of this project is a deepened understanding of the hydrological cycle.

Task 1. Describe the components of the hydrological cycle.

Task 2. Identify anthropogenic activities affecting the hydrological cycle.

Task 3. Describe one or more locations where there is increasing competition by users of fresh water.

PROJECT 2: The Effect of Global Climate Change on Water Issues

The objective of this project is a deepened understanding of the possible effects of global climate change on water issues.

Task 1. Describe the vulnerability of one or more regions to extremes in the hydrological cycle.

Task 2. Describe the vulnerability of one or more regions to a rise in sea level.

PROJECT 3: The Growing Importance of Water Resources Management in Sustainable Development

The objective of this project is to develop, practice, and refine skills for incorporating water resources management into planning for sustainable development.

Task 1. Identify a region with a sustainable development initiative. Describe the initiative.

Task 2. Determine how water resources management practices are incorporated into the initiative.

Task 3. Suggest solutions to observed problems for water resources management.

Task 4. Summarize your findings.

Acknowledgment

The author acknowledges the contribution of Jonathan Patz for draft text and references on the effects of sea level rise.

References

Bordas MP, Canali GE. 1980. The influence of land use and topography on the hydrological and sedimentological behavior of basins in the basalt region of South Brazil. In *Symposium on the Influence of Man on the Hydrological Regime with Special Reference* to Representative and Experimental Basins, Helsinki. IAHS Pub. No. 130. IAHS Press, Wallingford, U.K., pp. 55–60.

Chapman D, ed. 1996. *Water Quality Assessments: A Guide to the Use of Biota, Sediments, and Water in Environmental Monitoring,* 2d ed. Published on behalf of UNESCO, WHO, UNEP. E & FN Spon, London.

Donoso MC, de Groot NJPM. 1996. CATHALAC: A search for better understanding the humid tropics of Latin America and the Caribbean. In *Hydrology in the Humid Tropic Environment* (Johnson AI, Fernandez-Jauregui C, eds.). IAHS Publ. No. 253. IAHS Press, Wallingford, U.K., pp. 213–21.

Falkenmark M, Chapman T, eds. 1989. *Comparative Hydrology: An Ecological Approach to Land and Water Resources.* UNESCO, Paris.

Fetter CW. 1993. *Contaminant Hydrogeology.* Macmillan, New York.

Food and Agriculture Organization. 1995. *Planning for Sustainable Use of Land Resources: Towards a New Approach.* Background paper to FAO's Task Managership for Chap. 10 of U.N. Conference on Environment and Development's Agenda 21. FAO, Rome.

———. 1996. *World Food Summit Fact Sheets.* FAO, Rome.

Han M, Wu L, Hou J, Liu C, Zhao G, Zhang Z. 1995. Sea-level rise and the North China Coastal Plain: A preliminary assessment. *J Coastal Res* 14:132–50.

Intergovernmental Panel on Climate Change. 1996. *Climate Change 1995: Impacts, Adaptations, and Mitigation of Climate Change: Scientific Technical Analyses. Contribution of Working Group II to the Second Assessment Report for the Intergovernmental Panel on Climate Change* (Watson RT, Zinyowera MC, Moss RH, eds.). Cambridge University Press, Cambridge.

Lindh G. 1983. *Water and the City.* UNESCO, Paris.

Lundqvist J, Falkenmark M. 1997. World freshwater problem—call for a new realism. In *Comprehensive Assessment of the Freshwater Resources of the World.* WMO, Geneva.

Mackenzie FT. 1998. *Our Changing Planet: An Introduction to Earth System Science and Global Environmental Change,* 2d ed. Prentice Hall, Upper Saddle River, N.J.

Meybeck M, Chapman D, Helmer R, eds. 1989. *Global Freshwater Quality: A First Assessment.* Blackwell Reference, Oxford, U.K.

Nicholls RJ, Leatherman SP. 1995. Global sea-level rise. In *As Climate Changes: International Impacts and Implications* (Strzepek KM, Smith JB, eds.). Cambridge University Press, Cambridge, Chap. 4.

Peck AJ. 1983. Response of groundwaters to clearing in Western Australia. In *Australian Water Resources Council Conference Series No. 8: International Conference on Groundwater and Man.* Australian Government Publishing Service, Canberra, Vol. 2, pp. 327–35.

Schickendanz K, Ackermann WC. 1977. Influences of irrigation on precipitation in semi-arid climates. In *Arid Land Irrigation in Developing Countries: Environmental Problems and Effects. International Symposium, February 16–21, 1976, Alexandria, Egypt* (Worthington EB, ed.). Pergamon University Press, Oxford, pp. 185–96.

Shiklomanov IA. 1996. *Assessment of Availability of Water Resources in the World.* State Hydrological Institute, St. Petersburg, Russia.

Todd DK. 1980. *Groundwater Hydrology.* J. Wiley & Sons, New York.

United Nations Educational, Scientific and Cultural Organization. 1971. *Scientific Framework of World Water Balance, Technical Papers in Hydrology 7.* UNESCO, Paris.

United Nations Environment Programme, World Meteorological Organization. 1997. *Common Questions about Climate Change.* Brochure. UNEP, New York, and WMO, Geneva.

Vereniging voor Landinrichting. 1992. *Cultuur Technisch Vademecum.* Brouwer Offset Utrecht B.V., Utrecht, The Netherlands.

Wang B, Zhang K, Shen J. 1995. Potential impacts of sea-level rise on the Shanghai area. *J Coastal Res* 14:151–66.

World Health Organization. 1984. *Guidelines for Drinking-Water Quality.* World Health Organization, Geneva.

World Meteorological Organization, United Nations Educational Scientific and Cultural Organization. 1997. *The World's Water—Is There Enough?* WMO, Geneva, and UNESCO, Paris.

Worthington EB, ed. 1977. *Arid Land Irrigation in Developing Countries.* Pergamon Press, Oxford, U.K.

Electronic References

Food and Agriculture Organization. 1996. World Food Summit Fact Sheets. Water and Food Security. Irrigation's Contribution. http://www.fao.org/wfs/fs/E/WatIrr-e.htm (Date Last Revised 11/13/1996).

United Nations Development Programme. 1999a. Human Development Indicators: Statistics from the 1998 Human Development Report. Population Trends. http://www.undp.org/hdro/population.htm (Date Last Revised 4/26/1999).

———. 1999b. Human Development Indicators: Statistics from the 1998 Human Development Report. Growing Urbanization. http://www.undp.org/hdro/urban.htm (Date Last Revised 4/26/1999).

CHAPTER 10

Ecology and Infectious Disease

Mark L. Wilson, Sc.D.

Recognition that the environment affects risk of infectious diseases is not novel. What has changed over the past few centuries is our understanding of the mechanisms that underlie such interactions. Because parasitic microorganisms that are necessary for the occurrence of infectious diseases share many fundamental biological properties with humans, an "ecological" perspective on their life cycles should improve understanding and increase control. An ecological perspective recognizes that fundamental physical and biological processes affect the survival and reproduction of all living organisms, including the infectious agents that cause diseases. Similar evolutionary processes influence the development, behavior, diversity, and distribution of plant and animal species, some of which may live inside us and occasionally cause harm. Advances in ecological understanding and theory during the past few decades, mostly developed through studies of vertebrates, insects, or plants, have led to paradigms that are relevant to microorganisms as well. In addition, human social, economic, and political activities have become a major influence on these ecological and evolutionary processes. This chapter addresses the various ways in which physical and biological environments, mediated through human activities, influence the ecology of survival, reproduction, and transmission of these microorganisms.

The relationship among microorganisms, their hosts, and the environmental conditions under which they interact has developed through a long history of coevolution (Ewald 1993). Thus, one might expect that an equilibrium has been reached between infectious agents and their hosts, resulting in a low incidence of disease and only rare appearance of new diseases. In fact, the diversity, distribution, and effects of infectious diseases are constantly in flux, both locally and globally, with new diseases appearing, well-recognized diseases spreading to new areas, and the severity of symptoms changing. Such "emerging" infectious diseases recently have been characterized as an important threat to public health (Lederberg et al. 1992; Centers for Disease Control and Prevention [CDC] 1994a, 1998; Wilson et al. 1994) and now constitute a major focus of research and policy development. The factors underlying the emergence of disease include various ecological changes such as deforestation, irrigation, or urban crowding, but also the rapid, long-distance movement of people or animals and the development of resistance to antibiotics. Understanding how these and other factors influence the emergence of infectious disease has taken on an urgency and importance rivaling the pressing issues of environmental conservation, natural resource utilization, population growth, or economic development.

Infectious diseases are fundamentally different from noninfectious diseases by virtue of the fact that a process involving contagion determines who becomes ill. In the absence of appropriate contact with an infected person, animal, or object, transmission of an infectious agent cannot occur. Thus, spatial proximity often plays an important role in infectious diseases. Unlike most noninfectious diseases, for which a component of the environment serves as the actual exposure (e.g., toxins, temperature extremes, and radiation), the role that the environment plays in infectious diseases is to mediate the extent of contagion by altering the abundance of pathogens or the frequency and nature of infectious contacts. Different environmental variables are more or less important depending on the mechanism of transmission of each infectious agent. In general, the transmission of agents that are waterborne, foodborne, vector-borne, or airborne or have an animal reservoir tends to be more strongly influenced by environmental variables.

One goal of this chapter is to provide students, policymakers, and educators with a basic understanding of the complex and multidisciplinary environmental interactions that are part of our "human ecology," particularly those that affect our health. Although these environmental factors represent only one of many kinds of influences on microbes and their associated infectious diseases, they often have important effects on public health. Despite this, however, knowledge of environmental influences usually is inadequately or inappropriately applied to prevention, planning, and public health education. Thus, another objective of this chapter is to develop a framework within which to view, analyze, and act upon diverse kinds of environmental changes as they influence the risk of disease. Examples of environmental changes include the results of dam construction and irrigation, intensification of agriculture, elevated precipitation, ocean warming, new mining activities, urbanization and urban crowding, deforestation, housing and road construction, and reforestation (Table 10.1).

Table 10.1 Examples of Environmental Changes and Recognized Effects on Infectious Diseases

Environmental Changes	Example Diseases	Pathway of Effect
Dams, canals, irrigation	Schistosomiasis	↑ snail host habitat, human contact
	Malaria	↑ breeding sites for mosquitoes
	Helminthiases	↑ larval contact due to moist soil
	River blindness	↓ blackfly breeding, ↓ disease
Agricultural intensification	Malaria	Crop insecticides and ↑ vector resistance
	Venezuelan hemorrhagic fever	↑ rodent abundance, ↑ contact
Urbanization, urban crowding	Cholera	↓ sanitation, hygiene; ↑ water contamination
	Dengue	Water-collecting trash, ↑ *Aedes aegypti* mosquito breeding sites
	Cutaneous leishmaniasis	↑ proximity, sandfly vectors
Deforestation and new habitation	Malaria	↑ breeding sites and vectors, immigration of susceptible people
	Oropouche	↑ contact, breeding of vectors
	Visceral leishmaniasis	↑ contact with sandfly vectors
Reforestation	Lyme disease	↑ tick hosts, outdoor exposure
Ocean warming	Red tide	↑ toxic algal blooms
Elevated precipitation	Rift Valley fever	↑ pools for mosquito breeding
	Hantavirus pulmonary syndrome	↑ rodent food, habitat, abundance

Ecological Characteristics and Processes Influencing Transmission

The classic disease triad of agent, host, and environment, usually portrayed simplistically as a static triangle (Fig. 10.1*a*), may be more appropriately conceived of as a process involving dynamic interactions (Fig. 10.1*b*). Not only does each of the three components include many variables and effects, but their interactions may produce indirect and nonlinear effects with time lags and various feedbacks over many temporal scales. Envisioned thusly, the notion of "environment" acquires a richer and more realistic meaning. (Chapters 4 and 5 provide additional discussion of concepts from dynamic systems.) This section summarizes how fundamental environmental factors determine various ecological interactions and how to gauge their influence on the dynamics of infectious disease.

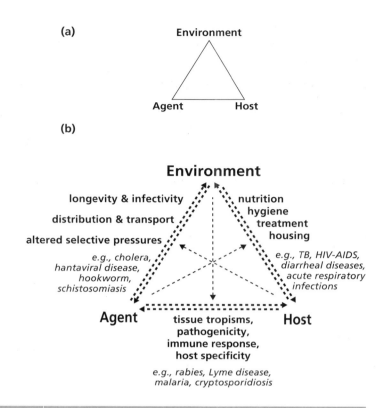

Figure 10.1 Three components of infectious diseases representing the agent, host, and environment in their classic, static conceptualization (*a*) and a more dynamic representation that considers directions of effect, interaction among entities, and possible conflicting effects (*b*).

The Interaction of Abiotic and Biotic Factors

Environmental variables are often grouped into two broad categories: abiotic and biotic. Abiotic factors are primarily physical or chemical and include elements such as temperature, water or precipitation, relative humidity, atmospheric gases, wind, and solar radiation. Biotic factors involve living plants and animals and can be grouped phylogenetically (species and higher taxa) or functionally (e.g., predators, competitors, nutritional resources, or guilds). While conceptually useful, the distinction between abiotic and biotic is blurred and can be misleading because extensive interdependence exists. Abiotic changes such as unusual precipitation may, for example, result in greater abundance of rodents through increased production of grain, or larger mosquito populations due to improved breeding habitat, or increased survival of certain aquatic bacteria if marsh water salinity is reduced. Many other similar effects on biotic processes are the direct result of changes in temperature, solar radiation, or other abiotic factors.

Conversely, biotic processes involving the presence, growth, or abundance of plants and animals also affect the amount or distribution of physical or chemical components of their environment. At the local or microclimatic level, for example, nests or houses result in temperatures that are warmer or cooler than ambient temperature, forest regeneration usually reduces evapotranspiration, irrigation often increases local relative humidity, and the creation of large urban areas may increase atmospheric particulates and generate warmer air temperatures than those found in the surrounding countryside. Thus, abiotic factors influence the broad distribution patterns of plants and animals, which in turn affect the manner in which the physical or chemical conditions of the environments are modified and experienced.

Environment may be modified and created by human activities. Such "anthropogenic" changes, whether intentional behaviors (house construction, dams, deforestation) or the unintended by-products of the consumption and production of resources, are largely determined by economic and cultural forces. Industrial activities now have global effects, as demonstrated by the emissions of chlorofluorocarbons (CFCs) that deplete the earth's protective layer of stratospheric ozone and the combined emissions of CFCs and other anthropogenic greenhouse gases that can affect patterns of global climate (see Chapter 7). Thus, another component of environment is that which is socially determined by human actions (see Chapter 6).

Temporal and Spatial Scale and Variation

Ecological changes may occur at spatial scales ranging from centimeters to continents and over time periods of days to millennia. The manner in which such temporal and spatial variation affects the ecology of infectious disease depends on characteristics of each agent and its mode of transmission. Epidemiological analyses that solely describe existing patterns of disease by person, place, and time may miss important features of the temporal or spatial scale over which ecological changes influence transmission of an infectious agent. The field of medical geography has explored spatial patterns of disease for decades, sometimes linking environmental patterns to those of cases. In addition, environmental changes, whether over days to weeks, months to seasons, or years to decades, can affect the dynamics of transmission. The duration of such effects generally corresponds to the time span over which differences occur. For example, a few weeks of unusually warm and wet weather may lead to a short-term increase in the abundance of vector mosquitoes. Other sudden ecological changes, such as fires, hurricanes, or the invasion of new species, however, may produce longer-term consequences.

The spatial scale over which ecological effects occur may be local (hectares to square kilometers), regional (counties to states), or large scale (biomes, continents, or the entire planet). For example, shifting the grazing location of a herd of cattle to a new field might increase the risk of waterborne disease in a downstream community, whereas widespread deforestation in the humid tropics could affect the transmission of malaria across thousands of square kilometers. Interannual climatic variability, such as that produced by the El Niño phenomenon (see Chapter 8), can affect climate-sensitive diseases worldwide. Our ability to understand, prevent, and ameliorate the consequences of ecological changes that influence infectious diseases depends on careful evaluation of the temporal and spatial distribution of events.

Ecological Zones and the Global Distribution of Diseases

The evolution of parasitic microorganisms has been occurring over millions of years, resulting in assemblages of parasites and hosts that until recently tended to be restricted to particular regions or biomes (see also Chapter 8). Although the global distribution of plants and animals has resulted from long-term evolutionary change, short-term events have affected the distributions of diseases on a global scale. Until the last millennium, most important infectious agents causing disease in humans were restricted to and separated by regional-scale macrogeographic distances. Typically, these distributions were constrained by ecological or host-determined factors that originally defined their development. The advent of rapid, long-distance movement of people and goods during the past few centuries has allowed certain infectious agents to be quickly transported over great distances, destroying some of the ecologically defined macrogeographic isolation that had developed over evolutionary time. Thus, contemporary shifts in the presence or intensity of transmission of microorganisms can occur at meso- to macrogeographic spatial scales and at annual to decadal temporal scales. Regional changes, for example, in the presence of particular host species, the salinity and turbidity of standing water, or the existence of a vegetation class might permit or limit the spread of an infectious agent. The resulting spatial heterogeneity could lead to regional differences or to the establishment of foci over the scale of kilometers.

Over very short time periods and small distances, microgeographic variation and hourly to seasonal fluctuations can result from population shifts, extremes in temperature or precipitation, local anthropogenic effects, or stochastic events. In such cases, the appropriate spatial measures would involve microhabitats covering tens of meters or less, sometimes changing during days or seasons.

Population Ecology and Behavior

The presence of a pathogenic microbe in a region is necessary but not sufficient for the occurrence of disease. Many factors influence which and how many people become infected and whether or not they become ill. Exposure is a function of the population density and distribution of infectious organisms and may occur through the vertebrate or human hosts that naturally harbor the pathogen or the invertebrates that transmit the agent. In general, the number of infectious animals is positively related to the probability of transmission, but other factors may be important. The behavior of infected hosts and of the humans who are at risk of exposure also influences how many potential infectious contacts are eventually realized. Not all contacts result in transmission, and this may differ according to the behavior of various ages or other groups. Thus, population and behavioral characteristics of infected hosts and susceptible people will mediate transmission, and these represent another set of variables to consider when evaluating the health repercussions of environmental change.

Immunity and Transmission

Because only some infections produce disease, an analysis of the ecology of health should consider how humans respond to infection, including factors such as previous exposure, nutrition, age, and immune status. Some of these factors have direct links to environmental change, while others are more a function of historical or genetic processes. For any population, the relative abundance of and contact patterns among susceptible, infected, infectious, and immune individuals largely determine the intensity and stability of transmission. The magnitude of an immune response, characterized by the production of humoral or cellular defenses to these foreign organisms, also is influenced by many variables, including the site, timing, and quantity of inoculum. Immunity may be complete, partial, or nonexistent, in turn influencing how humans shift among being susceptible and uninfected, being infected, or being infectious.

The basic reproduction ratio, R_0, is a useful conceptual tool for analyzing and comparing factors that contribute to the maintenance of transmission. This theoretical value represents the average number of new infections that arise during the period of infectiousness of one case that enters a population of totally susceptible people. Values of R_0 may range from zero to infinity, with a value less than one indicating that the infection will not persist in the population and a value greater than one indicating continued transmission at a rate that is positively related to the magnitude. A basic reproduction ratio of one is therefore a threshold value, which marks the transition from a system in which infection will not propagate to a system in which infection will propagate. Various versions of the basic reproduction ratio have been suggested, but the simplest is represented as the product of three components:

$$R_0 = \beta C D$$

where β is the probability of transmission given appropriate contact; C is the contact rate (per unit time) of an infectious person with susceptible people; and D is the duration of infectiousness (in the same units of time as C). These fundamental characteristics describe the spread of infection to be expected when an infectious agent is transmitted among susceptible people. A low probability of transmission (β), infrequent contact (C), or a short duration of infectiousness (D) all tend to reduce the basic reproduction ratio of an infectious agent. Similar considerations influence the dynamics of animal populations harboring infectious microorganisms that are pathogenic to humans; however, the relationship between such animal infections and human disease is complex (see under "The Diversity and Complexity of Transmission Cycles," below).

The concept of herd immunity defines the proportion of a population that is in the immune category. In one sense, it represents how much of the population is no longer at risk for continued transmission of an infection. In general, the larger the proportion of a population that is immune to a particular agent, the less likely it is that the infection will persist or spread. Typically, a population's herd immunity increases either as an infectious disease spreads or through vaccination. Herd immunity is reduced, however, when those who are already immune die or new susceptible people enter the population through births or in-migration. Such turnover among susceptible, infectious, and immune people in the population represents an important demographic component of the maintenance of infectious disease (Anderson and May 1991). Environmental changes that affect any of these human characteristics may influence the risk of transmission. At or above a certain level of herd immunity, here designated as the percentage P_H, transmission within that population cannot be maintained, as there are insufficient susceptible human hosts. This level of herd immunity necessary to interrupt transmission is related to R_0 as

$$P_H > 1 - (1/R_0)$$

This relationship shows that, to interrupt the transmission of diseases with a greater basic reproduction ratio, greater control effort generally will be required. In some contexts, the term *herd immunity* applies specifically to a population whose proportion immune exceeds that threshold P_H, which is equivalent to the basic reproduction ratio dropping below its threshold value of one after control is introduced (Aron 2000).

The Diversity and Complexity of Transmission Cycles

The manner in and extent to which ecological changes affect the spread of infectious microbes depend on the number of components in the transmission system, as well as their relationship to the environment and to each other. These components include the agent and humans, but also may involve nonhuman reservoirs and physical or biological mechanisms of transmission. Infectious diseases have been classified according to these characteristics as either anthroponotic or zoonotic and as either directly transmitted or indirectly transmitted. Anthroponotic diseases are caused by microorganisms that normally are transmitted solely from person to person. Zoonotic diseases, however, typically result from infectious agents that are transmitted in nature among nonhuman hosts, with accidental and occasional transfer to humans. The transmission mechanism may be direct (through physical contact or droplets) or indirect (via aerosol suspensions, food, soil, water, or a "vector" arthropod or snail that bites or otherwise contacts the host). The physical objects that aid indirect transmission are called *vehicles* or *physical vectors.*

Most diseases can be unambiguously categorized by these criteria into four kinds of transmission cycles, thereby permitting examination of common attributes that may be used to evaluate how environmental effects influence the distribution of each disease and its relative incidence (Table 10.2 and Fig. 10.2). *Directly transmitted anthroponoses* involve two components (agent and human), whereas *directly transmitted zoonoses* entail three (agent, animal reservoir, and human). *Indirectly transmitted anthroponoses* also involve three components (agent, vector or vehicle, and human), whereas *indirectly transmitted zoonoses* encompass four (agent, vector or vehicle, animal reservoir, and human). Depending on the nature of ecological changes and the responses of one or more of these components of transmission, shifts in the rate of spread, severity, and distribution of infections will result.

Directly Transmitted Anthroponoses

Infectious disease agents that normally are spread among people through contact or a physical medium generally have the least complicated associations with the environment. Nevertheless, impacts may be profound. Some changes directly alter the infectiousness of the microbe, but more often environmental impacts occur indirectly through the behavior of people. Natural abiotic factors such as precipitation, solar radiation, or relative humidity, for example, might affect the survival, persistence, and

ANTHROPONOSES

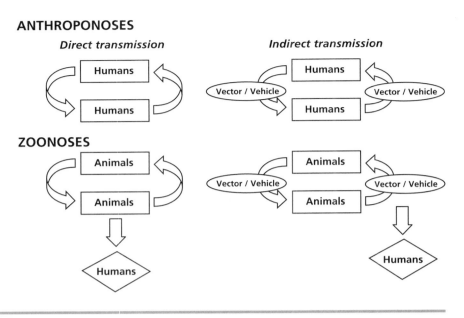

Figure 10.2 Four main types of transmission cycles of infectious diseases.

transmission of polio virus or streptococci bacteria in the environment. Most directly transmitted anthroponotic agents, however, are communicated by body fluids, sexual contact, or exposure to physical objects in a manner largely protected from these environmental influences. As examples of the latter, the agents that cause acquired immune deficiency syndrome (AIDS; human immunodeficiency virus), herpes (herpes simplex viruses 1 and 2), syphilis (*Treponema pallidum*), leprosy (*Mycobacterium leprae*), and gonorrhea (*Neisseria gonorrhoeae*) reside in and typically are transmitted by tissue or body fluids that tend to be shielded from the external environment. In these examples, ecological effects are most likely to affect transmission if they influence human behavior or responses to infection but are unlikely to affect survival or infectiousness of the agent directly.

Somewhat intermediate are other directly transmitted anthroponoses caused by microbes that may be spread via droplets, fecal material, or other particulates. Examples include diverse agents that cause measles (measles virus) and poliomyelitis (polio virus). These microbes are exposed to potentially harmful environmental conditions during short periods (minutes to hours) spent outside of the bodies of infected people. Nevertheless, most ecological impacts on directly transmitted anthroponoses do not occur by simple effects on the survival or spread of pathogens. Rather, changes in human behavior are more likely to affect transmission of these agents. Depending on the types of ecological changes associated with each particu-

Table 10.2 Examples of Environmental Effects on Infectious Disease Categorized by Cycle of Transmission

System Type	Transmission Characteristics			Environmental Impact	
	Components	Complexity	Focus of Effects	Examples	Probable Outcomes
Anthroponoses					
Direct	Agent Human	Low to moderate	Human body fluids or excrement, droplets	Proximity, sexual contact	Little change
Indirect	Agent Human Vector/vehicle	Moderate	Food, soil, water, aerosols, animal vector, human behavior	Agriculture growth, hygiene, irrigation, urbanization	Variable, large changes likely
Zoonoses					
Direct	Agent Reservoir Human	Moderate	Animal reservoir, human behavior, contact with body fluids	Deforestation, water impoundment	Variable, small changes possible
Indirect	Agent Reservoir Human Vector/vehicle	High	Animal reservoir, animal vector, human behavior	Deforestation, reforestation, climate change	Uncertain, multiple impacts may conflict

Note: Transmission may be more complex and varied than the basic patterns shown. For example, humans as well as dogs, cats, pigs, cattle, and other animals may serve as the reservoir of *Schistosoma japonicum*, a cause of schistosomiasis in East Asia (Benenson 1995, 419). Some zoonotic diseases with indirect transmission may be foodborne. For example, trematodes causing paragonimiasis cycle from snails to freshwater crabs and crayfish that are eaten by humans, in whom eggs are coughed up, swallowed, and excreted (Benenson 1995, 343–44).

lar agent, factors such as shifts in human population density, diminished housing conditions, increased crowding, reduced hygienic practices, or sharing of utensils, for example, are more likely to influence indirectly the transmission of these disease agents. In general, ecological effects are likely to be greater on pathogens spread from animals or by other indirect means.

Directly Transmitted Zoonoses

Zoonotic diseases result when microbes that are normally transmitted among nonhumans occasionally and accidentally spread to people. Since we are not part of the normal transmission cycle, we become "dead-end" infections to the parasite. For these agents, a vertebrate species, but not a human, serves as the natural reservoir host. People typically play no role in the transmission cycle of these agents because their persistence in nature depends on other processes involving the reservoir host and its environment. Thus, factors such as the rate of transmission, development of host immunity, and rate of reproduction or death in the nonhuman host population are critical to the continued existence of such microorganisms. Enzootic transmission of these zoonotic disease agents among animal hosts allows their perpetuation in nature, a situation analogous to that of endemic transmission in human populations. Correspondingly, epizootic transmission means that new infections in these vertebrates are occurring more frequently than normal, a situation analogous to that of epidemic transmission in human populations.

The risk to humans of acquiring zoonotic diseases and the roles that changing ecological conditions play in such risk depend in part on the mode of transmission of the infectious organism. Directly transmitted zoonoses are spread among reservoir hosts via media not unlike those of directly transmitted anthroponoses (e.g., direct contact with body fluids or droplets). Thus, similar environmental conditions influence the probability of contamination. Because humans usually are infected in the same manner as the natural vertebrate host, environmental characteristics affecting enzootic transmission function analogously during such spillover transmission to humans. Additional ecological effects, however, operating through the presence, abundance, and condition of the nonhuman reservoir host species, must be considered for directly transmitted zoonoses.

Environmental Factors Determining Transmission

The presence of, for example, soil-inhabiting *Clostridium tetani*, the anaerobic bacterium associated with human tetanus, may vary with environmental conditions at the ground surface. This common intestinal symbiont is normally transmitted by a spore stage that is frequently found in soil and fomites contaminated with animal feces. Infection by puncture wound may lead to unusual subcutaneous growth and production of tetanus toxin, which produces severe muscular symptoms. Although immunization with tetanus toxoid results in long-term protection, many cases occur in developing countries with inadequate vaccination and poor sanitation. Thus, proper treatment of human sewage and separation of animal waste from areas of human activity are likely to reduce risk of tetanus.

Other vertebrate host infections depend on ecological characteristics of the natural water supply or associated vegetation. Cryptosporidiosis, for example, is a waterborne intestinal infection caused by protozoa of the genus *Cryptosporidium* (Fayer and Ungar 1986). Potentially a life-threatening human illness in immunocompromised people, mild to chronic diarrhea occurs in most other cases. In 1993, contamination of the water supply in Milwaukee resulted in an outbreak in which more than 400,000 people became ill and 47 died (MacKenzie et al. 1994). Although considered an emerging disease, human cryptosporidiosis still tends to be underdiagnosed (Gallaher et al. 1989; Skeels et al. 1990). A study of patients with acute infectious diarrhea in England found that *Cryptosporidium* infection was almost as common as *Salmonella* infection and nearly three times more common than that of *Shigella* (Public Health Laboratory Service Study Group 1990).

Many vertebrates, but particularly domestic and wild ungulates, seem to be naturally infected by and frequently infectious with *Cryptosporidium* spp. (Navin and Juranek 1984). Rainfall, vegetation surrounding watersheds, concentration of potential reservoir hosts, and sources of drinking water represent environmental characteristics that are likely to affect transmission. Furthermore, the quantity of fecal shedding of *Cryptosporidium* oocysts or spores and, hence, the extent to which surface water will be infected probably are proportional to the abundance of infected hosts. While domestic farm animals may be important reservoirs, deer also harbor this agent (Simson 1992; Tzipori et al. 1981) and may be implicated in transmission to humans. Thus, the density of vertebrate reservoirs, including domestic animals, and their relationship to watersheds used for drinking water influence the probability of transmission among these hosts and to people.

Reservoir Population Ecology and Transmission

For certain other directly transmitted zoonoses, the agents occur in body fluids or as relatively protected encysted stages that experience minimal direct environmental effects. In such cases, different ecological effects on reservoir abundance and human contact nevertheless may affect risk of human infection. Directly transmitted zoonoses, such as rodent-borne viral hemorrhagic fevers and rabies, are examples that illustrate some basic principles.

Various hantaviruses (Johnson 1989) and arenaviruses (Childs and Peters 1994) are naturally maintained in particular rodent species through transmission in excreta or saliva via aerosol or bite. Hantavirus-associated zoonoses such as Korean hemorrhagic fever and nephropathia epidemica (NE), now grouped as "hemorrhagic fever with renal syndrome" (HFRS), have been known for decades throughout much of Asia and Europe (LeDuc et al. 1994). A new member of this group of hantaviruses, sin nombre virus (Elliott et al. 1994), has been identified as the agent of a recently recognized, life-threatening disease in the United States, hantavirus pulmonary syndrome (HPS) (Duchin et al. 1994). The causative viruses of each of these syndromes are related taxonomically; however, each is maintained through transmission in different rodent reservoir host species. Hantaan and Seoul viruses, causes of HFRS, circulate among field mice (*Apodemus* spp.) and Norway rats (*Rattus norvegicus*), respectively (Lee et al. 1978, 1982). Puumala virus, the causative agent of NE, is maintained in bank voles (*Clethrionomys glareolus*) and occurs primarily in Europe and western Russia (LeDuc 1987; Gavrilovskaya et al. 1990). Although details of the cycles of these hantaviruses differ among the various combinations of virus and reservoir, infection in the primary reservoir species typically produces chronic infection that persists for life (Gavrilovskaya et al. 1990; Lee et al. 1981; Yanagihara et al. 1985).

Over broad geographic regions, the abundance of rodents has been linked to outbreaks of hantaviral disease in humans. In the southwestern United States, unusually abundant rainfall produced lush vegetation and increased food resources for rodents during the spring and summer of 1993. It was then that the first recognized outbreak of HPS occurred. Data from the Sevilleta research station in Socorro County, New Mexico (south of the geographic location of most HPS cases), indicated that populations of the deer mouse (*Peromyscus maniculatus*) in some locations increased 10-fold between May 1992 and May 1993 (Parmenter et al. 1993). Deer mouse populations of 30 mice per hectare in May 1993 during the HPS epidemic declined to fewer than 3 mice per hectare by August (Parmenter and Vigil 1993), when the outbreak of HPS was waning; this suggested that increased abundance of rodents may have contributed to the large number of HPS cases during the spring and summer (Childs et al. 1994). Several studies in Sweden have convincingly linked the abundance of rodents to patterns in the incidence of NE among humans (Nyström 1982). At a finer spatial scale, local abundance of rodents may also influence the risk of human acquisition of hantaviral disease. One of the principal environmental variables that distinguished those households in which patients with HPS resided from selected control dwellings was a significantly greater abundance of rodents, as indicated by trapping success (Childs et al. 1995).

Arenaviruses, such as Junin virus from the corn mouse (*Calomys mus-*

culinus) in Argentina, Machupo virus in the related *Calomys callosus* from Bolivia, and Lassa virus from multimammate rats (*Mastomys* spp.) in Africa, also establish chronic infections in their respective rodent hosts. Transmission also occurs via contact with urine, feces, or rodent tissues. There seems to be a direct link between the density of rodent reservoirs and arenaviral disease in humans. A longitudinal study of *C. musculinus* populations in the area of Argentine hemorrhagic fever documented a dramatic increase in the density of rodents during the third year of the study. The number of human cases of Argentine hemorrhagic fever reached a 20-year high during the epidemic season following the increase in the density of rodents (Mills et al. 1992). During an outbreak of Bolivian hemorrhagic fever in San Joaquín, nearly 3,000 *C. callosus* (about 10 per household) were removed during a three-week period (Mercado 1975), apparently contributing to the rapid decline in new cases of Bolivian hemorrhagic fever. Again, a combination of changes in resources and predators that affected the abundance of rodents, combined with patterns of agricultural production and land use, seem to be the major determinants of the risk of arenaviral disease in the human population in South America. In particular, studies have shown that the grassy habitat on the edges of corn fields in Argentina represents ideal habitat for *C. musculinus* rodents that serve as reservoirs for Junin virus (Mills et al. 1992, 1994).

Rabies is a well-known directly transmitted viral agent that, unlike most zoonotic diseases, is highly pathogenic in virtually all vertebrate reservoir species that become infected. By killing most infected hosts, rabies virus creates a conundrum surrounding its persistence in nature. Several variants of rabies virus have co-evolved with particular mammalian carnivore or bat reservoir species that are considered natural hosts (Smith et al. 1992), although all variants are pathogenic in most mammals. Based on practical experience and simulation modeling, the ecology of rabies virus depends largely on the density of reservoir hosts and is considered critical to understanding the temporal and spatial dynamics of transmission (Bacon 1985). Studies of red foxes (*Vulpes vulpes*) in Germany, for example, indicate that the incidence of disease is positively associated with the density of this dominant reservoir species (Steck and Wandeler 1980). After the appearance of epizootic rabies, fox populations are diminished and reports of animal rabies in a given locale decline precipitously; evidence of disease may disappear from given regions for some time. A threshold density of hosts (K_T of Anderson et al. 1981) is apparently required for rabies to perpetuate in fox populations; below this threshold, infectious contacts seem too few to maintain transmission. As fox density increases, models suggest that an increasingly larger percentage of the fox population would have to be vaccinated or removed to locally eliminate rabies (Anderson et al. 1981).

Human exposure to rabid animals in the United States also occurs primarily through contact with terrestrial wildlife, although bat-associated variants of rabies have been the source of most human deaths from rabies during the last decade (Krebs et al. 1993, 1994). The density of wildlife, such as that of raccoons, seems related to the risk of human exposure (Winkler and Jenkins 1992). Postexposure treatment of humans with rabies immunoglobulin and vaccine is a measure of human contact with potentially rabid animals. In New York State, for example, the number of humans receiving antirabies treatment increased from 1,125 to 2,905 from 1992 to 1993 (CDC 1994b) at the same time that the number of rabid raccoons rose from 1,355 to 2,320 (Krebs et al. 1993, 1994). A reduction of the population of reservoir hosts may reduce wildlife rabies in certain specific applications but is unlikely to be effective as the sole control method (Debbie 1991). Thus, ecological conditions that alter the density and distribution of hosts will affect the maintenance of rabies virus as well as human exposure.

Unlike most anthroponoses, transmission of these directly transmitted zoonoses usually is not influenced by the density or the immune status of human populations. Except for a few agents against which vaccines have been developed (rabies, tetanus toxoid), risk and prevention are associated with ecological conditions influencing the reservoir.

Indirectly Transmitted Anthroponoses

Indirect transfer of pathogenic microorganisms may occur by physical processes (waterborne, soilborne, aerosols) or by biological organisms (mosquitoes, ticks, sandflies, snails, or other animals capable of "vector-borne" transmission). For pathogens that normally are spread from human to human by these physical or animal vectors, the resulting disease is considered to be a vector-borne anthroponosis. As with the directly transmitted zoonoses just discussed, three factors comprise the transmission cycle of indirectly transmitted or vector-borne anthroponoses: human hosts, the infectious agent, and the vector. Unlike directly transmitted zoonoses, however, the pathogens of vector-borne anthroponoses are not dependent on another vertebrate for their life cycle, but instead require a physical or invertebrate vector to mediate infection of humans. Environmental changes typically influence the risk of human disease by altering exposure to the soil or water medium or by influencing the abundance and distribution of vectors, their behavior, or their rate of infection (Lake et al. 1993).

Most animal vectors are blood-sucking arthropods. These insects or ticks normally must feed on vertebrate blood for their sustenance and reproduction. In the process they acquire and transfer infectious microorganisms. Various disease agents have evolved the capacity to employ particular characteristics of the arthropod's life cycle, thereby permitting reproduction and transfer from human to human. Although blood-sucking arthropods are the vectors associated with most vector-borne anthroponoses, snails are also considered vectors in transmitting the agents that cause human schistosomiasis.

Environmental Factors Determining Transmission

Numerous environmental factors determine the geographic distribution, abundance, longevity, activity, habitat-associations, or biting behavior of arthropod vectors. These, in turn, influence the ability of a vector to contact humans and transmit infectious agents. In addition to the inherent species-specific *vector competence* of a blood-sucking arthropod to support the development and transfer of a pathogen, factors such as the frequency of biting, rate of pathogen development, and longevity of the vector have been integrated into the risk of transmission. The measure termed *vectorial capacity* summarizes the contribution of such variables to spread of a vector-borne infection (Dye 1992). This ensemble of factors determining vectorial capacity (VC) has been combined into a simple formula originally developed for malaria by MacDonald (1957) and subsequently reformulated by others such as Garrett-Jones (1964) as

$$VC = (ma)\,(P/F)e^{-n/E}E$$

where

ma = number of blood meals on humans taken by the vector population per person per day (person-biting rate)
P = proportion of blood meals taken by a vector on humans
F = time interval between blood meals taken by a vector (days)
e = base of natural logarithm (≈ 2.71828)
E = vector's life expectancy (days)
n = time for parasite development or sporogony in a vector (days)

VC is the potential number of secondary parasite inoculations by the vector that would be produced from an individual infective human host per day of infectivity. The secondary parasite inoculations typically occur many days after an infective host transmits the parasite. VC can be thought of as part of an R_0 for vector-borne diseases, in which the contact rate (C) has many components involving feeding behavior and longevity of the vector and the probability of transmission given contact (β) is related to the development of the parasite in the vector. To complete the calculation of R_0, one multiplies the daily VC by the average number of days that a human host is infective (D). The equation for VC incorporates the important variables that influence the transmission of malaria and, with some modification, most other vector-borne anthroponotic agents as well. Although many assumptions underlie the definitions of and relationships among variables, most essential features of the risk of transmission are included, thereby permitting comparison of relative values of risk among places, conditions, or time.

Environmental effects are not explicitly identified in the formulation of VC, but it should be obvious that they might affect the magnitude of components of the VC in diverse ways. For example, temperature may alter the duration of parasite development inside the mosquito vector (n), wind speed or relative humidity may affect the frequency of feeding by vectors (F), and surrounding vegetation or alternative sources of blood meals (e.g., cattle) may alter the rate of blood feeding on humans by vectors (ma). To contribute to transmission of parasites, an individual vector needs to obtain two blood meals from humans, one to acquire and a second to transfer the infection. The likelihood that the transfer occurs depends on the proportion of blood meals taken on humans (P) and the overall frequency of feeding (F), both of which might be affected by alternative sources of blood meals or weather. Predators and a complex of other abiotic factors will influence the life expectancy of the vector (E), perhaps the most important single variable that influences the extent of transmission.

In addition to quantitative comparisons, the equation for VC offers more abstract and generally heuristic value by explicitly identifying the relative importance of different factors as they affect transmission. Indeed, many of the variables are difficult to estimate without considerable error or among-site variation. Such error may compromise quantitative comparisons, but the relative magnitude of the effects of various factors is illustrated.

Human Ecology and Transmission

Characteristics of the human population also are important in the transmission of vector-borne anthroponoses. Since vectors acquire agents by feeding on infected humans and in turn must feed on another nonimmune human to transmit the agents, the population density and infectious or immune status of human hosts affect transmission and the expression of disease. Social or economic processes that increase human population or crowding, reduce herd immunity, or modify human contact with vectors are likely to increase the risk of spread of a vector-borne anthroponosis. Crowding influences how frequently vectors contact human hosts, and its effect may be less straightforward than for a directly transmitted anthroponosis.

One important vector-borne anthroponosis that illustrates these principles is dengue fever, which is caused by dengue virus. The widespread mosquito *Aedes aegypti* is the primary vector of this agent (Gubler 1988). When dengue-infected humans with adequate virions circulating in their blood system are fed upon by *Ae. aegypti* mosquitoes, these mosquitoes efficiently acquire the virus and, after a period of viral replication, become infectious to other humans. Various environmental factors influence the abundance of this mosquito, particularly via egg laying (e.g., water in discarded cans or tires) and hatching, larval development, or adult survival (e.g., rainfall, temperature, or relative humidity). In addition, the ability of these mosquitoes to transmit dengue virus may be influenced by ambient temperatures that affect the *extrinsic incubation period* in the mosquito. This period represents the time needed for viral multiplication and invasion of the mosquito salivary glands, which must occur before the next infectious blood meal may take place.

Schistosomiasis is another type of indirectly transmitted and primarily anthroponotic infection normally caused by trematode worms, or blood flukes, of the genus *Schistosoma*. The transmission cycle involves specific snail species, primarily of the genera *Biomphalaria, Bulinus,* and *Onchomelonia,* that serve as vectors of schistosome parasites. Human disease is principally the result of eggs that lodge in various organs when they are produced by adult worms in people who are chronically infected. Thus, both people and snails play a vital role in *Schistosoma* biology, and environmental influences penetrate the ecology of both. Initially, human infection results from working or bathing in water where snails release the infectious cercarial stage of this parasite. The cycle of transmission is completed when these infectious people urinate or defecate in snail-infested water, thereby infecting other individual snails of these species that can eventually perpetuate the cycle. The role of environmental factors in the epidemiology of schistosomiasis is considerable, as determinants such as precipitation, water salinity and flow, vegetation, and water use all affect risk. Environmental changes such as the creation of dams, the diversion of water, canalization and irrigation, or diminished hygiene also may influence the risk of schistosomiasis.

Malaria represents another example of an indirectly transmitted anthroponosis. By nearly all measures the world's most important vector-borne disease, malaria is discussed in detail in Chapter 12. The variety and magnitude of environmental influences on this vector-borne disease are enormous. Not only do abiotic elements such as precipitation or temperature affect the abundance of mosquito vectors and the development of parasites in those vectors, but also biotic factors operating through vegetation, deforestation, agricultural activity, and housing construction may influence mosquito survival and vectorial capacity. In general, because of influences on the abundance and survival of the vector, vector-borne anthroponoses are more strongly affected by environmental influences than are directly transmitted anthroponotic agents.

Various types of filariasis represent yet another example of arthropod-borne anthroponoses (see Chapter 2). The nematode species that cause this infection, most dramatically seen as elephantiasis, are all transmitted by various species of mosquitoes. Two main genera of filarial nematode worms, *Wuchereria* and *Brugia,* are the principal human parasites, with both groups being transmitted by particular species of *Culex, Anopheles,* and *Aedes* mosquitoes. Primarily a disease of warmer regions of Asia, Africa, the Pacific Islands, and South America, microfilaria circulating in the bloodstream appear many months after initial infection, producing a range of symptoms from none to recurrent fever, lymphadenitis, elephantiasis, and even pulmonary disease. Interestingly, the intensity of *Wuchereria* spp. microfiliaria in the blood varies in a cyclical pattern throughout the day, with the greatest number occurring either at night or during the day, depending on species. Although *Wuchereria* and *Brugia* species differ in their biology and disease manifestations, worms of both genera depend on mosquitoes as vectors of transmission. Thus, various environmental factors may affect the abundance of vectors and transmission in areas where people are currently infected. Although some species of *Brugia* also infect other vertebrates, humans seem to be the principal reservoirs in most settings. Thus, many different genera of mosquitoes may harbor and transmit infective larvae of both genera of filaria, making control of transmission complicated and often ineffective.

Indirectly Transmitted Zoonoses

As judged by the number and diversity of organisms involved in transmission, vector-borne zoonoses tend to be the most ecologically complex infectious diseases. Accordingly, environmental change may have the greatest number and diversity of effects, but in possibly contradictory ways. In addition to humans, a nonhuman reservoir, an arthropod vector, and the pathogen all may be influenced by ecological variables. Many of the factors that affect the abundance and survival of vectors also might have influences, possibly dissimilar, on the abundance and distribution of reservoir hosts. Risk of infection in humans is, thereby, a function of not only vectorial capacity, but also a more complex relationship involving the density and proximity of competent reservoir hosts (Wilson 1994b).

Examples of some North American vector-borne zoonoses include tick-borne Rocky Mountain spotted fever (*Rickettsia rickettsii*), mosquito-transmitted viral diseases such as Saint Louis encephalitis or eastern equine encephalitis, flea-vectored plague (*Yersinia pestis*), or sandfly-borne leishmaniasis (*Leishmania* spp.). Arthropod vectors acquire the infectious organism while feeding on an infected reservoir host and subsequently transmit during a later blood meal. Transmission to humans rather than the vertebrate reservoir represents a dead end for the microorganism and possibly an atypical source of blood for the arthropod vector. Thus, a contradiction exists between efficient vector-borne maintenance of an infectious agent among reservoirs and zoonotic transmission to humans (Spielman and Kimsey 1991). Human blood meals that produce such dead-end infections deny new infections in the vertebrate reservoir. Alternatively, efficient vector-borne transmission in the reservoir population would rarely spill over to humans if the vector were not partially anthropophilic. In this case, the presence of nonhuman hosts suitable to the vector's needs would simultaneously provide a measure of zooprophylaxis by absorbing infectious spillover blood meals from the vector. Nevertheless, zoonotic transmission still may occur if additional "bridge" vector species that feed both on the reservoir and on humans are sufficiently abundant. This may partly explain the transmission of urban yellow fever (Monath 1989). Yellow fever virus seems to be maintained among nonhuman primates in a jungle cycle but can be introduced into urban areas where other mosquito species can efficiently transmit from human to human.

Environmental Factors Determining Transmission

The environment's effects on the transmission of vector-borne zoonoses may operate through the reservoir, vector, interactions between the reservoir and vector, or contact between the human and vector. Efficient enzootic transmission requires that certain arthropod species that permit survival and development of particular pathogens take an infective blood meal on vertebrate reservoir hosts that normally favor the development of those pathogens. Ecological factors that affect the vector's abundance, longevity, activity, or feeding behavior may influence enzootic transmission. Similar consequences may alter the reproduction, survival, and abundance of vertebrate reservoirs. These factors are multiple, diverse, and interacting, including weather, natural or agricultural vegetation, housing construction, land use patterns, surface water, and other variables.

Still other circumstances affect the risk of transmission to humans by altering the proximity of people to vectors or reservoirs or influencing the extent and timing of outdoor activities that expose people. For certain conditions, the strength and duration of the human immune response is a function of age. In addition, poor nutrition, excessive exposure to sunlight and pesticides, or other concomitant infections may serve to weaken the immune response. The environmental conditions that affect humans may be less obvious but generally include outdoor behavior, housing in proximity to vector habitats, quality and quantity of food, and agricultural or other uses of the environment.

The Ecology of Reservoir-Vector-Human Interactions

Lyme disease is an example of an indirectly transmitted zoonosis with multiple environmental determinants. The principal vector in North America is the deer tick, *Ixodes scapularis,* which, to survive and reproduce, must take blood meals as larva, nymph, and adult. Typically, various small mammals are fed upon by larvae and nymphs, while white-tailed deer and other large mammals serve as hosts to adult ticks. Larvae or nymphs may acquire infection with the bacterial agent of Lyme disease, *Borrelia burgdorferi,* which then can be transmitted during subsequent blood feeding by nymphs or adults. Environmental factors such as changes in habitat that affect the distribution and abundance of these vertebrate reservoirs or their behavior and response to infestation would in turn affect this tick's feeding, survival, and ability to transmit. Furthermore, during the weeks to months of quiescence between periods of host-seeking, deer ticks are vulnerable to desiccation, drowning, predators, and other influences of weather and microhabitat conditions.

Finally, for people to become infected with Lyme disease, they must encounter the vector. People whose lifestyles involve intimate interaction with nature through housing in proximity to woodland environments or outdoor occupational and recreational activities will experience increased contact with vector ticks and greater risk of acquiring Lyme disease. These factors, which led to the emergence of Lyme disease in 1975, are also responsible for cases of human babesiosis caused by the protozoan *Babesia microti* and the 1994 emergence of human granulocytic ehrlichiosis (HGE) caused by *Ehrlichia phagocytophila* or a closely related rickettsia (Bakken et al. 1994; Walker et al. 1996).

A very different example of an arthropod vector-borne zoonosis is that of Rift Valley fever. Caused by a mosquito-borne virus of the same name, Rift Valley fever is widespread throughout much of the African continent. Although details of the circulation of Rift Valley fever virus are incomplete, enzootic transmission involving mosquitoes (probably *Aedes* species) may be transovarial, yet also can involve infection of domestic ungulates (animals such as cattle, sheep, and goats). Most infected ungulates experience mild illness, but abortion of fetuses and even death of young individuals frequently occur. These vertebrates, in turn, may infect other "bridge" mosquitoes (perhaps *Culex* species), which are also known to feed on people, thus transmitting the virus (Wilson 1994a).

Various environmental factors seem to determine the intensity of enzootic transmission of Rift Valley fever virus. Rainfall sufficient to fill shallow depressions temporarily with water stimulates hatching of the "rain-pool" *Aedes* mosquito vectors. Other ecological factors, such as the local abundance of ungulates, their movement while searching for forage, and their proximity to human populations, also may be important. Other "bridge" mosquito species with different habitat requirements introduce additional environmental factors in whether humans become infected. Thus, a complex of ecological factors affect where, when, and with what intensity Rift Valley fever occurs within its range in sub-Saharan Africa. Occasionally, the virus has escaped its normal range where its transmission is stable. Two recent outbreaks in Egypt along the Nile River (in 1978 and 1993) probably can be explained either by winds having blown infected mosquitoes northward from Sudan or Kenya or by transport of inadequately quarantined domestic animals that introduced the virus to the region.

Plague is yet another indirectly transmitted vector-borne zoonosis with strong links to ecology. Fleas, particularly of the genus *Xenopsylla*, are responsible for infecting vertebrates during blood feeding. These fleas typically transmit infection by the causative bacterium, *Y. pestis*, among reservoir vertebrates. The reservoir species of importance vary with region but tend to be rodents or other small mammals. Urban transmission involving principally rats of the genus *Rattus* occurs rarely today but may appear where wild mammals such as ground squirrels, rabbits, or wild carnivores contact these peridomestic pests. Most often, people are exposed when in contact with rodents in forested or other heavily vegetated sites. Two forms of human disease are recognized: bubonic and pneumonic plague. In bubonic plague, transmission by the bite of infected fleas produces enlarged lymph glands (buboes) near the site of the initial infection. In pneumonic plague, which represents an unusual form of human-to-human transmission of this normally vector-borne zoonosis, respiratory droplets serve as the mechanism of pathogen transfer. Numerous environmental factors that affect the abundance of rodents, the survival or feeding behavior of fleas, and contact with humans or domestic pets can alter risk of disease in humans.

African trypanosomiasis, commonly known as sleeping sickness, represents yet another example of a vector-borne infection that is primarily a zoonosis dependent upon particular environmental factors. Two subspecies of the protozoan *Trypanosoma brucei* produce disease in people as well as wild ungulates and cattle throughout much of tropical Africa. Transmitted by at least six species of tsetse flies (genus *Glossina*), infection with *T. brucei* hemoflagellates produces fatal disease in humans if untreated. A complex of environmental factors affect the longevity and survival of tsetse fly vectors, the abundance of nonhuman vertebrate reservoirs, and the ability of flies to find human hosts and transmit the infectious agent. In west and central Africa, transmission by flies of the *Glossina palpalis* group is most frequent in forested riverine habitats, whereas *Glossina morsitans* flies, which predominate in dry savanna regions of East Africa, are responsible for most infections there. Not only do changes in the abundance of reservoirs influence the risk of human disease, but large-scale shifts in land cover and habitat may produce changes in the incidence and distribution of sleeping sickness.

Ecological Change and Emerging Diseases

Infectious diseases have recently been grouped not by their usual habitat, mode of transmission, or type of reservoir, but by virtue of being newly recognized or reappearing with increased incidence or severity. Scientific and popular attention has been directed toward these so-called emerging diseases, which are newly recognized syndromes, diseases appearing in new areas, or infections with more severe or less easily treated symptoms. Since the warning sounded by the Institute of Medicine report on emerging disease (Lederberg et al. 1992), many articles, conferences, and books have pointed out the present and future dangers of new and resurgent infectious diseases (CDC 1994a; Morse 1994; Wilson et al. 1994; Krause 1998). Most reports recognize that ecological changes, including those that result directly from human social and economic behavior, play a major role in the emergence of infectious diseases (Lake et al. 1993; Roizman 1995).

Foundations of the Emergence of Disease

Infectious diseases may be considered emerging or reemerging if they conform to one or more of the following categories:

1. newly described disease or syndrome recognized within the past few decades;
2. expanding distribution of a familiar disease into a new region or habitat;
3. increased local incidence of a disease; or
4. increased severity or duration of a disease or increased resistance to treatment.

Environmental changes affect these different categories of emerging diseases through various pathways (see also Chapters 6, 7, 8, and 9).

Newly Recognized Diseases

Most truly new diseases appear when enough people are exposed, often suddenly, to a pathogenic microorganism that it attracts the attention of health providers, epidemiologists, or researchers. Such diseases are new, but the causative agents usually are not. These microorganisms often have existed in the same environment with humans for long periods, occasionally infecting people without being recognized. At some point, a shift in the type or extent of contact may occur, thus leading to more frequent disease. Alternatively, people who move into environments that have been sparsely inhabited may suddenly become exposed to an already extant pathogen. In either case, the infectious agent had previously existed in the area, usually unrecognized in a silent, zoonotic cycle of transmission involving little or no human contact. Both scenarios may occur as people migrate to new areas and exploit these environments in a manner that enhances the abundance of a pathogen.

One example of such a newly recognized emerging disease is legionellosis (Legionnaire disease), caused by bacteria in the genus *Legionella*. Although an outbreak occurred in the late 1950s, only within the past two decades has the worldwide distribution of this agent and disease been recognized. Normally residing in water or soil, where it is particularly stable, the transmission of *Legionella* species to humans apparently occurs primarily via aerosol. With increased construction of water towers used in central air conditioning and heating units of large buildings, transmission to greater numbers of occupants began to occur. In this example, ecological changes involved the creation of a highly suitable, sheltered environment in which the pathogen was able to multiply, thereby exposing numerous people in the vicinity.

Alternatively, humans may move into environments where infectious microorganisms naturally circulate, thereby becoming exposed in a manner that had not occurred previously. Among the most publicized of such examples is Ebola hemorrhagic fever. The highly virulent Ebola virus, which has been recognized in various parts of sub-Saharan Africa, causes severe, often fatal, hemorrhagic disease in humans. Although the reservoir of this virus remains unknown, evidence suggests that natural infection of forest animals somehow results in spillover human infection, often producing catastrophic disease. Because of hemorrhaging, health care providers and family members may be infected accidentally, producing focal epidemics that usually persist for only weeks or months.

Another example is Venezuelan hemorrhagic fever, an emerging disease recognized in the late 1980s in the plains of central Venezuela, where intense agricultural development was ongoing (Salas et al. 1991). Caused by Guanarito virus, a newly recognized member of the arenavirus family, Venezuelan hemorrhagic fever is a severe, often fatal disease found solely among residents of this region. As with other arenaviruses, which cause Lassa fever in Africa or Argentine and Bolivian hemorrhagic fevers in South America (see under "Directly Transmitted Zoonoses," above), Guanarito virus normally circulates among wild rodents that naturally occur in the region (Tesh et al. 1993). Although investigations are incomplete, it seems that *Zygodontomys brevicauda* rodents, which thrive in nonforested fields throughout the area, are serving as the vertebrate reservoir for Guanarito virus. A closely related but less pathogenic arenavirus from the region, Pirital virus, recently has been distinguished from Guanarito virus and seems to have a reservoir in field-inhabiting *Sigmadon alstoni* rodents (Fulhorst et al. 1999). Thus, intense agricultural production of grain crops seems to have encouraged dense populations of both granivorous rodent species, thereby augmenting enzootic transmission through more frequent contact. In addition, human settlements and increased population in the area have created suitable rodent shelter in grain depositories or the modest dwellings where people live, thereby elevating the number of people being exposed.

Thus, newly recognized emerging diseases such as these result from human intrusion into and modification of environments where zoonotic agents normally circulate unnoticed. Ecological changes may either intensify enzootic transmission or produce spillover infection of humans, or both, thus leading to the emergence of newly recognized diseases (Table 10.3). Other examples include Legionnaire disease, human monocytic ehrlichiosis and human granulocytic ehrlichiosis, Lassa fever, and Ebola hemorrhagic fever. The recognition of such agents has been made possible by improved surveillance, careful diagnosis, better communication among health workers, and new techniques for detecting agents. Among these techniques are the polymerase chain reaction (PCR), which is used to amplify genomes of suspected agents, and more sensitive and specific techniques for identifying antibodies indicating recent infection.

Expanding Distributions

Long-recognized diseases that appear in new regions or habitats or whose foci are locally expanding also may be considered emerging. The distant transport of infectious agents into suitable habitats hundreds or thousands of kilometers away, for example, may extend the distribution of associated disease. Historically, such long-distance movement has occurred progressively over short distances (e.g., infectious contacts and gradual expansion of range) or more rarely during catastrophic environmental events (e.g., storm-disturbed migrating birds or upper atmospheric drift of arthropods or aerosol). Today, however, the extensive and rapid transport of people and commercial goods throughout the world has multiplied the risk and reality of disease conveyance worldwide. In effect, such long-distance transport, whether via physical, mechanical, or living means, represents a population ecological shift in a pathogen's demography that often has serious effects on public health.

The 1978 emergence of Rift Valley fever into the Nile (see under "Indirectly Transmitted Zoonoses," above) represented the sudden expansion of this viral zoonosis from a region of enzootic transmission hundreds of kilometers to the south. Because ecological conditions are not conducive to continued transmission in Egypt, Rift Valley fever disappeared two years later, only to return during 1993 in a similar outbreak (Arthur et al. 1993).

Transport of other infectious agents into areas where suitable habitats, reservoirs, or vectors already exist may lead to more permanent expansion and emergence of disease. The growing epizootic of raccoon rabies in the

Table 10.3   Examples of Emerging Diseases Newly Recognized since 1947

Disease	Year First Recognized	Pathogen	Environmental Link
Legionnaire disease	1947	*Legionella pneumophila*	Air conditioning in large buildings
Argentine hemorrhagic fever	1958	Junin virus	Agriculture practices favoring rodent reservoirs
Bolivian hemorrhagic fever	1959	Machupo virus	Rodent abundance near houses, agriculture
Ebola hemorrhagic fever	1966	Ebola virus	Contact with natural reservoir
Lassa fever	1969	Lassa virus	Rodent abundance near houses
Lyme disease	1975	*Borrelia burgdorferi*	Increased habitat favoring reservoir and tick vector
Cryptosporidiosis	1976	*Cryptosporidium parvum*	Animal contamination of drinking water
Acquired immunode-ficiency syndrome (AIDS)	1981	Human immunodeficiency virus (HIV)	Contact with primate (?) reservoir
Escherichia coli O157:H7	1982	*E. coli* O157:H7	Fecal contamination of food, pathogen evolution
Venezuelan hemorrhagic fever	1989	Guanarito virus	Rodent abundance in houses, cultivated fields
Hantavirus pulmonary syndrome (HPS)	1993	Sin nombre virus	Rodent population increases
Human granulocytic ehrlichiosis (HGE)	1994	*Ehrlichia phagocytophila*	Increased habitat favoring reservoir and tick vector

northeastern United States is believed to have resulted from translocation of rabies virus–infected raccoons from Georgia to northern Virginia during the 1970s. Subsequently, transmission has expanded among the dense populations of raccoons found in the vast wooded and suburban areas of the northeastern states (Krebs et al. 1993). The reforestation and suburban development of much of the region, combined with decreased hunting and increased tolerance of wildlife living near human dwellings, created the ecological precondition for abundant raccoon hosts. Transmission among them has become intense, producing considerable risk of rabies to humans living nearby (Wilson et al. 1997).

Yet another example involves dengue fever. Its mosquito vector, *Ae. aegypti,* was spread throughout the world on transoceanic ships during the nineteenth century (Gubler 1989). For dengue virus to be spread over such great distances, however, infected people were needed, as they serve as the reservoir host, with any individual remaining infectious to the vector only for a few days. The worldwide tropical distribution of dengue seems to have occurred during the past few decades largely due to the rapid movement of infected people over vast distances; they have been able to introduce the virus into regions where *Ae. aegypti* populations were already present, thus permitting local transmission in new settings.

In the fall of 1999, the sudden appearance of severe and fatal encephalitis in people and many dying birds in and around New York City gave yet another example of the expanding range of an emerging disease. Studies have revealed that West Nile virus, heretofore known only in the Old World, had been introduced, apparently quite recently, into North America. Whether transatlantic movement was in infected birds, vector mosquitoes, or viremic humans is unknown, but it seems likely that rapid travel via aircraft would have allowed introduction. Molecular epidemiological studies indicate that the West Nile virus strain in the New York area is very closely related to a strain recently isolated in Israel (Lanciotti et al. 1999). Whether West Nile virus will continue to persist in North America and expand its range is presently unknown.

An alternative mechanism by which emerging diseases may increase their distribution involves gradual expansion, usually made possible by human-induced changes in environmental conditions. Various ecological changes such as deforestation, modification of river flow, or dense human crowding may locally influence the geographic pattern of a disease focus. Such changes over larger areas can have correspondingly large-scale effects on dispersal of the agent and disease. For example, increased regional distribution and incidence of schistosomiasis in Africa have been attributed to the impoundment of river water for irrigation. Altered patterns of water flow, nutrient loads, and vegetation have permitted expansion of competent snail vectors; human contact with water, including contamination with excrement, may increase as well, setting the stage for a larger area and greater human population affected by emerging schistosomiasis.

Some evidence and considerable speculation indicate that global climate change may be altering the distributions of diseases that are particularly sensitive to climate. The distribution of malaria, for example, may be gradually expanding in response to an increase in average temperatures. Not only are vector mosquitoes able to survive and better reproduce under warmer conditions, but the development time for the *Plasmodium* parasites in their cold-blooded vectors would be shortened with warmer temperatures, thereby increasing the amount of transmission. This may partly explain why malaria is being observed at higher elevations (Lindsay and Martens 1998). Warmer oceans may increase the risk of toxic algal blooms or of cholera epidemics if plankton that harbor *Vibrio* become more abundant (Harvell et al. 1999; see Chapter 11). Gradual expansion of or shifts in the potential distribution of many infectious diseases whose transmission is temperature-sensitive may occur if forecasts of global warming are correct (see Chapter 7).

Increased Local Incidence

Recognized infectious diseases that are increasing in incidence within a region also may be considered to be emerging. Causes for increasing local incidence are many, including migration of susceptible people into an area, environmental changes that enhance local transmission, or changes in pathogen virulence. In general, no single factor is responsible for increasing local incidence. Under certain circumstances, more aggressive surveillance or reporting may make it seem that incidence has increased, thus providing the impression that a disease is reemerging.

People who immigrate into a region where infectious diseases are already endemic generally are at risk if they have not been previously infected. Furthermore, ignorance of simple prevention practices may further result in increased transmission to such newcomers. Once exposed, lack of protective immunity is likely to result in more severe disease among individuals or greater incidence among the population, or both. Increased transmission may even indirectly affect long-term residents as the expanding population results in lowered herd immunity and more transmission. An example illustrating these processes involves the Brazilians who have migrated into previously heavily forested regions of the Amazon River basin where rapid deforestation for agriculture or gold mining is occurring (see also Chapter 12). The influx of people seeking employment into the area not only has augmented ecological changes in land use that favored *Anopheles* mosquito breeding, but also has exposed for the first time many persons who had not previously experienced malaria infection. What once was a feeble incidence of malaria in these regions quickly became epidemic. This example also illustrates how increased incidence can result from environmental changes that enhance local transmission.

Increased Severity, Duration, or Resistance to Treatment

Syndromes in which symptoms have become more severe or are less effectively treated make up the final category of emerging diseases. For these diseases, the causative pathogens have been well recognized and the therapy previously had been successful. What has changed is the responsiveness of the pathogen to treatment (antibiotic resistance), its efficiency of transfer or virulence (genetic drift or natural selection), or the host's natural ability to fight the pathogen (people with compromised immune systems). Some examples illustrate these processes.

Resistance to antibiotics has become a major concern in the treatment

of several infectious agents (Cohen 1992), most notably the common
bacteria *Staphylococcus aureus* and *Streptococcus pneumoniae,* the complex of bacteria producing tuberculosis (*Mycobacterium tuberculosis*), and
the species of *Plasmodium* that cause human malaria. Although environmental changes play little direct role in this process, fundamental principles of natural selection and evolution are operating. Such diverse factors
as the extent of treatment, the nature and site of antibiotic action, or the
genomic complexity of the parasite all influence the rate at which resistance is likely to appear. Very simply, antibiotics will not select against genetic variants that tolerate such treatment, thus allowing these strains to
flourish and be transmitted to other hosts. Such genetic strains will eventually be found with increasing frequency in the host population. Most
concern is focused on those microbes that cause anthroponoses, since
treatment of people infected with zoonotic agents does not result in selective pressure on the population of infecting organisms.

Increased efficiency of transmission or greater virulence may result
from genetic changes that occur by chance mutation, drift, or recombination and are selected because they increase the basic reproduction ratio, R_0.
Here again, ecological factors permeate this process, but the roles of specific factors are difficult to delineate. Examples include the likelihood that
genetic drift has increased the virulence of certain strains of measles virus
or that more highly virulent strains of influenza virus appear periodically,
perhaps because of genetic recombination in pig or duck hosts. The genomic structure of certain infectious agents could increase the likelihood
of significant genetic changes, as may occur with human immunodeficiency virus (HIV).

Humans respond physically to infections by producing fever, phago-
cytes, and agent-specific molecules in the immune system that kill or prevent replication of agents. An individual's capacity to produce an immune
response is one of the most important factors influencing the dynamics of
transmission of any infectious microbe. As that immune response is compromised, not only might more transmission occur, but also symptoms
may be more severe, long-lasting, and life-threatening. Among environmental factors that affect immunity, nutrition is perhaps the most important and best understood. In addition, various kinds of toxins and electromagnetic radiation are suspected in reducing the body's ability to tolerate
and eliminate pathogenic microbes. Another factor that recently has become increasingly important, particularly in developed countries, involves
the side effects of specific therapies that reduce immune function. Each of
these conditions can lead to a state in which people's immune systems become compromised, thus enhancing the expression or extent of disease.
Together, these constitute yet another mechanism by which some diseases
are considered to be emerging.

Evaluation of the Emergence of Disease

The importance of, responses to, and future development in emerging diseases are likely to differ widely among agents, conditions, and causes. Each mechanism outlined in this section, "Ecological Change and Emerging Diseases," presents different forces that allow the appearance of disease. Each of the four general categories of emerging diseases (newly described, expanded distribution, increased local incidence, and elevated severity/antibiotic resistance) has multiple processes and mechanisms, some operating very differently for various agents and in different situations. For these reasons, it is difficult to anticipate or predict what diseases are likely to emerge or where, when, or at what rate (Wilson 1994b).

Many of the ecological and evolutionary processes just described are inherently uncertain, with the exception of some that are directly related to human activities. For example, rapid, long-distance air travel is likely to increase in the years to come, thus allowing introduction of pathogens into new areas. Movements of humans and animals or the transport of infected material may lead to the emergence of disease. Whether this is more likely for diseases with particular types of transmission is worth considering. Perhaps diseases transmitted by certain pathways are more prone to transport and establishment and hence more likely to emerge in this manner. Directly transmitted anthroponoses, such as measles or herpes, solely require the introduction of the pathogen into a previously unaffected region for disease to emerge. Other diseases exhibiting more complex transmission cycles usually offer a greater number of vehicles for transport, yet often exhibit greater constraints on the conditions necessary for their maintenance. For example, if malaria is to become established after introduction by a parasitemic human, competent, human-biting anopheline mosquitoes must already exist in the region. Similarly, diseases such as schistosomiasis will become established only if the proper snail species are present. The highly complex cycles of indirectly transmitted zoonoses, such as Rift Valley fever, typically require a variety of competent vertebrate hosts and particular mosquito vector species for the virus to become established. Although seasonal variations sometimes permit brief epidemics that do not persist, the emergence of vector-borne zoonoses leading to endemicity may be relatively rare. A framework such as this allows us to question how, when, and where transport or movement of existing agents is more likely to lead to the emergence of disease.

We also might anticipate increased emergence of those diseases affected by human alteration of the environment or the manner in which humans interact in certain ecological cycles. For example, diseases such as Ebola and Venezuelan hemorrhagic fever appeared after changes in the patterns of human use of the environment allowed the causative agents to infect people. Increased use and disruption of previously underexploited habitats may result in the appearance of more zoonoses such as these. The deterioration of living standards in some developing countries, sometimes brought on by war, drought, rapid urbanization, and the global economy, is reducing access to clean water or proper housing. One result often is greater incidence of endemic enteric diseases and vector-borne diseases. Many other mechanisms can be imagined, making accurate forecasting of the emergence of disease very difficult.

Other methods might also be considered as a framework search for patterns of disease emergence. By classifying agents according to the immune response they invoke in people (e.g., whether protection is partial or complete, lifelong or short-lived), we may be better able to anticipate which introduced pathogens are more likely to emerge. Similarly, pathogens whose antigenicity or pathogenicity is more variable generally should be considered more likely candidates to emerge. Alternatively, comparing diseases by the host specificity of the pathogen (e.g., number and variety of organisms that can be successfully parasitized) may be useful. Agents whose host range is very narrow may be less likely to emerge than those that can survive in a wide variety of organisms or media.

Many of the same factors that contribute to the emergence of human diseases also apply to diseases of other organisms. The prevalence of wildlife diseases may increase because of factors such as translocation or migration of wildlife species or shifts in population density or immune status. These factors can affect the appearance and severity of new infections in wildlife populations, with risk of transfer to humans (Stohr and Meslin 1997). Similarly, domestic animals are at risk of such infections (e.g., Cheneau et al. 1999), which may display increasing resistance to antibiotics or be less effectively treated by antiparasitic drugs. Many of the same issues also apply to agricultural activities, particularly with regard to dissemination of viruses of plants, as well as the distribution and abundance of insect pests of plants. Thus, the principles and mechanisms of the emergence of disease are relevant to many infections and infestations of diverse organisms, most of which play an important role in our environment and, indirectly, our well-being.

Implications for Control and Risk Reduction

The foregoing discussion illustrates various ecological factors that affect the risk of infectious diseases in diverse ways and suggests that careful analysis of associations should permit the design of environmental change that minimizes such risk. Research has demonstrated the multiple pathways by which ecological variables influence components of transmission cycles of infectious disease. Furthermore, experience has shown how environmental changes resulting from deforestation, population shifts, changing human behaviors, or unintended side effects of economic development projects have led to increased or, in some cases, decreased transmission. Most such effects can no longer be dismissed as unanticipated outcomes of an otherwise sound policy or development plan. Indeed, both theoretical analyses and the historical record suggest that what was once considered "unexpected" should now be viewed as very likely. The simple notion that environmental changes will affect solely the proximate or obvious must be discarded and replaced with a modern view that intentionally and carefully asks how changing biotic and abiotic factors will influence the pathways of disease transmission. Multiple, indirect, and nonlinear effects should be considered the rule rather than the exception (see also Chapters 4 and 5). From this perspective, the challenge to environmental and public health scientists is to undertake analyses based on an understanding of process and history that produces realistic predictions.

Ecological Pattern, Uncertainty, and Prediction

General outcomes of environmental change often can be predicted by evaluating historical patterns and applying general principles; simple predictions, however, can be deceptively difficult. As noted above, the complexity of ecological pathways is such that the directions of effects may be opposite on different components of a cycle of disease transmission. Thus, small changes may produce large and diverse consequences. Indirect feedback may lead to counterintuitive effects that are not easily anticipated or recognized. Because of such complex interactions and multiple variables, it may never be possible to predict completely and precisely the outcomes of infectious disease resulting from ecological change. Nevertheless, predictions that forecast probable directions of modification or likely new events can be accurate, even if not precise. Even accurate predictions will be disease- and context-specific and are unlikely to include the time at which or extent to which outcomes will appear. Despite such difficulty and uncertainty, serious efforts to anticipate how environmental changes will affect infectious diseases must become part of economic development theory, public health practice, and integrated assessment of global ecosystem change (see Chapters 4 and 5).

Treatment, Prevention, and Control

Public health practices are aimed at prevention of new cases of disease, as opposed to the treatment of ill people. For some anthroponoses, treatment of disease also reduces the probability of new infections. For many sexually transmitted diseases (STDs), such as gonorrhea and syphilis, as well as water- or arthropod-borne anthroponoses, such as cholera, malaria, and onchocerciasis, treating individuals reduces their infectiousness and thus the probability of transmission. In essence, individual treatment is simultaneously population-level control. Many other diseases, however, with important ecological underpinnings are zoonoses (e.g., Lyme disease, hantavirus pulmonary syndrome, leishmaniasis or sleeping sickness), for which treatment of individual cases does nothing to reduce future transmission. In such situations, attention must be directed toward creating environments where individual exposures are decreased. At the level of populations, reducing the risk of infectious disease transmission is equivalent to "control" as commonly conceived. Such population-based control must serve as the basis for future efforts aimed at reducing risk through altering environments.

The ecological principles underlying population-based control of disease usually involve the frequency of infectious contacts or alteration of host responses. Given the strong role that ecological factors play in the transmission dynamics of so many diseases, we should strive to develop and apply basic ecological principles to the design of population-based programs to control disease.

The importance of vertebrate density to human risk of zoonotic disease, for example, is a complex function of many factors that depend on the natural mode of transmission of each microorganism (Wilson and Childs 1997). Risk generally increases as reservoir density increases but not necessarily in a linear manner. Under certain circumstances, increased density may not alter human risk or may even decrease risk, since the actual risk depends on the transmission dynamics among reservoirs and the mode of transfer to humans.

As the complexity of transmission or reservoir interactions increases, the relationship between reservoir density and risk becomes more tenuous. No simple formula can predict how changes in the environment and in reservoir density will affect zoonotic diseases. The particular transmission characteristics of each microorganism and its usual interaction with the vertebrate reservoir(s) must be analyzed. Then processes involving vectors, immunity, alternative hosts, and other factors may lead to counterintuitive results (Wilson and Childs 1997).

Effective control is ultimately aimed at modifying new, infectious contacts. Without careful evaluation, however, the long-term outcome of control efforts may be contrary to that originally sought. Other, sometimes unintended consequences of control activities may contradict or even negate the desired effects, in part because of the complex web of causation underlying strongly environmentally linked transmission cycles. The challenge faced by research scientists and public health workers is to anticipate, predict, and design for the unintended and sometimes indirect effects of efforts that alter multiple parts of the complex web of causation.

A Conceptual Framework for Integrated Control
One approach to control may involve efforts using multiple mechanisms, sometimes directed at many components of transmission. In addition to

education about risky behavior, for example, partner tracing, case treatment, and distribution of prophylactics may all be used to reduce the incidence of STDs. For environmentally linked anthroponoses and most zoonoses, local or regional interventions can involve vegetation, water resources, vector control, or animal reservoir populations. By definition, such efforts are in the realm of public health and will require participation and accord from many citizens. To be successful, such community-wide efforts will require cooperation of and input from the community. Each program will need to consider not only the ecological factors that influence risk, but also whether population immunity, vaccination, or other circumstances might be important. Ultimately, the network of transmission must be sufficiently well understood that interference at the level of the community, in the complex web of causation, does no harm.

Knowledge and Research Needs

The special characteristics of infectious diseases that are strongly linked to environment create particular problems for epidemiological and ecological research and information gathering. Classical laboratory experiments aimed at demonstrating survival, multiplication, or transmission of such agents cannot fully replicate diverse natural conditions. Not only do aspects of the environment influence these agents, but living organisms such as reservoirs, vectors, and amplifying hosts also affect transmission and the distribution of disease. Our understanding of the ecology of such ailments requires field observations and research that complement laboratory experiments. Long-term data gathering and surveillance are especially important as seasonal and interannual variation often is observed. As the kinds of information needed change, so must the design of observations, the tools used to gather and analyze these data, and the analytical approaches that are used.

Ecoepidemiological Data

Historically, long-term surveillance has been indispensable in studying how environmental changes affect the distribution and incidence of disease (see also Chapter 2). Without such systematically gathered epidemiological records, we lack the basic information needed to track and retrospectively analyze changes in diseases after ecological or demographic changes. Unfortunately, shrinking research budgets and an increasingly narrow focus on simple experiments that produce rapid results have meant that the number of long-term prospective observations has declined. Particularly in underdeveloped countries, standardized surveillance has suffered, reducing our ability to compare patterns of disease with the environmental data increasingly available from ground and satellite sources (see also Chapter 3). We need a renewed commitment to and respect for naturalistic observations and ecoepidemiological experiments aimed at understanding mechanisms and processes of environmental influences on infectious diseases.

Although the focus is often placed on emerging diseases, the approach should also include diseases that are on the decline because the local disappearance of diseases may provide basic insights into the links between ecological change and the transmission of infection. For example, echinostomiasis, which is a foodborne parasitic disease endemic in parts of Asia (Graczyk and Fried 1998), virtually disappeared from Lake Lindu Valley in central Sulawesi, Indonesia, between the 1950s and the 1970s because of the introduction of a new species of fish that fed on the clams primarily responsible for transmission to humans (Carney et al. 1980). Both to understand decline and to anticipate appearance, prospective multi-year monitoring designed to test hypotheses concerning ecological and health patterns should be encouraged. Critical environmental measures associated with particular diseases need to be identified for more detailed microenvironmental studies. Without a shift in the value placed on long-term naturalistic experiments, scientists will continue to be dissuaded from undertaking such important research.

Directed Fundamental Research

Although certain effects of environmental processes on infectious diseases can be studied through statistical association and natural observations, more traditional laboratory or field experimental manipulations sometimes are needed to analyze mechanisms and quantify the magnitude of interactions. Thus, the responses of growth rates to changing temperature or a shift in population due to increased competition, for example, may not be obvious without experimental intervention designed to quantify or enumerate interactions. For this reason, we need fundamental research aimed at particular interactions that are poorly understood or require quantification. Rather than experiments on better-known species or model interactions that have previously received attention, research organisms and interactions should be selected to illustrate basic properties and for their relevance to important infectious diseases. This could give rise to a new analytical ecology aimed at the reduction of disease risk by applying concepts and tools of traditional ecological research to organisms of public health importance. Furthermore, the recognition that complex interactions among physical, biological, and socioeconomic variables determine the probability of infection argues that multidisciplinary studies, including system dynamics of multivariable interactions, be encouraged in the health and ecological sciences. These could involve, for example, predominantly theoretical studies of complex dynamic behavior, spatial statistical investigations of the ecology of disease during ecological change, or integrative modeling of the effects of socioeconomic development on pathogen transmission.

New Perspectives and Methods

The analysis of multivariable interactions that may have spatiotemporal fluctuations, nonlinear rates of change, thresholds, or time lags requires different conceptual foundations and nontraditional analytic tools (see also Chapters 4 and 5). The usual disciplinary training of students may be inadequate if it lacks breadth across the natural and social sciences. Most training in the analysis of data focuses on description, statistical differences, or fitting data to predictions based on temporal or spatial associations. Less often are students provided with opportunities for developing skills in the analysis of process or modeling of dynamic interactions. In particular, development of and training in methods for the integrated analysis of interactions among qualitatively different variables are needed to address the complex processes that involve elements from many disciplines.

Conclusion

Ultimately, epidemiology and ecology must develop stronger links and borrow from each others' methods and paradigms. Ecology is very much a science of complex interconnection, of population-level analysis, of environmental impacts on organisms, and of change. Infectious diseases can be thought of as part of a larger "human ecology" in which human social systems, economic activities, interactions with the environment, and lifestyles represent some of many domains of interaction that influence risk. To reduce risk, whether immediate or long-term, whether locally or globally, will require an ecological conceptualization that accompanies detailed understanding of pathology and physiology. The challenge for the next few decades is made even more urgent as large-scale anthropogenic environmental changes are increasing the incidence of many old and new diseases alike. Not only should an ecological perspective on these processes produce deeper insight into important processes, but also it increases the likelihood that solutions will reduce disease risk for longer periods and for more people.

SUGGESTED STUDY PROJECTS

Suggested study projects provide a set of options for individual or team projects that will enhance interactivity and communication among course participants (see Appendix A). The Resource Center (see Appendix B) and references in all of the chapters provide starting points for inquiries. The process of finding and evaluating sources of information should be based on the principles of information literacy applied to the Internet environment (see Appendix A).

The objective of these three study projects is to grasp the diversity of the effects of ecological changes on infectious disease.

PROJECT 1: Ecological Characteristics and Processes Influencing the Transmission of Infectious Agents

The objective of this study project is to develop an understanding of environmental changes that influence the transmission of infectious agents.

Task 1. Using information provided in Table 10.1, select two or three types of environmental changes. Outline the pathways by which these changes increase or decrease the spread of infectious disease.

Task 2. Identify a location for each of the environmental changes selected in task 1. In each location, describe the specific effects on the spread of infectious disease.

Task 3. Compare the impacts of your selected environmental changes based on your findings in task 2.

Task 4. Summarize your results.

PROJECT 2: Cycles of Transmission of Infectious Diseases

The objective of this study project is to develop an understanding of four main types of cycles of transmission—directly transmitted anthroponoses and zoonoses and indirectly transmitted anthroponoses and zoonoses.

Task 1. Using information provided in Table 10.2, identify one example for each of the four system types and describe its transmission characteristics.

Task 2. Describe the possible effects of environmental changes on one or more of the examples selected in task 1.

Task 3. Search current literature for one or more of the examples selected in task 1. Critique the description of the effects of environmental changes in terms of accuracy and comprehensiveness.

PROJECT 3: Interdisciplinary Study in Ecological Change and Emerging Infectious Diseases

The objective of this project is to develop, practice, and refine research skills required for interdisciplinary studies in ecological change and the emergence of infectious diseases.

Task 1. Select a geographic area and describe its characteristics.

Task 2. Select a research strategy, using background information from this chapter and Part I of the book.

Task 3. Design your research plan.

Acknowledgments

I thank Betsy Foxman, James Koopman, and Pia MacDonald for useful discussions of this material. This work was supported in part by a grant from the National Institute of Allergy and Infectious Diseases (AI34409).

References

Anderson RM, Jackson HC, May RM, Smith AM. 1981. Population dynamics of fox rabies in Europe. *Nature* 289:765–71.

Anderson RM, May RM. 1991. *Infectious Disease of Humans: Dynamics and Control.* Oxford University Press, New York.

Aron JL. 2000. Mathematical modeling: The dynamics of infection. In *Infectious Disease Epidemiology: Theory and Practice* (Nelson KE, Williams CM, Graham NMH, eds.). Aspen Publishers, Gaithersburg, Md., Chap. 6.

Arthur RR, El-Sharkawy MS, Cope SE, Botros BA, Oun S, Morrill JC, Shope RE, Hibbs RG, Darwish MA, Imam IZE. 1993. Recurrence of Rift Valley fever in Egypt. *Lancet* 348 (8880): 1149–50.

Bacon PJ, ed. 1985. *Population Dynamics of Rabies in Wildlife.* Academic Press, New York.

Bakken JS, Dumler JS, Chen SM, Eckman MR, Van Etta LL, Walker DH. 1994. Human granulocytic ehrlichiosis in the upper midwest United States. *JAMA* 272:212–18.

Benenson AS, ed. 1995. *Control of Communicable Diseases Manual,* 16th edition. American Public Health Association, Washington, D.C.

Carney WP, Sudomo M, Purnomo A. 1980. Echinostomiasis: A disease that disappeared. *Trop Geogr Med* 32 (2): 101–5.

Centers for Disease Control and Prevention. 1994a. *Addressing Emerging Infectious Disease Threats: A Prevention Strategy for the United States.* Centers for Disease Control and Prevention, Atlanta.

———. 1994b. Raccoon rabies epizootic—United States, 1993. *MMWR* 43:269–73.

———. 1998. *Preventing Emerging Infectious Diseases: A Strategy for the Twenty-first Century.* Centers for Disease Control and Prevention, Atlanta.

Cheneau Y, Roeder PL, Obi TU, Rweyemamu MM, Benkirane A, Wojciechowski KJ. 1999. Disease prevention and preparedness: The Food and Agriculture Organization Emergency Prevention System. *Rev Sci Tech Office Int Epizooties* 18:122–34.

Childs JE, Krebs JW, Ksiazek TG, Maupin GO, Gage KL, Rollin PE, Zeitz PS, Sarisky J, Enscore RE, Butler JC, Cheek JE, Glass GE, Peters CJ. 1995. A household-based, case-control study of environmental factors associated with hantavirus pulmonary syndrome in the southwestern United States. *Am J Trop Med Hyg* 52 (5): 393–97.

Childs JE, Ksaizek TG, Spiropoulou CF, Krebs JW, Morzunov S, Maupin GO, Gage KL, Rollin PE, Sarisky J, Enscore RE, Frey JK, Peters CJ, Nichol ST. 1994. Serologic and genetic identification of *Peromyscus maniculatus* as the primary rodent reservoir for a new hantavirus in the southwestern United States. *J Infect Dis* 169:1271–80.

Childs JE, Peters CJ. 1994. Ecology and epidemiology of arenaviruses and their hosts. In *The Arenaviridae* (Salvato M, ed.). Plenum Press, New York, pp. 345–401.

Cohen ML. 1992. Epidemiology of drug resistance: Implications for a post-antimicrobial era. *Science* 257:1050–55.

Debbie JG. 1991. Rabies control of terrestrial wildlife by population reduction. In *The Natural History of Rabies,* 2d ed. (Baer RG, ed.). CRC Press, Boca Raton, Fla., pp. 477–84.

Duchin JS, Koster FT, Peters CJ, Simpson GL, Tempest B, Zaki SR, Ksiazek TG, Rollin PE, Nichol S, Umland E, Moolenaar RL, Reef SE, Nolte KB, Gallaher MM, Butler JC, Brieman RF, and the Hantavirus Study Group. 1994. Hantaviral pulmonary syndrome: Clinical description of disease caused by a newly recognized hemorrhagic fever virus in the southwestern United States. *N Engl J Med* 330:949–55.

Dye C. 1992. The analysis of parasite transmission by bloodsucking insects. *Annu Rev Entomol* 37:1–19.

Elliott LH, Ksiazek TG, Rollin PE, Spiropoulou CF, Morzunov S, Monroe M, Goldsmith CS, Humphrey CD, Zaki SR, Krebs JW, Maupin G, Gage K, Childs JE, Nichol ST, Peters CJ. 1994. Isolation of the causative agent of hantavirus pulmonary syndrome. *Am J Trop Med Hyg* 51:102–8.

Ewald PW. 1993. The evolution of virulence. *Sci Am* 268:86–93.

Fayer R, Ungar BLP. 1986. *Cryptosporidium* spp. and cryptosporidiosis. *Microbiol Rev* 50:458–83.

Fulhorst CF, Bowen MD, Salas RA, Duno G, Utrera A, Ksiazek TG, de Manzione NM, Miller E, Vasquez C, Peters CJ, Tesh RB. 1999. Natural rodent host associations of Guanarito and Pirital viruses (Family Arenaviridae) in central Venezuela. *Am J Trop Med Hyg* 61 (2): 325–30.

Gallaher MM, Herndon JL, Nims LJ, Sterling CR, Grabowski DJ, Hull HF. 1989. Cryptosporidiosis and surface water. *Am J Public Health* 79:39–42.

Garrett-Jones C. 1964. The human blood-index of malaria vectors in relation to epidemiological assessment. *Bull World Health Organ* 30:241–61.

Gavrilovskaya IN, Apekina NS, Bernshtein AD, Demina VT, Okulova NM, Myasnikov YA, Chumakov MP. 1990. Pathogenesis of hemorrhagic fever with renal syndrome virus infection and mode of horizontal transmission of hantavirus in bank voles. *Arch Virol* Suppl. 1: 57–62.

Graczyk TK, Fried B. 1998. Echinostomiasis: A common but forgotten food-borne disease. *Am J Trop Med Hyg* 58 (4): 501–4.

Gubler DJ. 1988. Dengue. In *The Arboviruses: Epidemiology and Ecology,* Vol. 2 (Monath TP, ed.). CRC Press, Boca Raton, Fla., pp. 223–61.

———. 1989. *Aedes aegypti* and *Aedes aegypti*–borne disease control in the 1990s: Top down or bottom up. *Am J Trop Med Hyg* 40 (6): 571–78.

Harvell CD, Kim K, Burkholder JM, Colwell RR, Epstein PR, Grimes DJ, Hofmann EE, Lipp EK, Osterhaus AD, Overstreet RM, Porter JW, Smith GW, Vasta GR. 1999. Emerging marine diseases—climate links and anthropogenic factors. *Science* 285 (5433): 1505–10.

Johnson KM. 1989. Hantaviruses. In *Viral Infections of Humans: Epidemiology and Control,* 3d ed. (Evans AS, ed.). Plenum Press, New York, pp. 341–450.

Krause RH, ed. 1998. *Emerging Infections.* Academic Press, San Diego.

Krebs JW, Strine TW, Childs JE. 1993. Rabies surveillance in the United States during 1992. *J Am Vet Med Assoc* 203:1718–31.

Krebs JW, Strine TW, Smith JW, Rupprecht CE, Childs JE. 1994. Rabies surveillance in the United States during 1993. *J Am Vet Med Assoc* 205:1695–1709.

Lake JV, Bock GR, Ackrill K, eds. 1993. *Environmental Change and Human Health.* J Wiley & Sons, New York.

Lanciotti RS, Roehrig JT, Deubel V, Smith J, Parker M, Steele K, Crise B, Volpe KE, Crabtree MB, Scherret JH, Hall RA, MacKenzie JS, Cropp CB, Panigrahy B, Ostlund E, Schmitt B, Malkinson M, Banet C, Weissman J, Komar N, Savage HM, Stone W, McNamara T, Gubler DJ. 1999. Origin of the West Nile virus responsible for an outbreak of encephalitis in the northeastern United States. *Science* 286 (5448): 2333–3337.

Lederberg J, Shope RE, Oaks SC, eds. 1992. *Emerging Infections: Microbial Threats to Health in the United States.* National Academy Press, Washington.

LeDuc JW. 1987. Epidemiology of Hantaan and related viruses. *Lab Anim Sci* 37:413–18.

LeDuc JW, Glass GE, Childs JE, Watson AJ. 1994. Hantaan virus and rodent zoonoses. In *Emerging Viruses: Evolution of Viruses and Viral Diseases* (Morse SS, ed.). Princeton University Press, Princeton, pp. 149–58.

Lee HW, Baek LJ, Johnson KM. 1982. Isolation of Hantaan virus, the etiologic agent of Korean hemorrhagic fever, from wild urban rats. *J Infect Dis* 146:638–44.

Lee HW, Lee PW, Baek LJ, Song CK, Seong IW. 1981. Intraspecific transmission of Hantaan virus, etiologic agent of Korean hemorrhagic fever, in the rodent *Apodemus agrarius*. *Am J Trop Med Hyg* 30:1106–12.

Lee HW, Lee PW, Johnson KM. 1978. Isolation of the etiologic agent of Korean hemorrhagic fever. *J Infect Dis* 137:298–308.

Lindsay SW, Martens WJM. 1998. Malaria in the African highlands: Past, present and future. *Bull World Health Organ* 76 (1): 33–45.

MacDonald G. 1957. *The Epidemiology and Control of Malaria*. Oxford University Press, London.

MacKenzie WR, Hoxie NJ, Proctor ME, Gradus MS, Blair KA, Peterson DE, Kaxmierczak JJ, Addiss DG, Fox KR, Rose JB, Davis JP. 1994. A massive outbreak in Milwaukee of *Cryptosporidium* infection transmitted through the public water supply. *N Engl J Med* 331:161–77.

Mercado RR. 1975. Rodent control programmes in areas affected by Bolivian haemorrhagic fever. *Bull World Health Organ* 52:691–96.

Mills JN, Ellis BA, Childs JE, McKee KT Jr, Maiztegui JI, Peters CJ, Ksiazek TG, Jahrling PB. 1994. Prevalence of infection with Junin virus in rodent populations in the epidemic area of Argentine hemorrhagic fever. *Am J Trop Med Hyg* 51 (5): 554–62.

Mills JN, Ellis BA, McKee KT Jr, Calderon GE, Maiztegui JI, Nelson GO, Ksiazek TG, Peters CJ, Childs JE. 1992. A longitudinal study of Junin virus activity in the rodent reservoir of Argentine hemorrhagic fever. *Am J Trop Med Hyg* 47:749–63.

Monath TP. 1989. Yellow fever. In *The Arboviruses: Epidemiology and Ecology*, Vol. 5 (Monath TP, ed.). CRC Press, Boca Raton, Fla., pp. 139–231.

Morse SS, ed. 1994. *Emerging Viruses: Evolution of Viruses and Viral Diseases*. Princeton University Press, Princeton.

Navin TR, Juranek DD. 1984. Cryptosporidiosis: Clinical, epidemiological, and parasitologic review. *Rev Infect Dis* 6:313–27.

Nyström K. 1982. Epidemiology of HFRS (endemic benign nephropathy—EBN) in Sweden. *Scand J Infect Dis Suppl* 36:92.

Parmenter RR, Brunt JW, Moore DI, Ernest S. 1993. *The Hantavirus Epidemic in the Southwest: Rodent Population Dynamics and the Implications for Transmission of Hantavirus-associated Adult Respiratory Distress Syndrome (HARDS) in the Four Corners Region*. Sevilleta LTER Publication No. 41. University of New Mexico, Albuquerque.

Parmenter RR, Vigil R. 1993. *The HARDS Epidemic in the Southwest: An Assessment of Autumn Rodent Densities and Population Demographics in Central and Northern New Mexico, October 1993*. Sevilleta LTER Publication No. 45. University of New Mexico, Albuquerque.

Public Health Laboratory Service Study Group. 1990. Cryptosporidiosis in England and Wales: Prevalence and clinical and epidemiological features. *Br Med J* 300:774–77.

Roizman B, ed. 1995. *Infectious Diseases in an Age of Change: The Impact of Human Ecology and Behavior on Disease Transmission*. National Academy Press, Washington.

Salas R, de Manzione N, Tesh RB, Rico-Hesse R, Shope RE, Betancourt A, Godoy O, Bruzual R, Pacheco ME, Ramos B, Taibo ME, Tamayo JG, Jaimes E, Vasquez C, Araoz F, Querales J. 1991. Venezuelan haemorrhagic fever. *Lancet* 338 (8774): 1033–36.

Simson VR. 1992. Cryptosporidiosis in newborn red deer (*Cervus elaphus*). *Vet Rec* 130:116–18.

Skeels MR, Sokolow R, Hubbard CV, Andrus JK, Biasch J. 1990. Cryptosporidium infection in Oregon public health clinic patients, 1985–88: The value of statewide laboratory surveillance. *Am J Public Health* 80:305–8.

Smith JS, Orciari LA, Yager PA, Seidel HD, Warner CK. 1992. Epidemiologic and historical relationships among 87 rabies virus isolates as determined by limited sequence analysis. *J Infect Dis* 166:296–307.

Spielman A, Kimsey RB. 1991. Zoonosis. In *Encyclopedia of Human Biology*, Vol. 7. Academic Press, San Diego, pp. 891–900.

Steck F, Wandeler A. 1980. The epidemiology of fox rabies in Europe. *Epidemiol Rev* 2:71–96.

Stohr K, Meslin FX. 1997. The role of veterinary public health in the prevention of zoonoses. *Arch Virol* 13 (Suppl): 207–18.

Tesh RB, Wilson ML, Salas R, de Manzione NMC, Tovar D, Ksiazek TG, Peters CJ. 1993. Field studies on the epidemiology of Venezuelan hemorrhagic fever: 1. Implications of the cotton rat *Sigmodon alstoni* as the probable source of infection to humans. *Am J Trop Med Hyg* 49:227–35.

Tzipori SK, Angus W, Campbell I, Sherwood D. 1981. Diarrhea in young red deer associated with infection with *Cryptosporidium*. *J Infect Dis* 144:170–75.

Walker DH, Barbour AG, Oliver JH, Lane RS, Dumler JS, Dennis DT, Persing DH, Azad AF, McSweegan E. 1996. Emerging bacterial zoonotic and vectorborne diseases: Ecological and epidemiological factors. *JAMA* 275 (6): 463–69.

Wilson ME, Levins R, Spielman A, eds. 1994. *Disease in Evolution: Global Changes and the Emergence of Infectious Diseases.* New York Academy of Sciences, New York.

Wilson ML. 1994a. Population ecology of tick vectors: Interaction, measurement and analysis. In *Ecological Dynamics of Tick-Borne Zoonoses* (Sonenshine DE, Mather TN, eds.). Oxford University Press, New York, pp. 20–44.

———. 1994b. Rift Valley fever virus ecology and the epidemiology of disease emergence. In *Disease in Evolution: Global Changes and the Emergence of Infectious Diseases* (Wilson ME, Levins R, Spielman A, eds.) New York Academy of Sciences, New York, pp. 169–80.

Wilson ML, Bretsky PM, Cooper GH, Egbertson S, Van Kruiningen HJ, Cartter ML. 1997. Emergence of raccoon rabies in Connecticut, 1991–1994: Spatial and temporal characteristics of animal infection and human contact. *Am J Trop Med Hyg* 57 (4): 457–63.

Wilson ML, Childs JE. 1997. Vertebrate abundance and the epidemiology of human zoonoses. In *The Science of Overabundance: Deer Ecology and Population Management* (McShea WJ, Underwood HB, Rappole JH, eds.). Smithsonian Institution Press, Washington, pp. 223–48.

Winkler WG, Jenkins SR. 1992. Raccoon rabies. In *The Natural History of Rabies,* 2d ed. (Baer RG, ed.). CRC Press, Boca Raton, Fla., pp. 325–40.

Yanagihara R, Amyx HL, Gajdusek DC. 1985. Experimental infection with Puumala virus, the etiologic agent of nephropathia epidemica, in bank voles (*Clethrionomys glareolus*). *J Virol* 55:34–38.

CASE STUDIES

In Part III, the purpose is to present examples of public health issues that demonstrate interactions among the various themes of the book. The first case study is on cholera, one of the few bacterial diseases that continue to cause pandemics (Chapter 11). Recent research on the aquatic environment has shown that the causative agent of cholera multiplies in association with plankton independently of humans. A series of studies suggests that changes in the environment, rather than simply human migration and the fecal-oral route of transmission of the disease, may be a primary factor in the rapid pandemic spread of cholera. The case study on malaria also focuses on the relationship between an infectious disease and global ecosystem change (Chapter 12). Global changes in the growth and movement of populations, patterns of economic development, and climate are changing the risk of exposure of human populations to malaria. Examples from Zimbabwe, Gambia, Niger, Sri Lanka, Brazil, and the United States demonstrate complex interconnections between these global changes and the local situations. Another case study examines interactions among atmospheric changes and their implications for human health (Chapter 13). Traditional forms of air pollution and greenhouse gases interact in several ways: the use of fossil fuels serves as a common source, the concentrations of some air pollutants are expected to increase under projections of enhanced global warming due to anthropogenic emissions of greenhouse gases, and the effects of air pollutants and projected changes in heat stress due to enhanced global warming may possibly exhibit synergy. The primary focus on air pollutants is augmented by a discussion of other possible pathways of the effect of global climate change on threats to public health. The last case study examines the importance of water resources through four examples that illustrate the diverse ways in which too little or too much water can adversely affect human health (Chapter 14). Cholera in a refugee camp in Africa due to a shortage of clean drinking water contrasts with the serious health effects of flooding in Brazil. The consequences of diversions of water for agricultural or municipal use are shown in terms of massive ecological deterioration in the Aral Sea basin and, on a smaller scale, an increase in dusty air in California.

See Part I for approaches to research. See Part II for background on environmental changes on a global scale.

CHAPTER 11

Cholera and Global Ecosystems

Anwar Huq, Ph.D., R. Bradley Sack, M.D., Sc.D., and Rita R. Colwell, Ph.D., D.Sc.

Cholera is an ancient disease, widely assumed to be native to the Indian subcontinent and a scourge for thousands of years. Periodically, however, during the last 170 years, cholera has occurred worldwide as a pandemic disease, involving human populations geographically distributed over nearly the entire globe. Indeed, cholera is one of the few bacterial diseases that continue to cause pandemics. The properties that give *Vibrio cholerae,* the causative agent of cholera, such "dispersability" (Craig 1996) are still unknown, but the ecology of *V. cholerae* provides some clues. It is recognized that increased populations in urban areas, especially those with less than adequate water and sanitation, facilitated the spread of this disease. During the last 35 years, cholera has again emerged in areas outside its endemic home in Asia, causing pandemic disease in Africa and South America, geographic areas not experiencing the disease for almost 100 years. Although there is no facile explanation for these phenomena, it was widely assumed, until relatively recently, that the disease was spread only by infected humans to other susceptible persons via fecal contamination of water and food and that global movement of human populations accounted for global "movement" of the disease. Recent studies of the aquatic environment, however, have shown that *V. cholerae* is a normal inhabitant of surface water and survives and multiplies in association with plankton, quite independently of infected humans (Huq et al. 1983, 1990; Islam et al. 1990b). Since climate affects the growth of plankton, it is reasonable to assume that growth of vibrios associated with plankton also will be influenced. Results of studies conducted during the past 20 years on the ecology of *V. cholerae* and the recently recognized significance of global climate in human health suggest that the continuing presence of cholera in the Indian subcontinent and the reemergence of cholera in

327

other countries or continents may be significantly related to environmental factors.

This chapter reviews data suggesting that changes in the environment, rather than simply human migration and fecal-oral route of transmission of the disease, may be a primary factor in the rapid pandemic spread of cholera. Available information on relevant environmental parameters are discussed, with the goal of predicting and, ultimately, preventing the spread of cholera.

The History of Cholera

Since detailed accounts of the history of cholera are available (Barua 1988; Barua and Greenough 1991; Pollitzer 1959), only a brief summary is provided here.

The First Six Pandemics

The ancient history of cholera dates back thousands of years. Cholera has been endemic in the Ganges delta region from the time of recorded history in that part of the world. The clinical disease was well described in ancient Sanskrit writings, estimated to have been written about 400–500 b.c., which also included methods for its treatment (Barua 1988). However, the first large outbreak of cholera in India, according to written accounts, occurred in 1503 (Barua 1988). Because cholera occurred regularly in epidemic form and had a high mortality rate (up to 70%), it was a much-feared disease among both indigenous populations and visitors from abroad, particularly from Europe.

Between 1817 and 1925, six cholera pandemics were recorded in different parts of the world (Pollitzer 1959). The first pandemic began in 1817, when cholera occurred outside the Indian subcontinent in geographic regions as distant as the Middle East, China, and Japan. During the second pandemic, beginning in 1829, cholera occurred in Europe and the New World. Four more pandemics, each lasting 7–24 years and continuing through 1925, involved Africa, Australia, Europe, and all of the Americas. The causative agent, *V. cholerae,* was not identified until 1884, during the fifth pandemic (Koch 1884). From 1925 until the most recent pandemic (the seventh), which began in 1961, cholera was confined to the Asian subcontinent, except for sporadic outbreaks; one such outbreak occurred in Egypt in 1947, and another, smaller one occurred in Syria at approximately the same time.

Why these pandemics began and why they terminated are not known. Cholera, the disease, did not become endemic in any of the new geographic areas where it occurred, clearly unlike the situation in the Ganges delta, where cholera remains unabated to this day. Because of the large numbers of cases and deaths during these pandemics, the disease was viewed as a major public health problem to be dealt with by governmental control. Indeed, the epidemic of cholera in New York City was the primary reason that the first board of health in the United States was formed in 1866 (Duffy 1971).

The Seventh Pandemic and Early Signs of an Eighth Pandemic

The seventh pandemic now has involved almost the entire world, including Europe, Africa, and South America, continents that had been free of cholera for the past century (Fig. 11.1). This pandemic is unlike any of the previous ones in several respects. The pandemic began in Indonesia, rather than in the Ganges delta, as had previous ones, and the organism causing this pandemic was a biotype of *V. cholerae* serogroup O1 called El Tor, which differs in certain taxonomic characteristics from the "classical" biotype that caused the fifth and sixth pandemics (see Box 11.1). The first four pandemics are also believed to have been caused by the classical biotype, but, since *V. cholerae* was not identified until 1884, there is no information to confirm this.

El Tor vibrios were first isolated in 1906 in a quarantine station in the village of El Tor, Egypt, from pilgrims returning from the Haj. The pilgrims did not have clinical cholera, although some had a diarrheal illness (Tanamal 1959; Barua 1988). El Tor vibrios were not recognized as a cause of diarrheal disease, however, until they were isolated from patients with severe watery diarrhea in Sulawesi (Celebes), Indonesia, in 1937. During the next 20 years in Sulawesi, there were four outbreaks of cholera-like illness, often with a high (80%) mortality (Tanamal 1959). This cholera-like disease was originally called *paracholera* to differentiate it from classical cholera, but later investigations determined that El Tor and classical cholera were clinically identical. In 1960, for unknown reasons, El Tor began to spread around the world. This pandemic is now in its 39th year and still shows no signs of receding, making it longer in duration than any of the previous pandemics, the longest of which was 24 years. The disease is considered to be endemic in many places where it now occurs, particularly in Africa. It also seems to be establishing itself in Central and South America (having occurred there during the past eight years); endemicity did not develop after any of the previous pandemics.

In 1992, a newly described non-O1 serogroup of *V. cholerae*, subsequently designated as serogroup O139 (see Box 11.1 and Fig. 11.1), presumed to be the eighth pandemic in its infancy, dramatically appeared in India and, within a few months, in Bangladesh, causing widespread epidemics of cholera (Ramamurthy et al. 1993; Cholera Working Group 1993). There are presently more than 152 serogroups of *V. cholerae*, with new ones continuing to be identified. At the time of discovery of O139, the non-O1, non-O139 serogroups were also referred to as *noncholera vibrios*.

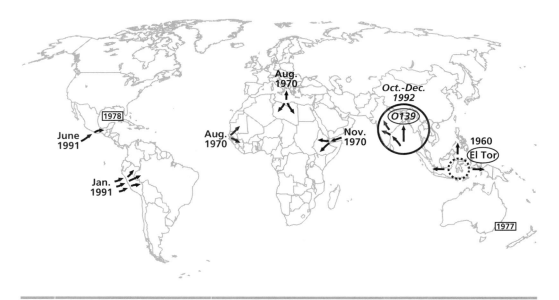

Figure 11.1 Selected aspects of the seventh and incipient eighth pandemics of cholera. *Arrows* show invasions by cholera; *boxes* show localized, distinct clones of vibrios that have not invaded. The origin of the seventh pandemic is shown in Sulawesi (Celebes), Indonesia. All designations refer to El Tor cholera except in the Bay of Bengal, where O139 cholera invaded in what may be the early stages of an eighth pandemic. A full map of the seventh pandemic of cholera can be found in Glass and Black (1992).

These strains produced cholera toxins and caused diarrheal disease, but none had been associated with epidemic disease, as had been O1 and now O139. Over the next several years after its first appearance, the O139 serogroup of *V. cholerae* accounted for large outbreaks of cholera, first in India, then Bangladesh, and subsequently in many surrounding countries, such as Thailand, Malaysia, China, and Pakistan. In many places, it temporarily displaced the O1 vibrios. It diminished in prevalence during the next two or three years but did not disappear. By the summer of 1997, it was again the predominant serogroup throughout India, while, in Bangladesh, the O1 serogroup remained predominant. At present, both serogroups are found in India and Bangladesh, with the O1 serogroup generally being more commonly isolated than the O139 serogroup.

Vibrio cholerae serogroup O139 is closely related to the epidemic serogroup of *V. cholerae* O1 El Tor but possesses a capsule, and the capsular layer is distinct from the lipopolysaccharide (LPS) antigen in the cell wall (Waldor et al. 1994; Waldor and Mekalanos 1994). Although *V. cholerae* serogroup O139 is distinctly different from O1 vibrios, changes from one serogroup of vibrios to another have been observed; cells of *V.*

Biological Classification of *Vibrio cholerae*

Vibrio cholerae, the bacterial organism that causes cholera, exhibits a great deal of biological diversity that is classified according to serogroups, serotypes, and biotypes. Serogroups and serotypes are based on O antigens of the cell wall, which are specific biological components affecting the immunological response to the organism. Biotypes are based on phenotypic differences, which are biological abilities or sensitivities of the organism. Biological variation is important in understanding the epidemic potential of cholera.

Approximately 150 serogroups of *Vibrio cholerae* have been identified. Only serogroups O1 and O139 cause epidemic cholera; both serogroups produce identical cholera toxin. Serogroup O1 has two major serotypes (Inaba and Ogawa), each of which has two biotypes (classical and El Tor). Serogroup O1 also contains a rare serotype, Hikojima. No serotypes have been identified for serogroup O139.

One aspect of biotype is whether enzymes from the organism cause the red blood cells to lyse. The El Tor biotype has this capability, while the classical biotype does not. Vibrios may also differ in their sensitivity to bacteriophages.

---

cholerae O1 were detected in pure cultures of non-O1 *V. cholerae* in a laboratory microcosm experiment using fluorescent antibodies (Colwell et al. 1995). This suggests that non-O1 or non-O139 *V. cholerae* cannot be ignored and need to be carefully and closely monitored in the environment.

Epidemics caused by O139 are a significant turning point in the history of cholera because the evidence points to this strain having arisen from genetic recombination, horizontal gene transfer, and acquisition of unique DNA (Waldor and Mekalanos 1996). This horizontal gene transfer refers to a naturally occurring process by which a bacteriophage (virus) transfers genetic material from one bacterium to another. It is, therefore, necessary that both O1 and O139 antigens be used in constructing diagnostic tests for cholera. Since a new serogroup is associated with cholera epidemics, at this time it can only be speculated as to whether *V. cholerae* O139 will eventually also cause a global pandemic. Because the O antigens of the two serogroups are not cross-reactive, previous infection with *V. cholerae* O1 does not confer protection against cholera caused by serogroup O139. Thus, the entire world population may, at least theoretically, be susceptible to this new form of cholera.

Old and New Beliefs about How Cholera Occurs

In ancient times, it was believed that cholera was a punishment from the gods. In the early 1880s in Calcutta, India, a goddess of cholera (Ola Bebee, "Our Lady of the Flux") was designated and worshiped (De 1961). In the times of the first pandemics of cholera in Europe and the New World, it was thought that cholera was a disease of prostitutes, the poor, and the lazy and that God was punishing people with this plague. When Pacini discovered the cholera vibrio in 1884, it was possible to determine why and how people became infected. Even before the documented discovery of the causative organism and its description by Robert Koch, however, John Snow showed conclusively that cholera in London was transmitted by contaminated water (Snow 1855). That fecal contamination of water and food could introduce the vibrios was later demonstrated microbiologically and, since no animal reservoirs were found and the only habitat of the vibrio then known was the human intestine, most public health practitioners assumed that the disease was transmitted only by fecal contamination originating from infected persons. This hypothesis, however, did not fit all observations about the behavior of cholera. In endemic areas, for instance, epidemics of cholera would often begin in multiple sites at the same time, without a clear chain of personal transmission. Also, during the recent pandemic spread of cholera in a "virgin" area, like Peru in 1991, cholera occurred almost simultaneously in several locations along the Pacific coast, suggesting something other than simply the dissemination of the disease by infected persons (Seas et al., in press). Furthermore, in Australia, transmission of cholera was traced to rivers that were not fecally contaminated (Desmarchelier et al. 1995).

An Update on the Ecology of *Vibrio cholerae*

Historically, most of the outbreaks and epidemics of cholera during the past two decades have originated in coastal areas (see Fig. 11.1 and Glass and Black 1992). In addition, cholera has been detected in a specific geographic location only mysteriously to "disappear" and remain absent for a few years to more than 100 years (Barua 1988). This erratic behavior of cholera is unlikely to be explained simply by dense populations, poor hygiene, or social and cultural peculiarities of the regions.

Vibrio cholerae O1 and the newly recognized serogroup O139, both causative agents of cholera, are the most studied members of the genus *Vibrio* because of their ability to cause epidemic disease of public health importance. The prevailing view of the ecology of cholera was that *V. cholerae* was associated only with humans, until Colwell hypothesized and later demonstrated that *V. cholerae* is an indigenous, autochthonous member of brackish and estuarine environments (Colwell et al. 1977). The autochthonous nature of *V. cholerae* has proven to be an important factor in understanding the epidemiology of cholera, promoting new insight leading to new knowledge and possible approaches to prevention of cholera. Eradication of this organism is very unlikely because of its natural habitat and interactions in estuarine ecosystems.

Vibrio cholerae O1 has been detected or isolated from almost everywhere in the world where there is water. Prior to 1970, nearly all of the methods used for isolation and characterization of this organism had been developed for clinical diagnosis of cholera in hospital laboratories (Finkelstein 1973). During the last decade, new methods based on molecular biology were developed or optimized to examine environmental samples and, as a result, *V. cholerae* O1 has been isolated or detected in ponds, rivers, lakes, and seawater samples collected from all five inhabited continents.

Bacteriological culture methods have not always been successful in isolating *V. cholerae,* and for many years microbiologists were frustrated as they sought an explanation for the "mysterious" disappearance of *V. cholerae* O1 from the environment during interepidemic periods, notably in countries like Bangladesh, where cholera is endemic. Discoveries of the past decade have revealed the existence of a dormant (i.e., viable but nonculturable) state, into which *V. cholerae* O1 enters in response to nutrient deprivation and other environmental conditions (Colwell et al. 1985). This finding led to the demonstration of the presence of *V. cholerae* O1 throughout the year in the aquatic environment in Bangladesh (Huq et al. 1990), a finding later confirmed by other studies (Islam et al. 1994). *V. cholerae* O1 in the viable but nonculturable state maintains its virulence and is capable of producing disease, as demonstrated in a rabbit model and in experiments using human volunteers (Colwell et al. 1990, 1996).

The association of *V. cholerae* with changing environmental conditions, particularly changes in the pH of water where phytoplankton are abundant, was first addressed by Cockburn and Cassanos in 1960. They demonstrated a correlation between the incidence of cholera and the presence of increased numbers of chlorophyll-bearing organisms in water. This correlation was further explained as being the result of elevated pH and increased dissolved oxygen caused by photosynthesis of blue-green algae, conditions supporting growth of *V. cholerae.* However, these investigators failed to note the significance of zooplankton blooms that follow phytoplankton blooms (see below). *Vibrio cholerae* produces an enzyme, mucinase, that actively degrades mucin and mucinlike substances present in plant cells (Schneider and Parker 1982). Silvery and Roach (1964) demonstrated that, when a large number of algae begin to disintegrate after the peak bloom, the gram-negative heterotrophic bacterial population, including *V. cholerae,* begins to increase. Heterotrophic bacteria obtain energy from one or more organic compounds and use nutrients released by disintegrating phytoplankton. The authors further noted, from careful observation, that the gelatinous cover of the blue-green filaments contained high concentrations of gram-negative, rod-shaped bacteria (Silvery and Roach 1964). The mucilaginous sheath of blue-green algae (*Anabaena variabilis*) was found to serve as a site of attachment for *V. cholerae* cells (Pearl and Keller 1979; Islam et al. 1990a, 1990b).

A direct detection method revealed that cells of *V. cholerae* persisted for about 15 months in the mucilaginous sheath of *A. variabilis* (Islam et al. 1990a, 1994). Thus, bacterial association with algae may be important in persistence of the organism but not in its amplification. In any case, *V. cholerae* is directly dependent on environmental conditions. However, specificity of the association of blue-green algae with *V. cholerae* has not been observed. Silvery and Roach (1964) detected only one large peak of an algal bloom occurring during an annual cycle, suggesting that, once an algal bloom disintegrates, the bacterial cells require another mechanism of survival and multiplication in the environment to cause later epidemics during a given year. Epidemics of cholera in Bangladesh, for example, peak twice a year.

The association of *V. cholerae* with microscopic crustaceans, a major component of zooplankton, was found to be important because of the chitin composition of the carapace of these animals. Attachment of different species of vibrios, including *V. cholerae* O1, to chitin particles and their protection from lethal actions of an acidic environment were demonstrated by Kaneko and Colwell (1973) and Nalin et al. (1977), respectively. Such protection can be important to *V. cholerae* exposed to gastric acid or in water with an acidic pH in the natural environment. In addition, chitin was found to exert protection for *V. cholerae* O1 against the effects of low temperature, suggesting extended survival of cholera vibrios in the environment, especially at freezing temperatures (Amako et al. 1987). In laboratory microcosm experiments, *V. cholerae* O1 and O139 were also resistant to the effects of alum and maintained culturability for longer periods when attached to copepods (Chowdhury et al. 1997) or to chitin particles (Huq and Colwell 1996). Earlier, in a field study in Bangladesh, viable but nonculturable cells of *V. cholerae* O1 and O139 were found to be attached to chitin particles (Huq et al. 1995).

According to a recent study, surface proteins of *V. cholerae* play an important role in attachment by vibrios to chitin particles (Renato and Pruzzo 1999). Since vibrios, including *V. cholerae* (Colwell 1970), produce chitinase, an enzyme that degrades chitin and, thereby, supplies nutrients needed for survival (Hood and Meyers 1977), enhanced survival of *V. cholerae* O1 was demonstrated in laboratory experiments by Huq et al. (1983) in the presence of tiny crustacean copepods. Other investigators have confirmed this finding, namely, that *V. cholerae* O1 survived significantly longer in laboratory microcosms when attached to copepods (Dumontet et al. 1996). However, these investigators did not observe differences in attachment between live and dead copepods, as reported by Huq et al. (1983).

Besides the natural occurrence of *V. cholerae* in aquatic environments, mechanical transportation has been hypothesized to play a significant role in the spread of the vibrio in the environment. Ship ballast has been a factor in the rapid distribution of marine vertebrates and invertebrates globally (Carlton and Geller 1993). Examination of ballast water revealed transmission of a wide variety of phytoplankton and zooplankton in the marine environment, as well. Copepods are usually the most abundant group of zooplankton (Miller 1984; Parsons et al. 1984), and copepods have been found in ballast water in large numbers (Carlton and Geller 1993). Ruiz et al. (2000) have observed that more than 90 percent of the plankton in ballast water is copepods, especially in the larval and juvenile stages. Within the last 15 years, six species of Asian copepods have been introduced to the Pacific coast of North America (Carlton and Geller 1993). Since copepods are hosts for *V. cholerae* O1, transport of the cholera vibrios can occur.

The seasonality of cholera in most areas of the world is well documented, usually at low levels in the winter and peaking during the summer when the temperature increases (Tauxe et al. 1994) (Fig. 11.2*a*). In Bangladesh, there are two distinct seasons of cholera epidemics, one major and the other minor. The major peak is during the months of September through November, and the minor peak is from January through April or May (Fig. 11.2*b*). According to Oppenheimer et al. (1978), zooplankton populations decrease during the monsoon (rainy) season in June and July because of reduced levels of nutrients; during August and September, after the monsoon season, the levels of nutrients significantly increase and blooms of phytoplankton followed by zooplankton occur. Kiorboe and Neilsen (1994) published a fascinating report on an extensive study of copepod production, demonstrating two distinct seasons for the production of several species of copepod eggs. One is February through April, and the other is during the months of August and September. This fits very well into the original hypothesis of Huq et al. (1983) that copepods play an important role in the survival, multiplication, and transmission of *V. cholerae* in the natural aquatic environment.

Scanning electron microscopy (SEM) showed that the egg cases and the oral regions of copepods were sites of significantly enhanced attachment by *V. cholerae* (Huq et al. 1981, 1983; Dumontet et al. 1996). The findings of Kiorboe and Neilsen (1994) on production of copepod eggs also fit the hypothesis of the role of zooplankton in *V. cholerae* ecology and are particularly important because, as freely swimming cells, *V. cholerae*, even in the nonculturable state, can attach to copepod eggs and multiply rapidly (Huq et al. 1981, 1983). Once the number of *V. cholerae* O1 cells is large enough to constitute an infectious dose (i.e., 10^4–10^6 cells per milliliter, depending on the state of health of the victim) (Hornick et al. 1971) and this dose is ingested by humans while bathing, swimming, or drinking untreated raw water from ponds, rivers, and lakes of cholera-endemic countries, notably Bangladesh, epidemics can be triggered (Huq and Colwell 1996).

zooplankton move in ballast of ships

seasonality of cholera corresponds to zooplankton blooms

(a)

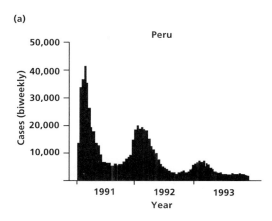

(b)

Figure 11.2 *a:* Seasonal cholera in Peru, 1991–93. The warm months of the year are December through March. *Source:* Reproduced with permission from Koo et al. 1996, Figure 2. *b:* Seasonal cholera (classical and El Tor) in Matlab, Bangladesh, 1981–91. Figures are the average number of cases per month for this 11-year period. *Source:* Data from the International Centre for Diarrhoeal Disease Research, Bangladesh (ICDDRB).

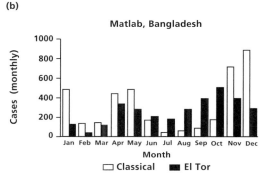

According to the Continuous Plankton Recorder survey, a survey of the northeast Atlantic established by Sir Alister Hardy in the 1930s and carried out each month from 1948 to the present, phytoplankton and zooplankton populations declined from 1948 to 1980 (Taylor 1991). The population of copepods, the dominant component of zooplankton in terms of biomass, diminished sixfold. However, since 1980, the trend has reversed (Taylor 1991). Gelbspan (1991) reported that algal blooms surpassing the usual spring bloom were observed during the spring months at various points worldwide, including North Carolina, California, Finland, Iceland, Tasmania, Japan, and Thailand. Phytoplankton are the "primary producers" of organic compounds (Hardy 1991), and the size of zooplankton populations increases significantly after algal blooms because zooplankton graze on phytoplankton.

Changes in climate may directly and indirectly affect the ecology of all living creatures. Fluctuations in weather conditions have minor to major effects on the environment that may bring ecological changes such as soil erosion, deforestation, drought, and human migrations from rural to urban areas. A hypothetical model, proposed by Huq et al. (1988) and subsequently modified by Colwell and Huq (1994a), suggests a role for the various physicochemical and biological factors of the aquatic environment in the transmission of *V. cholerae* O1. This is the first model to suggest that phytoplankton blooms and subsequent zooplankton blooms have direct or indirect effects on physicochemical parameters that play a role in the survival, multiplication, and transmission of *V. cholerae* in the natural environment. Subsequently, Islam et al. (1994) proposed a model describing the relationship of different environmental parameters to the formation of phytoplankton blooms and the multiplication of *V. cholerae*. The direct influence of sunlight and dissolved chemical nutrients on aquatic plants and phytoplankton (increases in pH and the concentration of dissolved oxygen) (Oppenheimer et al. 1978) provides a favorable environment for *V. cholerae* O1, since this bacterium prefers an elevated pH for growth and metabolism but is commensal to zooplankton species (e.g., copepods).

prefers ↑pH

Studies have shown that some strains of *V. cholerae* spontaneously produce CTX phage, a filamentous bacteriophage that contains structural genes for cholera toxin (Faruque et al. 1998). CTX phage may be responsible for the formation of new toxigenic strains of *V. cholerae* by lysogenic conversion, which allows the phage to replicate with the bacterial cell (Waldor and Mekalanos 1996), and its propagation may depend on environmental factors (Faruque et al. 1998). Increased production of phytoplankton provides additional nutrient for the next level of the food chain, the zooplankton (Kiorboe and Neilsen 1994). A pH of 8.5 positively influences the attachment of *V. cholerae* O1 to copepods (Huq et al. 1984a) and *Vibrio parahaemolyticus* to chitin (Kaneko and Colwell 1973).

Huq and Colwell proposed that cells of *V. cholerae* attached to zooplankton, and those in the gut of zooplankton are protected from the external environment to some degree and begin to proliferate, taking advantage of the increased surface area and improved conditions of nutrition, the latter deriving from the disintegration of phytoplankton and the release of nitrogenous products into the water (Huq et al. 1988; Colwell and Huq 1994a, 1994b). Dumontet et al. (1996) found that unidentified bacterial cells attached to copepods in the natural environment were often found to be detached when SEM was used. However, cells of *V. cholerae* remained attached to copepod surfaces during the vigorous process of SEM preparation (Dumontet et al. 1996), confirming previous findings by Huq et al. (1983), who reported firm attachment between *V. cholerae* and copepods. Most likely only nonvibrio cells detach, suggested by the observation of Huq et al. (1983). Furthermore, the gut contents of copepods comprise vibrios, including *V. cholerae* (Sochard et al. 1979).

We further suggest that, during interepidemic monsoon seasons in Bangladesh, when there is a significant alteration of nutritional conditions arising from seasonal changes in the chemical parameters of the water

(Oppenheimer et al. 1978), *V. cholerae* O1 may become nonculturable (Huq et al. 1984b; Huq and Colwell 1995). Conditions triggering an increase in the number of *V. cholerae* in water may include increased numbers of zooplankton following blooms of phytoplankton. Copepods in water are, thereby, ingested by humans where untreated water is the source of drinking water.

Temperature has a direct effect on attachment of *V. cholerae* to copepods (Huq et al. 1984b). Warmer temperature, in combination with elevated pH and plankton blooms, can influence growth and multiplication of *V. cholerae* in the aquatic environment. Recently, sea surface temperature in the Bay of Bengal has been shown to exhibit an annual cycle related to that of cholera cases in Bangladesh (Lobitz et al. 2000). Furthermore, sea surface height is an indicator of incursion of water with plankton into tidal rivers and is also correlated with outbreaks of cholera. However, the specific factor(s) inducing *V. cholerae* to multiply, thereby producing a large population, and to convert to the culturable state remains to be elucidated.

During the past two decades, it has become evident that the accumulated new information concerning the ecology of *V. cholerae* can provide possible explanations for the epidemic behavior peculiar to cholera. *V. cholerae,* a normal inhabitant of brackish waters, can exist in a viable but nonculturable state in the environment, which may be related to its association with both zooplankton and phytoplankton. Thus, the cholera vibrio can survive and multiply entirely without the benefit of the human intestine. Large increases in the populations of plankton (and thus of the vibrios) may be responsible for the sudden appearance of cholera among persons who ingest a large number of vibrios, the increase in number being related to environmental factors causing abundant growth of plankton in the water. The period of time after a zooplankton bloom is highly correlated with increased numbers of *V. cholerae* and, subsequently, with cholera.

A Summary of the Descriptive Epidemiological Aspects of Cholera

A brief summary of some known aspects of the epidemiology of cholera permits an appreciation of the contrasts between the "human" and the "environmental" aspects of the disease. Other, more detailed reviews provide additional information (Barua and Greenough 1991; Wachsmuth et al. 1994).

Vibrio cholerae is efficiently spread from human feces to other susceptible humans via contaminated water and food. This mechanism can account for most of the spread of the disease in highly localized epidemics. Most infections in humans are either asymptomatic or associated with mild diarrhea; severe cholera occurs only in 1 of every 10–50 individuals infected with the organism. The infectious dose needed to produce clinical cholera is known to be very high in normal volunteers, in the range of 10^8 viable organisms; this dose decreases by at least three to four orders of magnitude when gastric acidity is neutralized (Cash et al. 1974). Infectious doses under normal field circumstances are not known. Cholera vibrios are extremely acid-sensitive; they are quickly killed in normal stomach acid, whereas they thrive in alkaline environments.

Although animals such as cows and chickens may serve as reservoirs of infectious agents that are pathogenic to humans, no animal has been identified as a reservoir of *V. cholerae*. Thus, infected but asymptomatic humans were thought to be the reservoirs. A chronic gallbladder carrier state has been identified in a few humans (Azurin et al. 1967) but clearly is not a major mechanism or route for the spread of the disease. In areas where cholera is endemic, the highest attack rates of disease are in children two to four years old, while in areas newly invaded by cholera, the illness is usually seen first in adult men. This suggests that susceptibility to the disease is a combination of both lack of protective immunity and exposure to contaminated food and water. Clearly, patterns of water use in different areas will affect the spread of the disease; in some cities in Peru, cholera vibrios spread through the municipal water system (Ries et al. 1992), which resulted in extremely high rates of infection in the urban population. Where rivers or open wells are used for drinking water, primarily in rural areas, cases tend to cluster among persons living close to and drinking the contaminated water.

Geographic Examples of the Behavior of Cholera

At present, cholera is once again a global disease, but its epidemiological and ecological behavior varies widely according to where it occurs. Following are some examples of the observed behaviors of cholera in different parts of the world, which indicates how diverse they can be.

Bangladesh

In its endemic "home," the Ganges delta, clinical cholera occurs year-round, but there are, in addition, two regular epidemic peaks each year (see Fig. 11.2b): one in the spring and one in the fall of the year (Glass et al. 1982). One epidemic occurs just before the monsoons, which usually begin in mid-June, and the other at the end of the monsoons, usually in late September. When the epidemics occur, infected persons begin coming to a treatment center simultaneously from different geographic areas surrounding the center, rather than from a single locality that widens as the epidemic progresses. The latter situation would be expected if cholera were spreading only from infected persons. There is nothing obvious about differences in human behavior during these times; people use the same

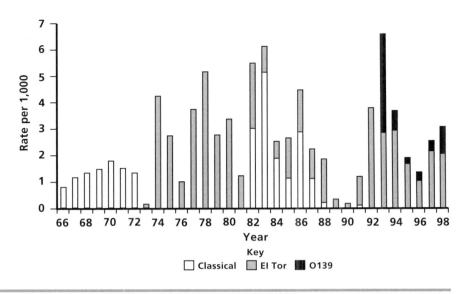

Figure 11.3 Annual incidence of cholera due to classical, El Tor, and O139 *Vibrio cholerae* in Matlab, Bangladesh, 1966–98. *Source:* Data from the International Centre for Diarrhoeal Disease Research, Bangladesh (ICDDRB).

sources of water and generally eat the same foods, with no observed change in sanitation practices. This is perhaps the best example of a seasonal occurrence of epidemic disease that might be explained on the basis of environmental changes and changes in plankton populations.

Although cholera appears every year, attack rates vary from year to year (Fig. 11.3). Increases in rates (with the exception of the last one in 1992) have coincided with the appearance of a "new" or reemergent cholera serotype. It is not known why these variations occur, but they may be related to the immunologic status of the population or to species variations in the zooplankton populations (e.g., copepods serving as hosts for *V. cholerae*). The emergence of new strains of vibrios could also be related to environmental factors, perhaps including changes related to the El Niño/Southern Oscillation climatic anomaly, such as warming of regions of the oceans (see Chapter 8). A large epidemic year in Bangladesh was 1991 (Table 11.1), when there were approximately 400,000 estimated cases of cholera and 8,000 deaths (Siddique et al. 1992). These large numbers of cholera cases occurred in the country as a whole, even though the attack rates at Matlab, Bangladesh, were relatively low that year (Fig. 11.3). As will be described later for other areas of the globe, 1991 was a major year for cholera, with large outbreaks also in Latin America and Africa, and a major El Niño event.

Bangladesh does not officially report cholera to the World Health Or-

Table 11.1   Major Outbreaks of Cholera, 1991

Region	No. Cases Reported	No. Deaths Reported
Latin America	400,000	4,000
Africa	150,000	14,000
Bangladesh	400,000 (215,000)[a]	8,000 (2,600)[a]

Source: Data for Latin America and Africa are based on reports to the World Health Organization. Data for Bangladesh are based on surveillance conducted by the International Centre for Diarrhoeal Disease Research, Bangladesh (ICDDRB).

[a]The numbers in parentheses are for three months only, September through November.

ganization (WHO) and, therefore, global figures published by the WHO are clearly an underestimate. In 1991, as an example, the number of cases of cholera in Bangladesh were approximately equal to the number of cases reported from all of Latin America, yet the official WHO figures do not reflect that.

Africa

In 1970, cholera invaded Africa after having been absent (except for Egypt) for more than a century. The invasion took place in three coastal areas (see Fig. 11.1). Guinea, on the west coast, was first to report cases and, within three months, cases were occurring in the northern part of the continent (Algeria) and in the eastern coastal region (Ethiopia and Somalia) (Goodgame and Greenough 1975). The epidemic moved along coastal countries and then inland. Within the first year of its introduction, 25 countries were reporting cholera, and, by 1991, 43 countries had reported cholera (Swerdlow and Issacson 1994). In many of these countries, cholera has subsequently become endemic.

Cholera does not, however, behave in these recently endemic regions with the regularity observed in the highly endemic area of the Ganges delta but is more often sporadic and less predictable. In 1991, a large epidemic of cholera occurred in 20 countries, with more than 150,000 cases and 14,000 deaths (Swerdlow and Issacson 1994) (see Table 11.1). In 1994, because of the extremely poor conditions of the refugee camps in Zaire, a massive cholera epidemic over a three-week period resulted in an estimated 70,000 cases and 25,000 deaths (Goma Epidemiology Group 1995). This seemed to be the result of a water source heavily contaminated with *V. cholerae* O1 and an almost complete lack of adequate treatment facilities (see Chapter 14). The country of Zaire has since been renamed the Democratic Republic of the Congo, but the name Zaire is retained in this chapter and Chapter 14 to describe events that occurred before 1997.

Cholera remains a major problem in Africa because of the lack of adequate sanitation and potable water. In addition, the mortality rates are high because adequate treatment is not routinely available.

Latin America
Somewhat comparable to its appearance in Africa, but 21 years later, cholera struck the coastal areas of Peru in January 1991, after being absent from Latin America for 100 years. Persons living in four Pacific coastal towns spanning a distance exceeding 1,000 kilometers developed cholera within a few days of each other, and the infection then quickly spread throughout the country, first to the jungles and later to the Andean regions. Within a year, cholera spread from Peru throughout the continent and affected all of the countries of South and Central America, with the exception of Uruguay (Koo et al. 1996). In several instances, municipal water supplies contained cholera vibrios and were efficient mechanisms of spread (Ries et al. 1992). It would have been useful to have plankton data for these water supplies as well.

In 1991, there were an estimated 400,000 cases in Latin America, with 4,000 deaths (Koo et al. 1996). In June 1991, about five months after the epidemic began in Peru, a second focus of cholera occurred on the Pacific Coast of Mexico and spread southward through Central America (Craig 1996). Thus, two widely separated coastal foci of cholera occurred within several months of each other and were most probably unrelated to any movement of persons.

There has been no reported extension of cholera to the Caribbean Islands.

The United States and Australia
In both the United States and Australia, there are foci of *V. cholerae* O1 in surface waters that are presumably unrelated to human fecal contamination. Since 1975, there have been sporadic cases and occasional clusters of cases of cholera along the Gulf Coast of the United States (see Fig. 11.1 and Mahon et al. 1996; Weber et al. 1994). Nearly all of the cases have been directly related to ingestion of undercooked shellfish, mostly crabs and, to a lesser extent, oysters. The cases almost always occur in the warm summer months (i.e., August and September). *V. cholerae* O1 could only rarely be cultured from the seawater in these incidents.

In Australia, some rivers in the southeast have been known to contain cholera vibrios since the late 1970s (Desmarchelier et al. 1995) (see Fig. 11.1). Here, cholera vibrios could be cultured from river water during all seasons of the year, although the few sporadic cases of clinical cholera all occurred during the warm season of the year, and all could be traced to drinking untreated river water. An extensive search for these vibrios in shellfish or water plants was unsuccessful. Plankton were not investigated, unfortunately.

In both the United States and Australia, the genotypes of *V. cholerae* O1 El Tor have been shown to be distinctive; they can be readily differentiated from cholera vibrios that caused the seventh pandemic in Asia, Africa, and Latin America (Wachsmuth et al. 1994). A recent study suggests that, although a single clone of pathogenic *V. cholerae* may be responsible for many cases of cholera in Asia, Africa, and Latin America during the seventh pandemic, other cases of cholera in those regions were caused by toxigenic *V. cholerae* strains that seem to have originated from local strains of *V. cholerae* O1 or non-O1 strains (Jiang et al. 2000b). In another study reported by Jiang et al. (2000a), isolates from single sampling sites in the Chesapeake Bay were genetically diverse, while genetically identical isolates were found at several of the sampling sites. So, their occurrence in the environment clearly suggests that a sole-source-of-contamination theory, as was first considered when the Peruvian epidemic began in Lima in 1991, can be ruled out.

Thus, in both of these geographic areas, *V. cholerae* O1 organisms are present and multiply in surface waters, independently of humans. In the United States, shellfish living in salt water have been hypothesized to be a reservoir. It is more likely that the shellfish, being filter feeders, concentrate the vibrios from the zooplankton on which they feed. In Australia, no reservoir, other than the fresh water itself, has been reported, but plankton populations have not yet been studied as a potential reservoir in that country.

The Control of Cholera

Cholera has been one of the most potent forces in bringing about the mobilization of public health resources. The occurrence of cholera epidemics places the burden of responsibility immediately on those political bodies that have not provided adequate safe water and sewage disposal for the affected population.

Since untreated, severe cholera has a very high mortality rate, up to 70 percent, a new cholera epidemic could produce a large number of deaths and, therefore, a great deal of panic in affected communities until adequate treatment facilities are established. In Zaire, as previously mentioned, the mortality rate nearly approached an "untreated" rate; it fell rapidly when adequate treatment was made available (Siddique et al. 1995). In contrast, the mortality rates in Latin America were less than 1 percent (Koo et al. 1996), exemplifying the importance of providing adequate treatment to those affected, in order to control cholera. The death rate from cholera is nearly always directly related to adequacy of medical therapy. In Latin

America, where oral rehydration had been widely used for treatment of children with diarrhea for at least a decade before the outbreak of cholera, physicians had a more adequate understanding of the physiology of the disease and the need for replacement fluids.

Adequate treatment of cholera involves (1) rapid replacement of fluid and electrolytes lost through passage of watery stools, (2) antimicrobial treatment to kill the cholera vibrios responsible for the illness, and (3) early resumption of feeding to prevent the worsening of malnutrition (Mahalanabis et al. 1992). Fluids and electrolytes can be replaced both intravenously and orally. Because intravenous fluids are scarce in developing countries, as well as expensive, and require administration by a trained health worker, oral rehydration salts (ORS) were developed initially to partially replace the need for intravenous fluids. The wide availability of these effective, yet inexpensive, oral fluids, which can be used by nonprofessional caregivers, has helped to make the treatment of cholera much simpler and more effective. ORS were later found to be effective for treatment of all diarrheal diseases in which dehydration is an important aspect, regardless of the causative agent or the age of the patient. ORS are today the mainstay of therapy for acute diarrheal illnesses the world over (Duggan et al. 1992; Centers for Disease Control and Prevention 1992).

Prevention of cholera ultimately depends on having adequate amounts of safe, potable water and a system for sanitary disposal of human feces. The long-term solution is, therefore, in the provision of bacteriologically safe water by water purification plants, with distribution via intact pipes to convenient collection points, and adequate latrines to collect and disinfect human fecal material. It is estimated that in Latin America alone this would cost about 200 billion dollars, clearly prohibitive as a solution to the problem (De Moredo 1991). Water purification and sewage treatment plants, as solutions, will clearly not be possible in the near future; therefore, less expensive and more immediate methods to accomplish these goals are needed. Such methods include (1) treatment of contaminated drinking water in the home by chlorination, boiling, or possibly filtration (Huq et al 1996); (2) storage of safe water in narrow-neck vessels to prevent contamination in the home; (3) improvements in personal hygiene (such as hand washing) to prevent direct fecal contamination of hands and foods; and (4) education to facilitate the implementation of these interventions. Presently, combinations of these interventions are being put into place in many developing countries.

Effective cholera vaccines, when available to the general public in areas where cholera is endemic, will also be important in the prevention of cholera, at least until long-term solutions are possible. Although parenteral cholera vaccines have been available for approximately 100 years, they are

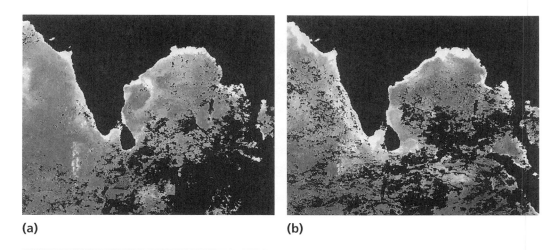

(a)   (b)

Figure 11.4 Concentration of chlorophyll-A in the Bay of Bengal and the adjacent marine environment in September 1997 (*a*) and September 1998 (*b*). Land and clouds show up as *black. White* represents the highest concentration of chlorophyll-A, which appears along coastal areas. Progressively darker shades of gray represent progressively lower concentrations of chlorophyll-A. Concentrations of chlorophyll-A were higher in 1998 than in 1997. *Source:* Courtesy of Byron Wood and Brad Lobitz of CHAART-NASA Ames, California, using SeaWiFS satellite.

poorly protective and of no use to public health. At present, oral vaccines (both killed vaccines and live attenuated vaccines), which can easily be administered, are being developed and tested. The only such vaccine to have been tested extensively in the field is a killed oral vaccine, which has been shown to be 50–70 percent protective for a period of three years (Clemens et al. 1990). The cost of these vaccines, however, has thus far limited their usefulness and prevented them from being used as public health tools.

If cholera is introduced and reintroduced into populations through aquatic reservoirs by plankton (see under "An Update on the Ecology of *Vibrio cholerae*," above), there may be new ways of preventing cholera from spreading, including inhibiting the growth of these organisms in surface waters or excluding plankton from drinking water by simple cloth or plastic net filtration. Furthermore, it may be possible to predict and recognize the onset of cholera epidemics by remote sensing from satellites circulating the globe that can measure chlorophyll concentrations (and phytoplankton and zooplankton populations indirectly) and sea surface temperatures (Figs. 11.4 and 11.5; Colwell 1996). The aim is to develop a forecasting model that integrates multiple sources of data. Being able to predict an epidemic may allow both mobilization of treatment supplies

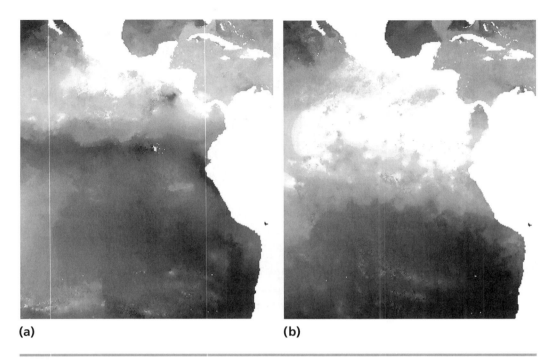

(a)   (b)

Figure 11.5   Sea surface temperature in the eastern Pacific Ocean in mid-January of 1997 (a) and of 1998 (b). Land shows up as *white*. Sea surface temperature is displayed in a gray scale for which *black* represents 19°C and *white* represents 29°C. Sea surface temperatures were warmer in 1998 than in 1997. *Source:* Courtesy of Byron Wood and Brad Lobitz of CHAART-NASA Ames, California, using NOAA-AVHRR satellite.

and perhaps even the rapid immunization of populations at particularly vulnerable times.

Conclusion

The advances made in the last 100 years concerning cholera have been both substantial and significant, especially in identifying the causative agent, *V. cholerae*, and its pathogenic properties and in developing simple, effective, and inexpensive means of treatment. However, until the environmental source of cholera was established and accepted, the disease remained enigmatic. With improved understanding of its ecology and especially its correlation with climate, the disease should become more predictable, based on those environmental parameters influencing cholera epidemics. With satellites and systems for monitoring the global climate, a new dimension of epidemiology and disease control is being charted, providing the possibility for prevention of outbreaks when conditions are conducive to the initiation and spread of the disease.

SUGGESTED STUDY PROJECTS

Suggested study projects provide a set of options for individual or team projects that will enhance interactivity and communication among course participants (see Appendix A). The Resource Center (see Appendix B) and references in all of the chapters provide starting points for inquiries. The process of finding and evaluating sources of information should be based on the principles of information literacy applied to the Internet environment (see Appendix A).

The objective of this chapter's study projects is to develop an understanding of the dynamics and control of cholera in relation to global ecosystem change.

PROJECT 1: Multiple Causes of Cholera Outbreaks

The objective of this study project is to develop an understanding of the dynamics of transmission of cholera outbreaks.

Task 1. Identify and describe various outbreaks of cholera in different locations (1817 to the present).

Task 2. Compare social and environmental factors leading to the outbreaks described in task 1.

PROJECT 2: The Framework for Ecological Change and Emerging Disease in Chapter 10

The objective of this study project is to develop an understanding of the framework for analyzing ecological change and emerging disease.

Task 1. Write a comprehensive paper on the relationship between this case study on cholera and the framework for ecological change and emerging disease discussed in Chapter 10.

PROJECT 3: Control Measures

The objective of this study project is to develop an understanding of ways to control cholera.

Task 1. Describe current public health practices to control cholera in a particular location.

Task 2. Describe prospects for improvements or deterioration of such practices in the future.

Acknowledgments

The authors acknowledge support received from National Institutes of Health Grants 1R01A139129–01 and R01NR04527–01 A1, Environmental Protection Agency Grant R824995–01, National Aeronautics and Space Administration Grant NA62–1195, and the Wallenberg Foundation. Authors also gratefully acknowledge the excellent collaboration of Byron Wood and Brad Lobitz, CHAART-NASA Ames, California, in providing satellite images and remote sensing data, as well as the helpful suggestions of Al Buck and the patient and skillful editing of Joan Aron.

References

Amako K, Shimodori S, Imoto T, Miake S, Umeda A. 1987. Effects of chitin and its soluble derivatives on survival of *V. cholerae* O1 at low temperature. *Appl Environ Microbiol* 53:603–5.

Azurin JC, Kobari K, Barua D, Alvero M, Gomez CZ, Daizon JJ, Nakano EI, Suplido R, Ledsma L. 1967. A long-term carrier of cholera: Cholera Dolores. *Bull World Health Organ* 37:745–49.

Barua D. 1988. Cholera during the last hundred years (1884–1983). In *Vibrio cholerae and Cholera* (Takeda Y, ed.). KTK Scientific Publishers, Tokyo, pp. 9–32.

Barua D, Greenough WB, eds. 1991. Cholera. In *Current Topics in Infectious Disease*. Plenum Medical Book Co., New York, Chap. 1.

Carlton JT, Geller JB. 1993. Ecological roulette: The global transport of non-indigenous marine organisms. *Science* 261:78–82.

Cash RA, Music SI, Libonati JP, Snyder MJ, Wenzel RP, Hornick R. 1974. Response of man to infection with *Vibrio cholerae:* I. Clinical, serologic and bacteriologic responses to a known inoculum. *J Infect Dis* 129:45–52.

Centers for Disease Control and Prevention. 1992. Update: Cholera—Western Hemisphere. *MMWR* 42:89–91.

Cholera Working Group. 1993. Large epidemic of cholera-like disease in Bangladesh caused by *Vibrio cholerae* O139 synonym Bengal. *Lancet* 342:387–90.

Chowdhury MA, Huq A, Xu B, Colwell RR. 1997. Effects of alum on free-living copepod associated with *Vibrio cholerae* O1 and O139. *Appl Environ Microbiol* 63:3323–26.

Clemens J, Sack DA, Harris JR. 1990. Field trial of oral vaccines in Bangladesh: Results from three-year follow-up. *Lancet* 335:270–73.

Cockburn TA, Cassanos JG. 1960. Epidemiology of endemic cholera. *Public Health Rep* 75:791–803.

Colwell RR. 1970. Polyphasic taxonomy of the genus *Vibrio:* Numerical taxonomy of *Vibrio cholerae, Vibrio parahaemolyticus,* and related *Vibrio* species. *J Bacteriol* 104:410–33.

———. 1996. Global climate and infectious disease: The cholera paradigm. *Science* 274:2025–31.

Colwell RR, Brayton PR, Grimes DJ, Roszak DR, Huq A, Palmer LM. 1985. Viable but nonculturable *V. cholerae* and related pathogens in the environment: Implication for release of genetically engineered microorganisms. *Bio/Technology* 3:817–20.

Colwell RR, Brayton PR, Herrington D, Tall B, Huq A, Levine MM. 1996. Viable but nonculturable *Vibrio cholerae* O1 revert to a cultivable state in the human intestine. *World J Microbiol Biotechnol* 12:28–31.

Colwell RR, Huq A. 1994a. Vibrios in the environment: Viable but nonculturable *Vibrio cholerae*. In *Vibrio cholerae and Cholera: Molecular to Global Perspectives* (Wachsmuth IK, Olsvik O, Blake PA, eds.). American Society for Microbiology, Washington, D.C., Chap. 9, pp. 117–33.

———. 1994b. Environmental reservoir of *Vibrio cholerae,* the causative agent of cholera, 1994. In *Disease in Evolution: Global Changes and Emergence of Infectious Diseases* (Wilson ME, Levins R, Spielman R, eds.). Annals of the New York Academy of Sciences, New York, Vol. 740, pp. 44–54.

Colwell RR, Huq A, Chowdhury MAR, Brayton P, Xu B. 1995. Serogroup conversion of *V. cholerae*. *Can J Microbiol* 41:946–50.

Colwell RR, Kaper JB, Joseph SW. 1977. *Vibrio cholerae* and *V. parahaemolyticus* and other vibrios: Occurrence and distribution in Chesapeake Bay. *Science* 198:394–96.

Colwell RR, Tamplin ML, Brayton PR, Gauzens AL, Tall BD, Herrington D, Levine MM,

Hall S, Huq A, Sack RB. 1990. Environmental aspects of *V. cholerae* in transmission of cholera. In *Advances in Research on Cholera and Related Diarrhoeas,* 7th ed. (Sack RB, Zinnaka Y, eds.). KTK Scientific Publishers, Tokyo, pp. 327–43.

Craig JP. 1996. Cholera: Outlook for the twenty-first century. *Caduceus* 12:25–42.

De SN. 1961. *Cholera: Its Pathology and Pathogenesis.* Oliver & Boyd, Edinburgh.

De Moredo CG. 1991. Presentation of the PAHO regional plan. In *Proceedings of the Conference: Confronting Cholera, the Development of the Hemispheric Response to the Epidemic.* Pan American Health Organization, Washington, D.C., pp. 39–44.

Desmarchelier PM, Wong FYK, Mallard K. 1995. An epidemiological study of *V. cholerae* O1 in the Australian environment based on the RNA gene polymorphism. *Epidemiol Infect* 115:435–46.

Duffy J. 1971. History of Asiatic cholera in the United States. *Bull NY Acad Med* 47:1152–68.

Duggan C, Santosham R, Glass I. 1992. The management of acute diarrhea in children: Oral rehydration, maintenance and nutritional therapy. *MMWR* 41:1–17.

Dumontet S, Krovacek K, Baloda SB, Grottoli R, Pasquale V, Vanucci S. 1996. Ecological relationship between *Aeromonas* and *Vibrio* spp. and planktonic copepods in the coastal marine environment in southern Italy. *Comp Immunol Microbiol Infect Dis* 19:245–54.

Faruque SM, Asadulghani, Alim ARM, Albert MJ, Islam KM, Mekalanos JJ. 1998. Induction of the lysogenic phage encoding cholera toxin in naturally occurring strains of toxigenic *V. cholerae* O1 and O139. *Infect Immunol* 66:3752–57.

Finkelstein RA. 1973. Cholera. *Crit Rev Microbiol* 2:553–623.

Gelbspan R. 1991. Alarm sounds at rise in algae blooms. *Boston Globe,* October 30, p. 10.

Glass R, Becker S, Huq I, Stoll B, Khan MU, Merson M, Lee J, Black R. 1982. Endemic cholera in rural Bangladesh, 1966–1980. *Am J Epidemiol* 116:959–70.

Glass R, Black RB. 1992. The epidemiology of cholera. In *Cholera* (Barua D, Greenough WB, eds.). Plenum Publishing Corp., New York.

Goma Epidemiology Group. 1995. Public health impact of Rwandan refugee crisis: What happened in Goma, Zaire, in July 1994? *Lancet* 345:339–44.

Goodgame RW, Greenough WB. 1975. Cholera in Africa: A message for the West. *Ann Intern Med* 82:101–6.

Hardy JT. 1991. Where sea meets sky. *Nat History* 591:55–59.

Hood MA, Meyers SP. 1977. Microbiological and chitinoclastic activities associated with *Penaeus stiferus. J Oceanogr Soc Jpn* 33:235–41.

Hornick RB, Music SI, Wenzel R. 1971. The Broad Street pump revisited: Response of volunteers to ingested cholera *Vibrio. Bull NY Acad Med* 47:1181–91.

Huq A, Chowdhury MAR, Felsenstein A, Colwell RR, Rahman R, Hossain KMB. 1988. Detection of *V. cholerae* from aquatic environments in Bangladesh. In *Biological Monitoring of Environment Pollution* (Yasuno M, Whitton BA, eds.). Tokai University Press, Tokyo, pp. 259–64.

Huq A, Colwell RR. 1995. Vibrios in the marine and estuarine environment. *J Marine Biotechnol* 3:60–63.

———. 1996. Vibrios in the marine and estuarine environment: Tracking of *Vibrio cholerae. J Ecosyst Health* 2:198–214.

Huq A, Colwell RR, Chowdhury MAR, Xu B, Moniruzzaman SM, Islam MS, Yunus M, Albert MJ. 1995. Co-existence of *V. cholerae* O1 and O139 Bengal in plankton in Bangladesh. *Lancet* 345:1249.

Huq A, Colwell RR, Rahman R, Ali A, Chowdhury MAR, Parveen S, Sack DA, Russek-Cohen E. 1990. Detection of *V. cholerae* O1 in the aquatic environment by fluorescent monoclonal antibody and culture method. *Appl Environ Microbiol* 56:2370– 73.

Huq A, Small EB, Colwell RR. 1981. A possible role of copepods in the ecology of *V. cholerae. Electron Microsc Central Facility Newslett Univ Maryland* 10:20–22.

Huq A, Small E, West P, Colwell RR. 1984a. The role of planktonic copepods in the sur- vival and multiplication of *Vibrio cholerae* in the aquatic environment. In *Vibrios in the Environment* (Colwell RR, ed). J. Wiley & Sons, New York, pp. 521–34.

Huq A, Small EB, West PA, Rahman R, Colwell RR. 1983. Ecology of *V. cholerae* with special reference to planktonic crustacean copepods. *Appl Environ Microbiol* 45:275–83.

Huq A, West PA, Small EB, Huq A, Colwell RR. 1984b. Influence of water temperature, salinity and pH on survival and growth of toxigenic *Vibrio cholerae* O1 associated with live copepods in laboratory microcosms. *Appl Environ Microbiol* 48:420–24.

Huq A, Xu B, Chowdhury MAR, Islam MS, Montilla R, Colwell RR. 1996. A simple filtra- tion method to remove plankton-associated *Vibrio cholerae* in raw water supplies in developing countries. *Appl Environ Microbiol* 62:2508–12.

Islam MS, Draser BS, Bradley DJ. 1990a. Survival of toxigenic *V. cholerae* O1 with a com- mon duckweed, *Lemna minor,* in artificial aquatic ecosystem. *Trans R Soc Trop Med Hyg* 84:422–24.

———. 1990b. Long-term persistence of toxigenic *V. cholerae* O1 in the mucilagenous sheath of a blue-green algae, *Anabaena variabilis. J Trop Med Hyg* 93:133–39.

Islam MS, Miah MA, Hasan MK, Sack RB, Albert MJ. 1994. Detection of nonculturable *V. cholerae* O1 associated with a cyanobacterium from an aquatic environment in Bangladesh. *Trans R Soc Trop Med Hyg* 88:298–99.

Jiang SC, Louis V, Choopun N, Sharma A, Huq A, Colwell RR. 2000a. Genetic diversity of *Vibrio cholerae* in Chesapeake Bay determined by amplified fragment length poly- morphism fingerprinting. *Appl Environ Microbiol* 66 (1): 140–47.

Jiang SC, Matte M, Matte G, Huq A, Colwell RR. 2000b. Genetic diversity of clinical and environmental isolates of *Vibrio cholerae* determined by amplified fragment length polymorphism fingerprinting. *Appl Environ Microbiol* 66 (1): 148–53.

Kaneko T, Colwell RR. 1973. Ecology of *Vibrio parahaemolyticus* in Chesapeake Bay. *J Bacteriol* 113:24–32.

Kiorboe T, Neilsen TJ. 1994. Regulation of zooplankton biomass and production in a temperate coastal ecosystem: 1. Copepods. *Limnol Oceanography* 39:493–507.

Koch R. 1884. An address on cholera and its bacillus. *Br Med J* 2:403–7, 453–59.

Koo D, Traverso H, Libel M, Drasbek C, Tauxe R, Brandling-Bennett D. 1996. Epidemic cholera in Latin America, 1991–1993: Implications of case definitions used for pub- lic health surveillance. *Bull Pan Am Health Organ* 30 (2): 134–43.

Lobitz B, Beck L, Huq A, Wood B, Fuchs G, Faruque ASG, Colwell RR. 2000. Climate and infectious disease: Use of remote sensing for detection of *Vibrio cholerae* by indirect measurement. *Proc Natl Acad Sci USA* 97 (4): 1438–43.

Mahalanabis D, Molla AM, Sack DA. 1992. Clinical management of cholera. In *Cholera* (Barua D, Greenough WB, eds). Plenum Medical Book Co., New York, pp. 253– 81.

Mahon BE, Mintz ED, Greene KD, Wells JG, Tauxe RV. 1996. Reported cholera in the United States, 1992–1994: A reflection of global changes in cholera epidemiology. *JAMA* 276 (4): 307–12.

Miller CB. 1984. The zooplankton estuaries. In *Ecology Estuaries and Enclosed Seas* (Kechum BH, ed). North-Holland, New York.

Nalin DR, Daya V, Ried V, Levine MM, Cisneros L. 1977. Adsorption and growth of *V. cholerae* to chitin. *Int Immunol* 25:768–70.

Oppenheimer JR, Ahmad MG, Huq A, Haque KA, Alam AKMA, Aziz KMS, Ali S, Haque ASM. 1978. Limnological studies in three ponds in Dhaka, Bangladesh. *Bangladesh J Fisher* 1:1–28.

Parsons TR, Takahashi M, Hargrave B. 1984. *Biological Oceanographic Processes.* Pergamon, Oxford, U.K.

Pearl HW, Keller PE. 1979. Significance of bacterial *Anabaena (cyanophyceae)* association with respect to N_2 fixation in fresh water. *J Phycology* 14:2.

Pollitzer R. 1959. *Cholera.* World Health Organization, Geneva.

Ramamurthy T, Garg R, Sharma SK, Nair GB, Shimada T, Takeda T, Karasawa T, Kuraziano H, Pal A, Takeda Y. 1993. Emergence of novel strains of *V. cholerae* with epidemic potential in southern and eastern India. *Lancet* 341:703–5.

Renato T, Pruzzo C. 1999. Role of surface proteins in *V. cholerae* attachment to chitin. *Appl Environ Microbiol* 65 (3): 1348–51.

Ries AA, Vugia DJ, Beingolea L, Palacios AM, Vasquez E, Wells JG, Baca NG, Swerdlow DL, Pollack M, Bean NH, Seminario L, Tauxe RV. 1992. Cholera in Piura, Peru: A modern urban epidemic. *J Infect Dis* 166:1429–33.

Ruiz GM, Rawlings TK, Dobbs FC, Drake LA, Mullady T, Huq A, Colwell RR. 2000. Global spread of microorganisms by ships. *Nature* 408:49.

Schneider DR, Parker CD. 1982. Purification and characterization of the mucinase of *V. cholerae. J Infect Dis* 145:474–82.

Seas C, Miranda J, Gil AI, Leon-Barua R, Patz J, Huq A, Colwell RR, Sack RB. New insights on the emergence of cholera in Latin America during 1991: The Peruvian experience. *Am J Trop Med Hyg* (in press).

Siddique AK, Salam A, Islam MS. 1995. Why treatment centers failed to prevent cholera deaths in Rwandan refugees in Goma, Zaire. *Lancet* 345:359–61.

Siddique AK, Zaman Z, Baqui AH, Akram K, Mutsuddy P, Eusof A, Haider K, Islam S, Sack RB. 1992. Cholera epidemics in Bangladesh, 1985–1991. *J Diarrheal Dis Res* 10:79–86.

Silvery JKG, Roach AW. 1964. Studies on microbiotic cycles in surface waters. *J Am Water Works Assoc* 56:60–72.

Snow J. 1855. *On the Mode of Communication of Cholera,* 2d ed. John Churchill, London.

Sochard MR, Wilson DF, Austin B, Colwell RR. 1979. Bacteria associated with surface and gut of marine copepods. *Appl Environ Microbiol* 37:750–59.

Swerdlow DL, Issacson M. 1994. The epidemiology of cholera in Africa. In *V. cholerae and Cholera: Molecular to Global Perspectives* (Wachsmuth K, Blake PA, Olsvik O, eds.). American Society for Microbiology, Washington, D.C., pp. 297–307.

Tanamal STW. 1959. Notes on paracholera in Sulawesi (Celebes), 1959. *Am J Med Hyg* 8:72–78.

Tauxe R, Seminario L, Tapia R, Libel M. 1994. The Latin American epidemic. In *V. cholerae and Cholera: Molecular to Global Perspectives* (Wachsmuth K, Blake PA, Olsvik O, eds.). American Society for Microbiology, Washington, D.C., pp. 321–44.

Taylor A. 1991. Plankton and the Gulf Stream. *New Scientist,* March, 52.

Wachsmuth IK, Blake PA, Olsvik O, eds. 1994. *V. cholerae and Cholera: Molecular to Global Perspectives* (Wachsmuth K, Blake PA, Olsvik O, eds.). American Society for Microbiology, Washington, D.C.

Waldor MK, Colwell R, Mekalanos JJ. 1994. The *Vibrio cholerae* O139 serogroup antigen includes an O-antigen capsule and lipopolysaccharide virulence determinants. *Proc Natl Acad Sci USA* 91:11388–92.

Waldor MK, Mekalanos JJ. 1994. *Vibrio cholerae* O139 specific gene sequences. *Lancet* 343:1366.

———. 1996. Lysogenic conversion of a filamentous phage encoding cholera toxin. *Science* 273:1910–14.

Weber JT, Levine WC, Hopkins DP. 1994. Cholera in the United States, 1965–1991. *Arch Intern Med* 154:331–36.

CHAPTER 12

Malaria and Global Ecosystem Change

Joan L. Aron, Ph.D., Clive J. Shiff, Ph.D., and
Alfred A. Buck, M.D., Dr.P.H.

Malaria, a potentially lethal parasitic disease once in decline and targeted for global eradication, afflicts hundreds of millions of people annually, and its impact is growing in many parts of the world. Malaria demonstrates the importance of the broad ecosystem concept of interactions between living organisms and nonliving elements. The transmission of malaria is directly affected by environmental conditions external to the human host because the infection is transmitted by mosquitoes. Environmental conditions, in turn, are affected by human activities. The purpose of this case study is to generate awareness of how global changes in the growth and movement of populations, patterns of economic development, and climate are changing the risk of exposure of human populations to malaria parasites. The forces underlying the spread of malaria are growing as populations with little or no immunity are infected more frequently by malaria parasites that are increasingly difficult to treat effectively (see Box 12.1).

Malaria is caused by a protozoan parasite that completes its complex cycle of development alternating between human hosts and mosquitoes of the genus *Anopheles;* the biting of human hosts by mosquitoes is the mode of contact that permits transmission from human to mosquito and back to human (see Chapter 10). Malaria reveals the relationship between ecosystem change and human health in several ways (see Fig. 12.1). Changes in land use, manipulations of water use, and variation in climate influence the distribution and abundance of the mosquito vectors of this disease. Also, the development of the malaria parasites within the mosquito vectors depends upon climatic conditions, particularly temperature. For example, ambient temperatures that are too cold or too hot inhibit the development of all *Plasmodium* species that infect humans; for *Plasmodium falciparum,* which is the species of malaria parasite that causes the most se-

BOX 12.1

The Emergence of Drug-Resistant Malaria

Chloroquine is a synthetic antimalarial drug that was made generally available in the late 1940s. Chloroquine became popular because it was cheap, relatively safe, and highly effective. However, its dual use, both for treatment after infection and for suppressive chemoprophylaxis before infection, and its uncontrolled distribution through private markets set the stage for the emergence of resistance to the drug. Doses that may be adequate to relieve symptoms are often not large enough to wipe out entire populations of parasites, so that resistant parasites are left to be transmitted to mosquitoes and then other people (see Chapter 10). During the 1950s, chloroquine-resistant *Plasmodium falciparum* became a major operational problem for malaria control in some areas; during the ensuing decades, chloroquine-resistant *P. falciparum* spread around the world (Wernsdorfer and Payne 1991; Wernsdorfer 1994). *P. falciparum* is of particular concern because it causes the most severe form of malaria.

Other antimalarial drugs have been introduced, but these alternative drugs are typically more expensive, more difficult to administer, and more likely to have adverse side effects than is chloroquine. Moreover, in some areas, *P. falciparum* parasites have already developed resistance to some of these drugs. There is evidence of cross-resistance in which resistance to one drug also creates resistance to other drugs, limiting the period during which new drugs can be effectively used (Olliaro et al. 1996). Approaches for dealing with cross-resistance include the administration of combinations of drugs and research for new drugs that target different biochemical mechanisms of parasite infection. Some new drugs are under development, although they are few in number (Olliaro et al. 1996). Vaccines are an additional line of active biomedical research, although the prospect of a safe and effective vaccine for general use seems unlikely in the near future.

---

vere illness in humans, the estimate of the minimum temperature required ranges from 16°C to 19°C and the maximum temperature for viable parasites is about 35°C (Lindsay and Birley 1996).

The growth and movement of human populations may bring more people into contact with mosquito vectors and malaria parasites. For example, resettlement of people without prior exposure to malaria may place them in malarious zones without the partial immunity acquired through years of exposure. In addition, genetic characteristics expressed in the blood cells of local populations may make them more or less susceptible to infection by malaria parasites (Livingstone 1984; Allen et al. 1992). Activities related to economic development—agriculture, irrigation, deforestation, and the construction of roads and buildings—modify the use of land and water, sometimes augmenting the number of possible aquatic breeding sites for the mosquito larvae and altering local or regional climate. Less directly, activities associated with economic development release into the atmosphere greenhouse gases that may accelerate global warming and affect patterns of global climate (see Chapter 7).

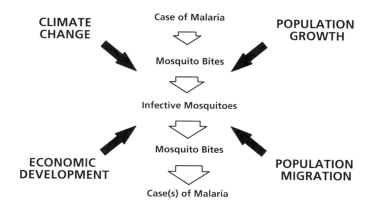

Figure 12.1 The transmission of malaria within the forces of global ecosystem change.

The relationship between ecosystem change and malaria at the global level is not divorced from conditions that operate on regional and local scales (see also Table 2.2). One particularly important issue is the diversity of anopheline species and their associated larval habitats, ranging from puddles in small, ephemeral depressions to swamps, marshes, and even brackish water (see Box 12.2). The geographic distribution of anopheline species is the main feature affecting the regional classification of the epidemiology of malaria (Rubio-Palis and Zimmerman 1997).

A better scientific understanding of the long-term changes in the global ecosystem could lead to the development of new strategies for the prevention of malaria. Decisions in the joint management of health issues and ecosystems involve complex tradeoffs (Wolman 1995; Graham and Wiener 1995). For example, the application of the insecticide DDT to kill vector mosquitoes is in conflict with a campaign to eliminate the dispersal of DDT and other persistent organic pollutants (*United Nations Environment Programme 2000; World Health Organization [WHO] 1999;* Hooper 1999). Any global program should also review successful approaches to the environmental management of malaria at local and regional scales. A traditional type of environmental management is the implementation of programs to manage water resources so as to eliminate the breeding sites of anopheline mosquitoes. Intermittent raising and lowering of water levels, alternately flushing and drying out breeding sites, has been used for larval control by the Tennessee Valley Authority in the United States and the Blue Nile Health Project in the Sudan (WHO 1995). Environmental management can also reduce the contact between humans and mosquito vectors by locating settlements away from breeding sites, by constructing housing with barriers to mosquito entry, and by the programmatic use of bednets

BOX 12.2

Swamp Fever

The diversity of aquatic habitats associated with malaria may be surprising to speakers of European languages, whose names for the disease reflect only the European and Mediterranean experience of an association with swamps and marshes. The English word derives from *mala aria* in Italian, meaning "bad air." *Paludisme* and *paludismo* in French and Spanish, respectively, derive from *paludis* in Latin, meaning "of the marsh." In traditional English usage, malaria has been called *swamp fever* or *marsh fever*, analogous to *Sumpffieber* in German.

treated with pyrethroid insecticides, which are less stable than DDT. Environmental management contributed to the successful control of malaria in some areas, such as the Pontine marshes in Italy and the Hula swamps in Israel.

This chapter provides examples of malaria as a public health problem in selected countries. In all examples, local factors interact with global changes in population growth and movement, economic development, or climate or some combination thereof. The examples also demonstrate that countries at very different levels of risk are affected by ecosystem change on a global scale.

Zimbabwe, Gambia, and Niger are selected as countries at high risk for malaria. Zimbabwe is in the southern portion of central Africa and is part of the Southern African Development Community, while Gambia and Niger are in West Africa. Variations in rainfall and temperature are major influences on the transmission of malaria in these countries. In Zimbabwe, the El Niño/Southern Oscillation (ENSO) climatic anomaly generates interannual variability in rainfall (see Chapter 8), and zones at higher altitudes are markedly cooler than zones at lower altitudes. In Gambia and Niger, fluctuation in rainfall from year to year is also important, although the pattern of climatic variability is different in West Africa. Global changes in climate are therefore likely to affect the transmission of malaria. Gambia and Niger illustrate how the pressures of population growth and associated economic development are creating new breeding sites for mosquito vectors and hence facilitating greater transmission of malaria. Migration is also an important regional factor, as demonstrated by the past movement of refugees from Mozambique to and from Zimbabwe.

Sri Lanka and Brazil are selected as countries at intermediate risk for malaria. The transmission of malaria in Sri Lanka is affected by major projects that manage water resources as well as interannual variability in rainfall subject to the influence of ENSO. In addition, the migration of population has contributed to malaria as a public health problem, since the development of water resources includes plans to resettle people from nonmalarious areas to malarious areas. The migration of population has been central to the resurgence of malaria in Brazil for similar reasons. Brazilians are encouraged for economic purposes to move into the malarious Amazonian region out of established cities and towns that are free of malaria. The transmission of malaria is further exacerbated by extensive deforestation associated with new settlement. Patterns of rainfall are also important as an environmental determinant of the transmission of malaria in Brazil. Amazonian deforestation is projected to alter regional climate, adding to the possible effects of global climate change.

The United States is selected as a country at low risk for malaria. Since there is no sustained transmission of malaria in the country, international movement of infected people is the critical factor in the small number of outbreaks that have occurred in recent years. Outbreaks have also been favored by weather conditions that were hotter and more humid than usual. Therefore, global warming would likely increase the risk of the transmission of malaria whenever the malaria parasite is introduced into the population.

Malaria in Africa

Malaria in Africa deserves special regional attention because over 90 percent of an estimated annual global total of 300–500 million clinical cases of malaria and 1.4–2.6 million deaths are among Africans (WHO 1995). The disproportionate burden of malaria in Africa can be largely attributed to the abundance of three species of mosquitoes that are very efficient at transmitting malaria—*Anopheles gambiae, Anopheles arabiensis,* and *Anopheles funestus* (Gillies and Coetzee 1987). The ability of malaria vectors in sub-Saharan Africa to breed in temporary pools without vegetation, such as a footprint filled with rainwater, is key to the exceptional persistence of malaria on the continent (Coluzzi 1994). The deadly importance of the African species of malaria vectors was demonstrated when *An. gambiae* from West Africa was accidentally introduced into Brazil, where South American vectors were already transmitting malaria. *Anopheles gambiae* is such an effective vector that, in 1938, only eight years after it had arrived in Brazil, it caused a major epidemic of malaria with 14,000 deaths (Soper and Wilson 1943; Walsh et al. 1993). Fortunately, Fred Soper and the Rockefeller Foundation responded and were able to eradicate *An. gambiae* from Brazil by intensive antimosquito operations. Unfortunately, eradication of *An. gambiae* from its native habitat in Africa would be infinitely more difficult or outright impossible.

Despite common regional concerns in Africa, understanding the relationship between ecosystem change and malaria requires an appreciation of local variation in ecology, geography, climate, and human activity. The countries selected illustrate the diversity of issues arising within the region.

Malaria in Zimbabwe

Malaria control has been in operation in Zimbabwe since the end of World War II, but, on April 17, 1996, the Panafrican News Agency reported that Zimbabwe had already experienced 828 deaths due to malaria and more than 300,000 cases of malaria since the beginning of the year. During the same period in 1995, the number of deaths had been reported to be 130 and the number of cases of malaria had been more than 100,000. The change in one year was clearly alarming and newsworthy. The primary cause was a significant increase in rainfall. Secondarily, the combination of rainfall and bad roads hindered the distribution of antimalarial drugs to affected areas, contributing to over a sixfold increase in the number of deaths due to malaria.

In southern Africa, the amount of rainfall is highly variable from one year to the next; heavy rainfall in 1996 had been preceded by five years of drought (Fig. 12.2a). Zimbabwe has a distinct rainy season characterized by monthly rainfall that begins to increase in October, peaks from December through March, and declines in April (Unganai 1996). The transmission of malaria peaks from February through May (i.e., toward the end of the rainy season) (Taylor and Mutambu 1986). Years with wetter rainy seasons are more favorable for transmission. In southern Africa, an increase in the incidence of malaria may also be one of the effects of a major flood episode (see Chapter 14). Years of heavy rainfall in the region tend to be associated with the La Niña phase of the ENSO climatic anomaly, which is the best-understood source of climatic variability on a seasonal to interannual time scale (see Chapter 8). Improvements in seasonal climate forecasting are gaining greater attention in Southern Africa Malaria Control (SAMC) and the larger public health community (*SAMC 2000; Jury 2000;* Kovats et al. 1999; *Liverpool School of Tropical Medicine 2000; National Oceanic and Atmospheric Administration [NOAA] 2000*).

A major uncertainty for the future is accelerated global warming as a consequence of emissions of carbon dioxide (CO_2) and other greenhouse gases (see Chapter 7). Projections of rainfall under conditions of accelerated global warming are not consistent from one climate model to another (see Chapter 9). Although historical analysis of the twentieth century's rainfall in Zimbabwe shows an average drop of 10–16 percent, climate scenarios projecting the effect of doubling CO_2 in the atmosphere allow for an increase or a decrease of 10–15 percent (Unganai 1996; Hulme 1996). The effect of accelerated global warming on ENSO fluctuations is also unclear (Unganai 1996; Hulme 1996). Increased interannual variability in precipitation would cause increased fluctuations in the transmission of malaria, very likely limiting the development of natural immunity to malaria that results from repeated exposure. The effect of environmental change is influenced by human action or inaction. In the epidemic of

(a)

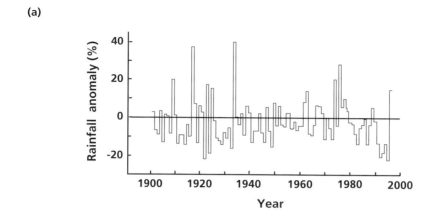

(b)

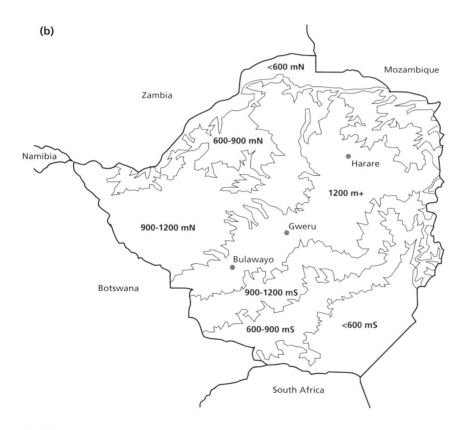

Figure 12.2 *a:* Rainfall in southern Africa, 1900/01–1995/96. The rainfall year is from July to June. *Source:* Redrawn with permission from Hulme 1996, Figure 8. *b:* Altitudinal classification of Zimbabwe. *mS,* meters elevation in the southern region; *mN,* meters elevation in the northern region; *1200 m+,* more than 1,200 meters elevation. *Source:* Redrawn with permission from Taylor and Mutambu 1986, Figure 1.

malaria in Zimbabwe in 1996, the death toll was probably increased because of shortages of antimalarial drugs in clinics in outlying areas as well as the unexpectedly high demand for treatment. One hopes that assessments of the consequences of climate change over several decades may be greatly reduced if the public health community is able to use seasonal to interannual climate forecasts to prevent morbidity and mortality due to malaria (see Chapter 5).

Temperature in Zimbabwe varies from the cooler central plateau, where the major population centers are located, to the warmer low-lying areas (Fig. 12.2b). Within Zimbabwe, malaria is much more common at lower altitudes (Taylor and Mutambu 1986). According to Leeson (1931), seasonal transmission of malaria at higher elevations is caused by the progressive migration of mosquito vectors from lower altitudes where transmission is perennial; seasonal transmission of malaria at higher elevations is interrupted when cool, dry weather kills off the populations of mosquitoes that grow during the warm, rainy weather. The potential for an increase in the transmission of malaria above an elevation of 900 meters is particularly sensitive to a projected increase of 2°C due to global climate change (Lindsay and Martens 1998).

Nevertheless, it is difficult to obtain clear empirical evidence of an effect of temperature changes alone on the incidence of highland malaria. An analysis of yearly variation in temperature in Zimbabwe suggests that, over the twentieth century, years of higher incidence of malaria have been associated with higher temperature (Freeman and Bradley 1996). However, that study is quite limited because it uses only the September temperature in Harare on the high central plateau, where malaria is not present, to compare to the incidence of malaria in the rest of the country. Over the twentieth century, the annual mean temperature for Harare has increased, probably because of the effect of an urban heat island, while the annual mean temperature for the country has decreased slightly (Unganai 1996).

Studies of Rwanda and Ethiopia, two East African countries at risk for malaria in highland areas, have examined an association between increases in the number of cases of malaria observed at higher elevations and increases in minimum temperatures (Loevinsohn 1994; Tulu 1996). However, multiple factors are responsible for the spread of malaria to highland areas (Mouchet et al. 1998; Lindsay and Martens 1998). For example, the El Niño phase of ENSO, which generates warmer and wetter conditions in East Africa, seemed to trigger an epidemic of highland malaria in Uganda in 1997–98, although only an association between excessive rainfall and the density of vectors proved to be statistically significant in the analysis of that epidemic (Lindblade et al. 1999). Interestingly, the rainfall of that same El Niño event was associated with a reduction in the incidence of malaria in highland Tanzania (Lindsay et al. 2000). Understanding these different effects will require a close examination of environmental differences in altitude, weather, and vector ecology as well as differences in study design that affect the measurements of increases and decreases in incidence.

A historical perspective also demonstrates the need for caution in interpreting the effect of climate change. The apparent burden of malaria in England during the Little Ice Age (a period of unusually cold weather starting in the middle of the sixteenth century and lasting approximately 150–200 years) underscores the importance of examining temperature in conjunction with other factors, including limitations in diagnostic accuracy (Reiter 2000; Martens 2000). Despite the fact that the central plateau of Zimbabwe is considered too cool to support the transmission of malaria, Leeson (1931), in a classic study of the anopheline vector mosquitoes in Zimbabwe, reported that "there are people still living who can remember that Salisbury [now Harare] was once no more free from malaria than any other place in the country." It is most likely that people with malaria in Harare actually acquired the infection when they traveled outside the city and that some of the supposed cases of malaria were other fevers of unknown origin; nevertheless, Leeson's statement reflects a reduction in malaria observed in Harare. Leeson attributed the absence of malaria in Harare to the development of improved housing, drainage, sanitation, and roads that prevented the breeding of mosquito vectors *An. funestus* and *An. gambiae*.

Human activities can modify the effect of environmental change in other ways. In the South Eastern Lowveld, Zimbabwe has major facilities for irrigating the agricultural production of sugar, wheat, and citrus. In theory, the risk of malaria could have increased because water provided for irrigation can create new habitat for breeding mosquitoes and allow mosquito populations to become less dependent on rainfall. However, for years, Zimbabwe has had an effective malaria control program for the irrigation projects. Human migration has also been important for the spread of malaria in Zimbabwe. Neighboring Mozambique, where the warmer and more humid conditions are more suitable for the transmission of malaria, was the source of refugees fleeing into Zimbabwe during Mozambique's civil war. The mixture of local and refugee populations near the border facilitated the spread of malaria. Movement of populations complicates any analysis linking reports of malaria to local environmental conditions.

Malaria in Gambia and Niger
Malaria is one of the most important causes of death among Gambian children; about 4 percent of children in rural areas die of malaria before reaching the age of five years (Greenwood and Pickering 1993). Given such a

high death toll, it is remarkable that the consequences of malaria were considerably worse in the 1950s. What happened even earlier is unknown because the 1950s are the earliest period in which systematic health surveys of the local (i.e., non-European) population were conducted. Reductions in rainfall have contributed to the reduction in malaria transmission since the 1950s. However, since drought is so harmful to natural resources, plans for economic development seek to combat the effects of drought. Unfortunately, these efforts often have the side effect of increasing the spread of malaria. Trends in malaria infection in West Africa reflect a tension between climatic forces tending to reduce transmission and human activity tending to increase transmission, underscoring the need for an integrated assessment of global change to consider multiple factors (see Chapter 5).

Rainfall is the major environmental determinant of the transmission of malaria in Gambia. A single, short rainy season starts in May or June and ends in October or November, followed by a long, dry season (Koram et al. 1995). Malaria is highly seasonal, roughly coinciding with the rainy season. Entomologists studying the mosquito vectors of West Africa are trying to identify exactly how and where transmission occurs in a limited fashion during the dry season; small foci of transmission are the source of expansion of transmission during the rainy season (Coluzzi 1993). Additionally, there is considerable variation in the amount of rainfall during the rainy season in West Africa (Tourre et al. 1999). Over the past 30 years or so, Gambia and surrounding areas have experienced a decline in precipitation (Greenwood and Pickering 1993). Thus, drought conditions, which have deleterious effects on agricultural and forest resources, have the beneficial effect of reducing the transmission of malaria. Some models of climate change predict even more drying in the region (Dixon et al. 1996).

Vegetation cover, which exhibits spatial and temporal variability, is an important predictor of the risk of malaria in West Africa and may be determined on the basis of data collected by remote-sensing satellites (see Chapter 3 and Thomson et al. 1997). However, bednets treated with the insecticide permethrin also affect the epidemiology of malaria in Gambia (Thomson et al. 1999). The use of bednets reduces exposure to infective mosquito bites; the effect of insecticide may be particularly important for preventing severe infections with high densities of parasites. The Gambian National Impregnated Bednet Programme has been working to treat all bednets found in all villages. A spatial model for predicting the risk of malaria in Gambia incorporates data on the use of bednets and estimates of vegetation cover (Thomson et al. 1999). The use of satellites for mapping environmental determinants of disease in Africa is an active area of investigation (*Liverpool School of Tropical Medicine 2000*).

Activities for economic development can promote the transmission of malaria even in dry areas. Poor practices of water management can create opportunities for year-round transmission of malaria where rainfall is virtually nonexistent. In Niger, in the oasis of Bilma in the Sahara Desert, only 0.8 mm of rain fell during a 12-month period that saw an increase in transmission of *P. falciparum*, the deadliest malaria parasite (Develoux et al. 1994). The prime breeding site for mosquito larvae was considered to be a large freshwater pond created during the prior decade by water leaking continuously from a broken borehole pipe. All of the other watering points were brackish and did not support the breeding of *An. gambiae*, the vector of greatest concern in the area.

Salinization of water is another environmental determinant of the transmission of malaria that also reflects a tension between climatic forces tending to reduce transmission and human activities tending to increase transmission. The entire country of Gambia is a narrow strip north and south of the River Gambia, which runs from east to west for about 300 miles and empties into the Atlantic Ocean. The mangrove swamps in the estuarine lower reaches of the River Gambia provide saltwater habitat for *Anopheles melas*. *Anopheles melas* is a malaria vector but is much less effective at transmitting malaria than *An. gambiae*, the freshwater breeder that dominates the transmission of malaria in the country. Some salinization of the ground water in Banjul, the capital city, has already occurred because of the years of drought and overexploitation of ground water; further salinization will occur with the accelerated rise in sea level expected under scenarios of climate change (Jallow et al. 1996). However, although a net increase in salinization should have the benefit of reducing the transmission of malaria, an integrated assessment must consider the tremendous cost associated with the loss of a supply of fresh water and agricultural capacity.

Human activities to counter salinization have the unintended consequence of increasing the transmission of malaria (Coluzzi 1994). Interventions to desalinize water have accelerated in recent years as a result of population growth, availability of new technologies, and international funding. As mangrove swamps are transformed into agricultural zones, often for the production of rice, *An. melas* is replaced by freshwater breeders *An. gambiae*, *An. funestus*, and *An. arabiensis*.

Growth in population also has mixed effects on the transmission of malaria. Urbanization can contribute to reduced transmission if proper land and water management prevents the creation of pools of stagnant water, which provide breeding sites for vector mosquitoes. But these ideal conditions are often not met. In Gambia, the periurban areas around the capital, Banjul, are experiencing a rapid pace of new construction with poor access roads and drainage (Koram et al. 1995). The new construction combined with rice farming along swamps provides excellent habitat for the breeding of *An. gambiae* during the rainy season. Travel to rural areas, where the transmission of malaria is even greater, also contributes to the risk of acquisition of malaria by urban residents. Overall, in the periphery of Banjul, malaria is associated with poor quality housing, crowding, and travel to rural areas. Urbanization may also increase the risk of the transmission of malaria by causing a shift from salt water to fresh water. In a study of Cotonou in Benin, the replacement of pile-dwelling traditional villages in lagoon areas with unplanned urban areas favored the replacement of a less effective vector, *An. melas,* by a more effective vector, *An. gambiae* (Coluzzi 1994).

Even in villages, changes in land use generated by the activities of an expanding population can create new opportunities for the transmission of malaria. In Niger and other countries, the excavation of clay for bricks for housing construction leaves large pits that fill with water, providing a breeding site for vector mosquitoes (Buck and Gratz 1990). Originally, these pits were at a considerable distance from human dwellings. As villages have expanded, people have moved in close proximity to the pits and have placed themselves at much greater risk of being bitten by mosquitoes. Moreover, the water in the pits allows transmission throughout the year, instead of the seasonal transmission associated with rainfall. An integrated assessment should consider changes in seasonal patterns of transmission, but such an approach is hindered in locations where a diagnosis of malaria is not routinely confirmed by an examination of parasites in a blood sample. In Niger, malaria parasites could not be found in the blood of many cases of malaria diagnosed on the basis of clinical symptoms alone; this was especially true in the dry season, when hardly any of the reported malaria cases had parasites (Buck and Gratz 1990).

Malaria in Sri Lanka

In 1963, after years of sustained control of malaria, only 17 cases of malaria were documented in the entire country of Sri Lanka. Of the 17, only 6 had acquired the infection in the country; the remaining 11 cases had been the result of infection from transmission outside the country (Wijesundera 1988). It seemed that the program to eradicate malaria would succeed in Sri Lanka. Eradication would be truly an amazing feat considering the long and devastating history of malaria on the island. Almost a thousand years ago, the kingdom of Yala in the southeast part of the island was deserted because of fever sickness, probably malaria (Jayawardene 1993). The worst Sri Lankan epidemic of malaria of the twentieth century occurred in 1934–35, when an estimated 80,000 people died in seven months (Jayawardene 1993).

Unfortunately, the program to eradicate malaria was not maintained (see also Box 12.3). The number of cases of malaria started to rebound in 1967. An important factor was the breakdown in the control of malaria due to logistical difficulties and the development of resistance to the insecti-

Failure of the Global Malaria Eradication Program

When the insecticide DDT became available after World War II, the United States and some other countries successfully interrupted the transmission of malaria by using DDT to kill adult biting mosquitoes that could transmit malaria. The approach was *indoor residual spraying,* which is a program of spraying on the interior surfaces of human dwellings, leaving a residue of DDT that kills infected or infective mosquitoes when they rest before or after biting. The eradication of malaria did not require the elimination of entire populations of mosquito vectors; mosquitoes that are capable of transmitting malaria may still be found in countries that eradicated malaria (see text under "Malaria in the United States"). That successful experience with eradication inspired the development of a global program based on the premise that an intensive schedule of indoor residual spraying over a limited period of five years could eradicate malaria, except in sub-Saharan Africa where transmission has been most intense. However, it soon became clear that the original plans were too optimistic and that maintenance of intensive vector control operations for an indefinite period of time was not possible. The major problem was the very high efficiency required of spraying operations; spraying had to include an adequate dose of insecticide on all surfaces in all housing on a regular schedule (Olliaro et al. 1996). In addition, mosquito vectors frequently developed resistance to DDT, certain vectors did not rest on walls inside houses, and many individuals refused to allow spraying in their homes. The collapse of the global program led to a resurgence of malaria in many countries around the world and, in 1978, a change in strategy from eradication to control (World Health Organization 1979).

---

cide DDT in the mosquito vectors. However, an examination of other factors provides a fuller picture. Since 1967, the numbers of cases have fluctuated, with the most recent upsurge in 1986 (Fig. 12.3*a*). Underlying this upsurge was an accelerated expansion of a major resettlement project for the development of the Mahaweli, Sri Lanka's longest river. The waters of the Mahaweli have been managed by a series of dams, reservoirs, and diversion tunnels for the purposes of irrigation, hydropower, and agricultural settlement (Wijesundera 1988). Malaria broke out in the upper reaches of the Mahaweli in 1986 and 1987, despite the fact that the transmission of malaria had been rare in that part of the island (see Fig. 12.3*b*). The explanation lies in an understanding of three factors that promote the transmission of malaria: changes in land and water use, immigration of people without any experience with malaria, and climatic conditions favorable for the transmission of malaria (Jayawardene 1993).

One of the great ironies of climate and geography in Sri Lanka is that water, which is necessary for the breeding of vector mosquitoes, is often too abundant to provide suitable breeding sites. Although too little rain can inhibit breeding by drying out pools of water, too much rain can also inhibit breeding by washing out pools of water. The upper reaches of the

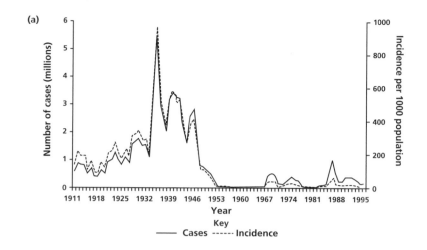

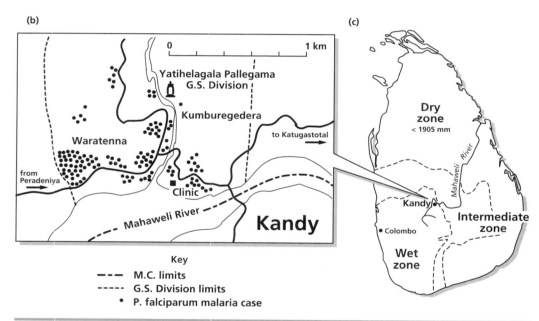

Figure 12.3   *a:* Annual incidence of malaria in Sri Lanka, 1911–96. *Source:* Redrawn with permission from Konradsen 1998, Figure 3. *b:* Concentration of falciparum malaria cases near Kandy, Sri Lanka, in May and June of 1987. *Source:* Wijesundera 1988, Figure 4. Reprinted with permission from Elsevier Science. *c:* Climatic zones of Sri Lanka. *Source:* Wijesundera 1988, Figure 2. Reprinted with permission from Elsevier Science.

**Table 12.1   Rainfall and River Discharge
in Upper Mahaweli, Sri Lanka, 1986–1987**

Year	Month	Rainfall (mm)	River Discharge (millions of cubic meters)
1986	May	242	126
	June	632	129
	July	670	133
	Aug	607	179
	Sept	947	208
	Oct	629	284
	Nov	295	216
	Dec	187	83
1987	Jan	44	44
	Feb	21	37
	Mar	85	8
	Apr	236	23
	May	223	39
	June	667	90
	July	47	68
	Aug	640	73

Source: Data from Wijesundera 1988, Table 1.

Mahaweli lie in the designated wet zone of the island, in the southwest, with annual rainfall of 2,000–5,000 millimeters (see Fig. 12.3*c*). Rainfall is present throughout the year, with one peak from the northeast monsoon between November and February and another peak from the southwest monsoon between May and September (Wijesundera 1988). Malaria is usually absent in the wet zone, since such heavy rainfall inhibits the breeding of vector mosquitoes by washing out pools along river and stream beds. However, in an unusually dry year, conditions in the wet zone may be suitable for an outbreak of malaria. A dry year in the wet zone makes the conditions more like those of the dry zone in Sri Lanka, where there is no southwest monsoon, annual rainfall is only about 1,500 mm, and malaria is perennial. The period from May 1986 to August 1987 was a period of relatively low rainfall in the upper Mahaweli, so the climatic conditions were favorable for an outbreak of malaria (see Table 12.1).

Changes in rainfall alone do not explain the role of water because the Mahaweli River is a managed system. An examination of the amount of water discharged into the river upstream of the outbreak of malaria shows low volumes of discharge in 1986 and 1987 (see Table 12.1), but this problem was not entirely because of low rainfall. In 1976, when the Polgolla dam was built in the area, transmission of malaria occurred downstream. The response was a program of intermittent flushing of the river, which interrupted transmission by washing out the sites suitable for vector breeding. However, water shortages forced the termination of this strategy to control vectors. The fundamental problem was that the priority for water use was hydropower and irrigation; only the excess was allowed downstream for inundating the breeding sites of vectors (Wijesundera 1988).

The resettlement of people is also part of the explanation for the outbreak. The inundation of land during the creation of reservoirs forced people from the local area with no experience of malaria to move to the dry zone where malaria is perennial. For social and economic reasons, migrants often traveled back and forth between malarious and nonmalarious areas. Nonimmune settlers picked up malaria more easily and were debilitated even as they were forced to continue subsistence activity (Jayawardene 1993). Movement from the malarious areas back to areas previously free of malaria further exacerbated the spread.

One study suggests that the climatic conditions favoring the transmission of malaria in Sri Lanka will become more common (Dhanapala 1998). Atmospheric models of a doubling of the atmospheric concentration of CO_2 predict warming (see Chapter 7) that will lead to more aridity in Sri Lanka because of increased evaporation and transpiration of water. Under model scenarios, the area of the dry zone of the island, where malaria is perennial, will increase by 45–65 percent while the area of the wet zone of the island, where malaria is rare, will decrease by 45–55 percent.

Temporal variation in rainfall will also continue to be important for the transmission of malaria. Historically, Sri Lanka has experienced variation in rainfall influenced by the ENSO pattern of atmospheric and oceanic circulation (see Chapter 8). The likelihood of an epidemic of malaria was increased by the failure of the southwest monsoon in the wet zone during a year in the El Niño phase of ENSO combined with the frequent failure of the northeast monsoon in the preceding year (Bouma and van der Kaay 1996). An unusually dry year in Sri Lanka plays the same role as an unusually wet year in Zimbabwe in terms of increasing the risk of malaria. As previously discussed for Zimbabwe, the swings in precipitation and malaria transmission might become more intense and present greater challenges for the control of malaria. On the other hand, assessments of the consequences of climate change may be greatly reduced if the public health community is able to use seasonal to interannual climate forecasts to prevent morbidity and mortality due to malaria (see Chapter 5 and Kovats et al. 1999; *NOAA 2000*). Strategies to manage water resources while promoting good health are the subject of active research (*International Water Management Institute 1999*).

Malaria in Brazil

The number of malaria cases has increased over 10-fold in Brazil since a low of 52,469 cases was reported in 1970 (Sawyer 1993). The official number of cases of malaria was 577,520 in 1990 (Mota 1992) and was only moderately reduced to 455,216 in 1997 (Ministério da Saúde [Brasil] 1998). The pattern of growth in the number of cases has been associated with a shift in the geographic distribution of cases. Malaria has been virtually eliminated from parts of the country with a long history of settlement by non-Indian populations (i.e., the major centers of population in the southeast and northeast near the Atlantic coast). As a public health problem, malaria is now concentrated in Brazil's frontier of development in the interior, the "Legal Amazon" region in the north and central-west. Each year, between 5 percent and 25 percent of the population along the Trans-Amazon Highway contracts malaria (Walsh et al. 1993).

The major contributor to the resurgence of malaria in Brazil is the migration of Brazilians from established cities and towns to the Amazonian frontier. Since the middle of the twentieth century, Brazil has encouraged agricultural settlement in the Amazon region as a means of distributing land to landless families (Sawyer 1993). The pressure for settlement is not simply a matter of a growing population. In Brazil, the ownership of land is concentrated in the hands of relatively few people, and the existing arrangement is politically and financially difficult to change (Sawyer 1993). Settlement in the Amazon region also serves a geopolitical strategy of surveillance of Brazil's vast borders with Peru, Bolivia, Colombia, Venezuela, and the Guyanas (Ascher 1999, 115–17).

Settlement of the Amazon region increases the opportunity for exposure to malaria because people who live in or near the forest are at greater risk of being bitten by malaria vectors than are those who live in established towns (Walsh et al. 1993). In particular, migrants who have not acquired immunity readily become ill when exposed to malaria. Immigration to the Amazon region has resulted in extensive contact between the migrants and the principal malaria vector, *Anopheles darlingi* (Zimmerman 1992). *Anopheles darlingi* breeds along interior forest streams and river backwaters. It can also breed in artificial ponds that have shaded edges. Deforestation has opened up areas, leading to an increase in the abundance of other vector species that previously were not considered important (Rubio-Palis and Zimmerman 1997). Environmental changes may have an indirect effect on the dynamics of malaria transmission by changing the composition of vector species and the predominant species of malaria parasite.

The pattern of settlement has established a social dynamic that fosters the spread of malaria in the Amazon region. Malaria has undermined the economic rationale for permanent settlement on family farms because of the burden posed by repeated bouts of malaria. Many young families choose to keep mothers and children away from malaria (Sawyer 1993). Temporary male employment in activities such as road construction and placer mining becomes more appealing. Temporary workers, especially migrant gold miners, have contributed to the spread of malaria. For example, the introduction of 40,000 gold miners into the Yanomami Indian area of the State of Roraima from 1987 to 1990 caused an epidemic of malaria that affected both migrants and Indians. Between 1991 and 1995, malaria was responsible for 25 percent of all deaths among the Yanomami (Confalonieri 1998). During that period, the annual incidence of malaria for Yanomami villages that had contact with immigrants was as high as 1,350 per 1,000 population; in contrast, the annual incidence in villages without such contact was around 20 per 1,000 (Confalonieri 1998).

Transient populations make it more difficult to provide appropriate health services. Treatment of malaria cases is often incomplete because of inadequate medical follow-up and possible side effects. Although incomplete treatment may relieve fever, the underlying infection persists. Since there is considerable movement of people within the Amazon region, infective individuals can be the source of transmission in new areas. Incomplete treatment also tends to generate resistance to the antimalarial drug used, so that future cases of malaria may not respond as well to therapy. The low quality of blood banking services provides a secondary mode of transmission through blood or blood products, which amplifies the primary mode of transmission through the mosquito vector. Transient populations also live in poor housing that increases exposure to mosquito bites at night and limits the opportunity to apply insecticides to interior walls. For those in occupations of extracting resources from forested areas, such as gold mining and rubber tapping, a dwelling might be as simple as a piece of canvas strung over four poles.

With adequate resources in the form of trained governmental and nongovernmental personnel at the local level, intersectoral cooperation, transportation, supplies, and equipment, malaria in the Amazon region can be controlled. However, if resources are lacking and cannot be sustained, even urbanization may not be protective. Transmission of malaria occurs regularly on the periphery of Manaus, a city of 1.5 million inhabitants situated along a tributary of the Amazon River. Manaus is experiencing tremendous growth due to immigration; many immigrants settle in outlying areas surrounded by small streams that have malaria vectors.

Climate change in the Amazon could affect the transmission of malaria. One analysis suggests that the region will become hotter and drier as a direct result of the replacement of the forest by degraded pasture (Shukla et al. 1990). Other global models of climate change related to global emissions of greenhouse gases into the atmosphere also show that tropical areas will become hotter and drier (Rind 1995). However, the nature of the consequences for malaria is unclear. On the one hand, greater temperatures could stimulate the development of the parasite in the mosquito vector. On the other hand, less rainfall could inhibit the breeding of mosquitoes, although uncertainty about the pattern of rainfall is critical, since most transmission occurs at the beginning and at the end of the rainy season. The uncertainties in the possible influence of climate change on malaria depend on uncertainties in the projections of climate change, especially the components related to patterns of precipitation.

Malaria in the United States

In two days in the summer of 1993, three residents of a single neighborhood of Queens in New York City experienced onset of fever caused by an infection of *P. falciparum,* the malaria parasite that causes the most serious illness and is potentially life threatening (Layton et al. 1995). Histories of travel ruled out the possibility that the infections were acquired in other countries. Histories of sharing needles or receipt of blood or blood products precluded the possibility of blood-induced transmission. Mosquitoes infected with malaria in other countries and transported by aircraft to the United States were also deemed unlikely sources because of the prevailing winds and the distances between the Queens neighborhood and the closest international airports. With other possibilities eliminated, the public health investigators concluded that the three cases of malaria had been caused by bites from infected mosquitoes that were local to the Queens neighborhood. Their analysis of how the cycle of malaria transmission operated that summer in Queens demonstrates the interplay of environmental conditions and movements of populations around the world that is increasing the vulnerability of the United States to a global resurgence of malaria.

Since malaria in the United States was officially eradicated in the 1950s (Zucker 1996), it may come as a surprise to many that mosquitoes capable of transmitting malaria can be found throughout the country. The most important mosquito vectors are *Anopheles quadrimaculatus* east of the Rocky Mountains, *Anopheles freeborni* west of the Rocky Mountains, and *Anopheles hermesi* in California (Zucker 1996). The United States has a history of endemic malaria at least since colonial times (Faust 1945). Although the transmission of malaria was more intense in the South, malaria was found throughout most of the country. The transmission of malaria was reduced even before World War II because of a combination of improvements in socioeconomic conditions, climatic conditions, increased drainage, better housing and nutrition, migration from rural to urban areas, greater access to medical services, and availability of quinine for treatment. The transmission of malaria was interrupted after World War II because of an additional program of applying the insecticide DDT on house walls to kill adult mosquitoes and in aquatic breeding sites to kill mosquito larvae (see Box 12.3 and Zucker 1996). Surveillance activities associated with the control of malaria allowed public health officials to focus on areas of transmission and treat people harboring malaria parasites. The goal was to break the cycle of malaria transmission, not to eliminate the mosquito vectors, and that goal was achieved. Thus, the investigators of the New York City malaria outbreak in 1993 knew that a mosquito vector was available, most likely *An. quadrimaculatus* (Layton et al. 1995).

Malaria parasites are constantly being reintroduced into the United States. For example, in 1992, the U.S. Centers for Disease Control and Prevention (CDC) received reports of 903 cases of malaria that were acquired outside the country (Zucker et al. 1995). These cases, which are designated as imported, reflect a mix of U.S. travelers, U.S. military personnel, and immigrants from countries where malaria is endemic. Every year, a small number of imported cases transmit malaria to others. For example, in 1992, seven cases arose from transmission in the United States (Zucker et al. 1995). As in 1992, the mode of transmission is usually congenital or via blood transfusion. Some imported infections may have been asymptomatic for years. In one instance of congenital transmission of *Plasmodium malariae*, the mother had emigrated from Laos in 1984 and had had no detectable malaria parasites in her blood. In another instance, a man who had donated blood to his brother was found to be the source of *P. malariae*, despite the fact that he had emigrated from Canton Province, China, in 1948 with no additional history of exposure through travel.

Imported cases who are harboring gametocytes, the infective stage of the parasite in humans, can provide a source of infection for anopheline mosquitoes, which in turn can transmit malaria to people. From 1957 to 1994, 74 cases of malaria acquired from mosquito-borne transmission were reported in 21 states (Zucker 1996). In one outbreak of *Plasmodium vivax* in San Diego County, California, transmission was sustained through a second generation of cases; that is, an imported case was the source of infection for more cases, who also became the source of infection for yet more cases (Maldonado et al. 1990). Before 1991, most of the outbreaks occurred in rural areas; from 1991 to 1994, three outbreaks occurred in densely populated urban and suburban areas, including New York City (Zucker 1996). The trend toward urbanization is worrisome because more people are at risk of acquiring infection. The trend toward urbanization is also puzzling, since urbanization is usually associated with the removal of breeding sites of malaria vectors and has contributed to the eradication of malaria in the United States. An explanation of the increased urbanization of malaria requires further research.

In the New York City outbreak of 1993, investigators could identify likely sources of gametocytes. The Queens neighborhood where the epidemic occurred had experienced a 31 percent increase in the number of foreign-born persons, as reflected in the 1990 census (Zucker 1996). Many of the immigrants were from countries where malaria is endemic, including parts of South America, Central America, Dominican Republic, and Haiti. In addition, more than 100 cases of imported malaria were reported in New York City in 1993 (Zucker 1996). It is clear that international population mobility plays an important role in the spread of malaria. The spread is likely to increase, as the global demographic changes causing population movements are likely to continue (see Chapter 6). However, proper planning can reduce the risk of importation of malaria. For example, a program for resettling East African refugees treated them for malaria before they left Africa (Slutsker et al. 1995).

Climate is also an important factor. Outbreaks of malaria in the United States, including the one in New York City in 1993, are associated with weather that is hotter and more humid than usual (Zucker 1996). Such an association should not be surprising, since a minimum ambient temperature is required to maintain the cycle of transmission in the mosquito vector, with the most virulent species of malaria requiring the warmest temperatures. In addition, the cycle of transmission is accelerated under warmer temperatures, that is, the parasite develops in a shorter period of time. Under scenarios of climate change in which the United States becomes warmer, the potential for the transmission of malaria is likely to increase (Patz et al. 1996). However, the resources of a wealthy country for surveillance and control of disease will probably prevent malaria from becoming an endemic problem again.

Conclusion

A better understanding of the global forces affecting the transmission of malaria should lead to improved models for the understanding of transmission and for the development of regional strategies of disease control. Also, it is well known that the transmission of malaria depends on local conditions, local perturbations, and local catastrophes. This means that local scenarios of the transmission of malaria must be analyzed in order to develop control strategies that are environmentally sound and socially desirable and that will function at the local level. The public health community will have to work with other sectors to control malaria as part of a larger effort to protect public health in a changing ecosystem.

SUGGESTED STUDY PROJECTS

Suggested study projects provide a set of options for individual or team projects that will enhance interactivity and communication among course participants (see Appendix A). The Resource Center (see Appendix B) and references in all of the chapters provide starting points for inquiries. The process of finding and evaluating sources of information should be based on the principles of information literacy applied to the Internet environment (see Appendix A).

The objective of this chapter's study projects is to develop an understanding of the dynamics and control of malaria in relation to global ecosystem change.

PROJECT 1: Global Ecosystem Change and the Transmission of Malaria

The objective of this study project is to develop an understanding of the effects of global ecosystem change on the transmission of malaria.

Task 1. Identify and describe periods of increased transmission of malaria in different locations (1950 to the present).

Task 2. Compare how population growth, population movement, economic development, and climate have contributed to the increases in transmission of malaria described in task 1.

Task 3. Describe how possible changes of the factors mentioned in task 2 could affect the transmission of malaria in the future.

PROJECT 2: Framework for Ecological Change and Emerging Disease in Chapter 10

The objective of this study project is to develop an understanding of the framework for analyzing ecological change and emerging disease.

Task 1. Write a comprehensive paper on the relationship between this case study on malaria and the framework for ecological change and emerging disease discussed in Chapter 10.

PROJECT 3: Control Measures

The objective of this study project is to develop an understanding of ways to control malaria.

Task 1. Describe current public health policies and practices to control malaria in a particular location.

Task 2. Describe current public health policies and practices to deal with potential changes in malaria control strategies.

Task 3. Describe options for malaria control that address one or more of the global forces affecting the transmission of malaria.

Acknowledgments

We thank Robert Zimmerman and Ulisses Confalonieri for comments and suggestions on the manuscript. We also appreciate Wim van der Hoek, who supplied a graphic on the incidence of malaria in Sri Lanka; Kim Lindblade, Ned Walker, and Steve Connor, who provided information on malaria in Africa; Simon Mason and Yves Tourre, who identified useful climate information for Africa; and the assistance of Jonathan Patz and Nirbhay Kumar in finding references.

References

Allen SJ, Bennett S, Riley EM, Rowe PA, Jakobsen PH, O'Donnell A, Greenwood BM. 1992. Morbidity from malaria and immune responses to defined *Plasmodium falciparum* antigens in children with sickle cell trait in The Gambia. *Trans R Soc Trop Med Hyg* 86 (5): 494–98.

Ascher W. 1999. *Why Governments Waste Natural Resources: Policy Failures in Developing Countries.* Johns Hopkins University Press, Baltimore, Md.

Bouma MJ, van der Kaay HJ. 1996. The El Niño Southern Oscillation and the historic malaria epidemics on the Indian subcontinent and Sri Lanka: An early warning system for future epidemics? *Trop Med Int Health* 1 (1): 86–96.

Buck AA, Gratz NG. 1990. *Niger: Assessment of Malaria Control. Niamey, January 24–February 15, 1990.* Report for United States Agency for International Development Vector Biology and Control Project, Arlington, Va.

Coluzzi M. 1993. Advances in the study of Afrotropical malaria vectors. *Parassitologia* 35 (suppl.): 23–29.

———. 1994. Malaria and the Afrotropical ecosystems: Impact of man-made environmental changes. *Parassitologia* 36:223–27.

Confalonieri U. 1998. Malaria in the Brazilian Amazon. In *World Resources, 1998–99* (World Resources Institute, ed.). Oxford University Press, New York, Box 2.4.

Develoux M, Chegou A, Prual A, Olivar M. 1994. Malaria in the oasis of Bilma, Republic of Niger. *Trans R Soc Trop Med Hyg* 88:644.

Dhanapala AH. 1998. Impact on incidence and distribution of malaria. In *Final Report of the Sri Lanka Climate Change Country Study* (Ratnasiri J, ed.). Ministry of Forestry and Environment, Colombo, Sri Lanka, pp. 28–32.

Dixon RK, Perry JA, Vanderklein EL, Hiol Hiol F. 1996. Vulnerability of forest resources to global climate changes: Case study of Cameroon and Ghana. *Climate Res* 6:127–33.

Faust EC. 1945. Clinical and public health aspects of malaria in the United States from an historical perspective. *Am J Trop Med* 25:185–201.

Freeman T, Bradley M. 1996. Temperature is predictive of severe malaria years in Zimbabwe. *Trans R Soc Trop Med Hyg* 90:232.

Gillies MT, Coetzee M. 1987. *A Supplement to the Anophelinae of the Africa South of the Sahara,* No. 55. South African Institute for Medical Research, Johannesburg.

Graham JD, Wiener JB (eds.). 1995. *Risk versus Risk: Tradeoffs in Protecting Health and the Environment.* Harvard University Press, Cambridge.

Greenwood BM, Pickering H. 1993. A malaria control trial using insecticide-treated bed nets and targeted chemoprophylaxis in a rural area of The Gambia, West Africa: 1. A review of the epidemiology and control of malaria in The Gambia, West Africa. *Trans R Soc Trop Med Hyg* 87 (Suppl. 2): 3–11.

Hooper K. 1999. Breast Milk Monitoring Programs (BMMPs): Worldwide early warning system for polyhalogenated POPs and for targeting studies in children's environmental health. *Environ Health Perspect* 107 (6): 429–30.

Hulme M. 1996. *Climate Change and Southern Africa: An Exploration of Some Potential Impacts and Implications in the SADC Region.* Climatic Research Unit, University of East Anglia, Norwich, U.K., and WWF International, Gland, Switzerland.

Jallow BP, Barrow MKA, Leatherman SP. 1996. Vulnerability of the coastal zone of The

Gambia to sea level rise and development of response strategies and adaptation options. *Climate Research* 6:165–77.

Jayawardene R. 1993. Illness perception: Social cost and coping strategies of malaria cases. *Soc Sci Med* 37 (9): 1169–76.

Konradsen F. 1998. Malaria Transmission and Control in an Irrigated Area of Sri Lanka. Ph.D. diss., Danish Bilharziasis Laboratory and International Irrigation Management Institute, Copenhagen.

Koram KA, Bennett S, Adiamah JH, Greenwood BM. 1995. Socio-economic risk factors for malaria in a peri-urban area of The Gambia. *Trans R Soc Trop Med Hyg* 89:146–50.

Kovats RS, Bouma M, Haines A. 1999. *El Niño and Health* (WHO/SDE/PHE/99.4). World Health Organization, Geneva.

Layton M, Parise ME, Campbell CC, Advani R, Sexton JD, Bosler EM, Zucker JR. 1995. Mosquito-transmitted malaria in New York City, 1993. *Lancet* 346:729–31.

Leeson HS. 1931. *Anopheline Mosquitoes of Southern Rhodesia, 1926–1928*. London School of Hygiene and Tropical Medicine, London.

Lindblade KA, Walker ED, Onapa AW, Katungu J, Wilson ML. 1999. Highland malaria in Uganda: Prospective analysis of an epidemic associated with El Niño. *Trans R Soc Trop Med Hyg* 93 (5): 480–87.

Lindsay SW, Birley MH. 1996. Climate change and malaria transmission. *Ann Trop Med Parasitol* 90 (6): 573–86.

Lindsay SW, Bodker R, Malima R, Msangeni HA, Kisinza W. 2000. Effect of 1997–98 El Niño on highland malaria in Tanzania. *Lancet* 355 (March 18): 989–90.

Lindsay SW, Martens WJM. 1998. Malaria in the African highlands: Past, present and future. *Bull World Health Organ* 76 (1): 33–45.

Livingstone FB. 1984. The Duffy blood groups, vivax malaria, and malaria selection in human populations: A review. *Hum Biol* 56 (3): 413–25.

Loevinsohn M. 1994. Climatic warming and increased malaria incidence in Rwanda. *Lancet* 343:714–18.

Maldonado YA, Nahlen BL, Roberto RR, Ginsberg M, Orellana E, Mizrahi M, McBarron K, Lobel HO, Campbell CC. 1990. Transmission of *Plasmodium vivax* malaria in San Diego County, California, 1986. *Am J Trop Med Hyg* 42:3–9.

Martens P. 2000. Letters to the editor: Malaria and global warming in perspective? *Emerg Infect Dis* 6 (3): 313–14.

Ministério da Saúde (Brasil). 1998. Consolidado anual da distribuição de casos notificados de agravos de doenças infecciosas e parasitarias [In Portuguese: Annual consolidation of the distribution of reported cases of infectious and parasitic diseases]. In *Informe Epidemiológico do SUS VII*. Ministry of Health (Brazil), Vol. 1, p. 109.

Mota EGE. 1992. Fatores determinantes da situaçao de malária na Amazônia [In Portuguese: Determinants of the malaria situation in Amazonia]. *Rev Soc Bras Med Trop* 25 (Suppl. 3): 27–32.

Mouchet J, Manguin S, Sircoulon J, Laventure S, Faye O, Onapa AW, Carnevale P, Julvez J, Fontenille D. 1998. Evolution of malaria in Africa for the past 40 years: Impact of climatic and human factors. *J Am Mosq Control Assoc* 14 (2): 121–30.

Olliaro P, Cattani J, Wirth D. 1996. Malaria, the submerged disease. *JAMA* 275 (3): 230–33.

Patz JA, Epstein P, Burke TA, Balbus JM. 1996. Global climate change and emerging infectious diseases. *JAMA* 275 (3): 217–23.

Reiter P. 2000. From Shakespeare to Defoe: Malaria in England in the Little Ice Age. *Emerg Infect Dis* 6 (1): 1–11.

Rind D. 1995. Drying out the tropics. *New Scientist* 146 (1976): 36–40.

Rubio-Palis Y, Zimmerman RH. 1997. Ecoregional classification of malaria vectors in the Neotropics. *J Med Entomol* 34 (5): 499–510.

Sawyer D. 1993. Economic and social consequences of malaria in new colonization projects in Brazil. *Soc Sci Med* 37 (9): 1131–36.

Shukla J, Nobre C, Sellers P. 1990. Amazon deforestation and climate change. *Science* 247:1322–25.

Slutsker L, Tipple M, Keane V, McCance C, Campbell CC. 1995. Malaria in East African refugees resettling to the United States: Development of strategies to reduce the risk of imported malaria. *J Infect Dis* 171:489–93.

Soper FL, Wilson DB. 1943. *Anopheles gambiae in Brazil, 1930–1940.* Rockefeller Foundation, New York.

Taylor P, Mutambu SL. 1986. A review of the malaria situation in Zimbabwe with special reference to the period 1972–1981. *Trans R Soc Trop Med Hyg* 80:12–19.

Thomson MC, Connor SJ, D'Alessandro U, Rowlingson B, Diggle P, Cresswell M, Greenwood B. 1999. Predicting malaria infection in Gambian children from satellite data and bednet use surveys: The importance of spatial correlation in the interpretation of results. *Am J Trop Med Hyg* 61 (1): 2–8.

Thomson MC, Connor SJ, Milligan P, Flasse SP. 1997. Mapping malaria risk in Africa: What can satellite data contribute? *Parasitol Today* 13 (8): 313–18.

Tourre YM, Rajagopalan B, Kushnir Y. 1999. Dominant patterns of climate variability in the Atlantic Ocean during the last 136 years. *J Climate* 12:2285–99.

Tulu A. 1996. Determinants of Malaria Transmission in the Highlands of Ethiopia: The Impact of Global Warming on Morbidity and Mortality Ascribed to Malaria. Ph.D. diss., London School of Hygiene and Tropical Medicine, University of London.

Unganai LS. 1996. Historic and future climatic change in Zimbabwe. *Climate Res* 6:137–45.

Walsh JF, Molyneux DH, Birley MH. 1993. Deforestation: Effects on vector-borne disease. *Parasitology* 106:S55–S75.

Wernsdorfer WH. 1994. Epidemiology of drug resistance in malaria. *Acta Trop (Basel)* 56:143–56.

Wernsdorfer WH, Payne D. 1991. The dynamics of drug resistance in *Plasmodium falciparum. Pharmacol Ther* 50:95–121.

Wijesundera M de S. 1988. Malaria outbreaks in new foci in Sri Lanka. *Parasitol Today* 4:147–50.

Wolman MG. 1995. Human and ecosystem health: Management despite some incompatibility. *Ecosystem Health* 1:35–40.

World Health Organization. 1979. *Expert Committee on Malaria,* 17th Report. WHO Technical Report Series 640. World Health Organization, Geneva.

———. 1995. *Vector Control for Malaria and Other Mosquito-Borne Diseases.* WHO Technical Report Series 857. World Health Organization, Geneva.

Zimmerman RH. 1992. Ecology of malaria vectors in the Americas and future directions. *Mem Inst Oswaldo Cruz* 87 (Suppl. 3): 371–83.

Zucker JR. 1996. Changing patterns of autochthonous malaria transmission in the United States: A review of recent outbreaks. *Emerg Infect Dis* 2 (1): 37–43.

Zucker JR, Barber AM, Paxton LA, Schultz LJ, Lobel HO, Roberts JM, Bartlett ME, Campbell CC. 1995. Malaria surveillance—United States, 1992. *MMWR CDC Surveil Summ* 44 (SS-5).

Electronic References

International Water Management Institute. 1999. Health and Environment.
http://www.cgiar.org/iwmi/health.htm (Date Last Revised 2/15/1999).

Jury MR. 2000. Latest Southern African Seasonal Forecast. Climate Impact Predictions
Group at the University of Zululand. http://weather.iafrica.com/regional/
forecasts/index.htm (Date Last Revised 3/30/2000).

Liverpool School of Tropical Medicine. 2000. MALSAT. Environmental Information Systems for Malaria. http://www.liv.ac.uk/lstm/malsat.html (Date Last Revised 2/14/
2000).

National Oceanic and Atmospheric Administration. 2000. The ENSO Experiment. Office
of Global Programs. http://www.ogp.noaa.gov/mpe/csi/appdev/health/
ensoexp.htm (Date Last Revised 3/20/2000).

Southern Africa Malaria Control. 2000. Monthly Malaria Update.
http://www.safrimal.org/status.htm (Date Last Revised 1/2000).

United Nations Environment Programme. 2000. Persistent Organic Pollutants.
http://www.chem.unep.ch/pops (Date Last Revised 3/27/2000).

World Health Organization. 1999. Roll Back Malaria: A Global Partnership.
http://www.who.int/rbm (Date Last Revised 1999).

CHAPTER 13

Global Climate Change and Air Pollution

Interactions and Their Effects on Human Health

Jonathan A. Patz, M.D., M.P.H., and
John M. Balbus, M.D., M.P.H.

As the twenty-first century begins, the concern about many local and regional air pollutants has become less acute in the United States, Canada, and other developed nations. The implementation of regulatory measures and the development of cleaner technologies have resulted in the improvement of some indicators of air quality. Many developing countries have a different experience because they have not yet developed or adopted many of the regulatory measures of the industrialized world to reduce the levels of local and regional air pollution (United Nations Environment Programme 1991; World Bank 1997). However, for both developed and developing nations, evidence is growing that global climate changes caused by the emission of greenhouse gases may exacerbate the impacts of air pollution on human health and the environment (see Chapters 7 and 8). The focus of this chapter is on the interactions between greenhouse gas emissions and conventional air pollutants and their effects on human health.

Ever since the hazards of breathing soot and fumes were recognized midway into the industrial revolution, engineers have striven to develop cleaner technologies. Theoretically, the ideal engine, defined in terms of the cleanest by-products, would emit only carbon dioxide (CO_2) and water, neither being directly toxic to humans or the environment. Only recently has CO_2 been recognized as an undesirable by-product through its function as a greenhouse gas. The atmospheric concentrations of CO_2 and other greenhouse gases have been rapidly increasing (see Fig. 7.11) beyond any conditions observed in the historical record and represent a new environmental challenge (Houghton et al. 1996).

This chapter contrasts the ramifications of increases in greenhouse gas emissions against the backdrop of improving trends in conventional air pollution in the United States and Canada. A historical perspective reveals a new dimension of air pollution issues that are more regional and global in scale. This chapter then focuses on pollutants that are dependent on climatic conditions; climate change may actually make it more difficult to control the formation of some pollutants, such as ozone. Finally, the broad scope of potential health effects of climate change worldwide is briefly reviewed.

A Historical Perspective on Air Pollution

From Local to Regional to Global

Historically, the developed nations have gone through a series of transitions with respect to air pollution problems. By the mid-1800s, the combustion of fossil fuels had fouled the urban environment in industrial cities. At that time, smoke, soot (particles), sulfur dioxide (SO_2), and nitrogen dioxide (NO_2) were the predominant emissions of concern. These pollutants caused respiratory irritation, hindered visibility, and blackened buildings. Later, the advent of the automobile introduced the problem of increasing concentrations of nitrogen oxides (NO_x) and volatile organic compounds (VOCs) in the urban environment. As chemical precursors to ground-level ozone, these pollutants resulted in a new type of "photochemical smog," consisting primarily of ozone whose formation was catalyzed by sunlight.

After the 1952 disaster in London, England, where a four-day period of smog caused an estimated 4,000 deaths (Logan 1953), it became strikingly clear that local accumulations of air pollution were hazardous. The adage, "the solution to pollution is dilution," was fervently followed, leading to the construction of higher smokestacks and the relocation of polluting industries farther away from urban centers. While local air quality benefited from the greater dispersion of pollution afforded by higher smokestacks, problems of regional air pollution, such as "acid rain," emerged or were exacerbated.

Acid rain demonstrates the shortcomings of addressing pollution from only a local perspective. The regional problem of acid rain was actually first described in the mid-1800s in reference to the effect of industrial emissions on precipitation in the British midlands. It has long been recognized that acid rain is widespread in northern Europe and eastern North America. By the mid-1980s, roughly half of the sulfuric acid rain falling in eastern Canada originated in the United States, just as much of the Scandinavian acid precipitation has been traced to industrialized areas of central Europe and the United Kingdom (Schindler 1988). Recent studies have led to discoveries of areas affected by acid rain in the western United States, Japan, China, the former Soviet Union, and South America.

Transboundary Air Pollution

Toxic air pollutants such as persistent organic pollutants (POPs) can travel long distances. POPs include pesticides, such as aldrin, chlordane, DDT, dieldrin, heptachlor, and toxaphene; industrial chemicals like polychlorinated biphenyls (PCBs); and industrial or incineration by-products such as dioxin and furans. POPs are highly fat soluble; they degrade very slowly in the environment and can bioaccumulate to high levels as they move up the food chain. For example, dioxin bioaccumulates through terrestrial food webs and can become concentrated in milk and other dairy products. Indigenous people near the Arctic, who rely on fatty foods high on the food chain, such as polar bear, seal, and fish, can receive considerable exposure to POPs, and the placentas of pregnant women in these areas are often found to have high levels of POPs that pose fetal risk. Some POPs can act as endocrine disrupters, mimicking the body's hormones. Estrogenic chemicals (including some organochlorines such as DDT, some PCBs, dioxin, and furans) may be responsible for declining sperm counts and the rising incidence of abnormalities in the human male reproductive tract (Jensen et al. 1995).

The transport of POPs can transcend international boundaries, as detailed in a report of the Commission for Environmental Cooperation (CEC), which was established under the North American Free Trade Agreement (*CEC 2000*). For example, 15–25 percent of the dioxin deposited in Lake Michigan comes from sources as far away as southern Texas. Up to 90 percent of POPs applied as agricultural pesticides are retained in the atmosphere or are revolatilized. Atmospheric deposition currently contributes the majority of the total yearly input of PCBs to Lake Superior (Environmental Protection Agency [EPA] 1996a). POPs eventually concentrate in water, soil, and wildlife in the cooler, northern latitudes because of normal atmospheric convection patterns and the tendency of POPs to revolatilize many times, termed the *grasshopper effect*. When the chemical is cooled, it condenses and precipitates out of the atmosphere, so that deposition tends to prevail over evaporation at high latitudes. POPs also degrade more slowly at cold temperatures.

Polluted air masses from industrial sources have also been tracked across the Atlantic Ocean and the Arctic by documenting unique trace metal content within these air masses (Schindler 1988). In 1998, some air pollution in the northwestern United States was even found to have originated in Asia, a transport process that would generally take 4–10 days (Monastersky 1998). Dust clouds from Asia, tracked by satellite in the spring of 1998, deposited detectable levels of arsenic, copper, lead, and zinc at the pristine site of Crater Lake, Oregon.

Global warming is the ultimate example of the global effect of air pollution, on a par with the global threat of stratospheric ozone depletion

(Chapter 7). The geographic point of emission of greenhouse gases has little relation to the effects of this phenomenon. Aggregated local and regional emissions are having a global impact whose full implications for the world's climate are still being learned.

Trends in Air Pollutants and Greenhouse Gases

In 1972, under provisions of the U.S. Clean Air Act, the newly established EPA was required to set national ambient air quality standards (NAAQSs) for current "criteria" pollutants deemed to be potentially hazardous to human health. The term *criteria* is an administrative designation by the EPA for carbon monoxide (CO), NO_2, particulate matter (PM), SO_2, ozone, and lead (Pb). The standards covering these pollutants were designed to be protective of the most vulnerable subgroups within the population, such as asthmatics or persons with emphysema. In 1997, EPA revised the original health-based standards for criteria air pollutants to add new standards for particles that were less than 2.5 microns in diameter ($PM_{2.5}$).

Since 1970, the emissions and concentrations of air pollutants in the United States have decreased substantially despite large increases in total population, vehicle miles traveled (VMT), and gross domestic product (GDP). From 1970 to 1996, the U.S. population increased by 29 percent, VMT increased by 121 percent, and the GDP increased by 104 percent, yet aggregate emissions of *criteria air pollutants* decreased by 32 percent (Fig. 13.1). Changes in the emissions of individual pollutants range from a 98 percent decrease for lead to an 11 percent increase for NO_2. Table 13.1 summarizes the percentage changes in national air quality concentrations and emissions. The experience in Canada has been similar (*Environment Canada 2000*). Nevertheless, direct exposure to air pollution continues to affect human health adversely in the United States and Canada. For example, a recent study of 11 Canadian cities showed an increase of 8 percent in daily rates of nontraumatic mortality associated with days of high levels of air pollution for NO_2, ozone, SO_2, and CO (Burnett et al. 1998).

While emissions of most criteria air pollutants have been decreasing, emissions of CO_2 from fossil fuel combustion in the United States grew by 9 percent between 1990 and 1996 (*EPA 2000a*). Figure 13.1 compares trends in the emission of criteria air pollutants (along with population, VMT, and GDP) with trends in the emission of CO_2. Thus, as cleaner technologies mandated under the U.S. Clean Air Act have reduced the levels of criteria pollutants (except for NO_2 and ozone in some regions), the emissions of greenhouse gases have been relatively unaffected by regulations. In essence, while the United States may be benefiting from cleaner air nationally, the country continues to emit substantial amounts of CO_2, exacerbating the problem of global climate change.

The catalytic converter developed for motor vehicles illustrates this

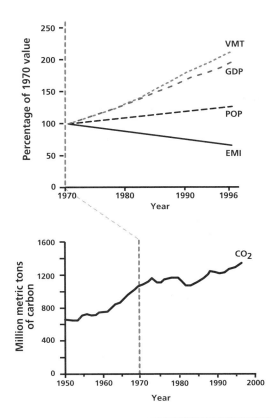

Figure 13.1 U.S. trends in emissions of criteria air pollutants, emissions of carbon dioxide, and related factors. Emissions of volatile organic compounds (VOCs) are included in the emissions of criteria air pollutants, since tropospheric ozone is created from VOCs rather than emitted directly from a source of ozone. *VMT,* vehicle miles traveled; *GDP,* gross domestic product; *POP,* population; *EMI,* emissions of criteria air pollutants. *Source:* Redrawn from *Environmental Protection Agency 1999a,* Executive Summary, p. 3, and *Marland et al. 1999.*

paradox of trading one air pollution problem for another. Vehicles now sold in the United States emit 96 percent less CO, 98 percent fewer hydrocarbons, and 90 percent less NO_x than vehicles sold in the early 1970s (*EPA 1999a*). Some of these reductions are the result of catalytic converter technology. As exhaust gases pass over the catalysts, chemical reactions convert pollutants (hydrocarbons, NO_x, CO) into harmless gases (CO_2, nitrogen) and water: (1) hydrocarbons combine with oxygen forming CO_2, (2) NO_x react with CO to produce nitrogen and CO_2, and (3) NO_x react with hydrogen to produce nitrogen and water vapor.

While greatly reducing emissions of CO, NO_x, and hydrocarbons, catalytic converters generate nitrous oxide (N_2O), a greenhouse gas that is presently unregulated and has 310 times the *global warming potential* of CO_2 (see Chapter 7). Combined emissions of greenhouse gases from stationary and mobile sources have increased by 21 percent from 1990 to 1996, primarily because of rising rates of N_2O generation in motor vehicles (*EPA 2000a*). Cars with catalytic converters emit up to five times the N_2O of cars without the devices, although the most current generation of converters have somewhat improved.

Table 13.1 Long-Term Change in U.S. National Air Quality Concentration and Emissions

Criteria Air Pollutant	Air Quality Concentration Percentage Change 1978–97	Emissions Percentage Change 1970–97
Carbon monoxide	−60	−32
Lead	−97	−98
Nitrogen dioxide	−25	+11
Ozone	−30 (1 hr)	−37
Particulate matter (<10 microns)	Not available	−75
Sulfur dioxide	−55	−35

Source: Adapted from *Environmental Protection Agency 1999b, 9.*

Fossil Fuels: The Common Source of Air Pollutants and Greenhouse Gases

Fossil fuel combustion is a common source of emissions of CO_2 and the six criteria air pollutants: CO, NO_2, PM, SO_2, ozone, and lead (Table 13.2; EPA 1997). (See also Box 13.1.) The use of fossil fuels, particularly gasoline, also releases air toxics, such as benzene, toluene, and VOCs (EPA 1997). The term *air toxic* is an administrative designation by the EPA for a variety of air pollutants, largely volatile compounds and heavy metals, that are not designated as criteria air pollutants.

Fossil fuel combustion produces nearly all of the CO_2 released into the atmosphere from human activities. As the most abundant greenhouse gas, CO_2 accounts for 81 percent of the total U.S. emissions of greenhouse gases, equaling 1,788 million metric tons of carbon equivalent (MMTCE) in 1996. This unit of measure is used because greenhouse gases differ in their global warming potential (Chapter 7). Although greenhouse gases such as methane (CH_4), N_2O, and chlorofluorocarbons (CFCs) have a greater warming potential than does CO_2, the magnitude of its emissions ensures that CO_2 contributes the most to enhanced greenhouse warming, and CO_2 levels are the focus of many projections of climate change. CO_2, CH_4, and N_2O occur naturally in the atmosphere, but human activities have substantially increased their concentrations. Since the mid-1800s, these greenhouse gases have increased by 30, 145, and 15 percent, respectively (Houghton et al. 1996). While the burning of fossil fuel is the largest source of CO_2, agriculture and decomposition of landfills are important sources of CH_4 and N_2O (*EPA 2000a*). Table 13.2 shows trends for all greenhouse gas emissions and sinks, presented in MMTCE units. Note the prominence of fossil fuel combustion.

BOX 13.1

Legacy of Gasoline Additives: Lead and Methyl Tertiary Butyl Ether

Lead interferes with brain development and has been associated with lower intelligence quotients (IQs) in children. Even relatively low blood lead levels (i.e., 10–25 micrograms per deciliter) have been shown to affect IQ in children. Elevated levels of lead in the blood of adults have been associated with high blood pressure. The use of tetraethyl lead as a gasoline additive in the United States resulted in widespread airborne dissemination of lead oxide, particularly in urban areas. After the phaseout of leaded gasoline in the United States, emissions of lead from transportation sources rapidly declined, and between 1977 and 1996, ambient concentrations of lead fell by 97 percent. This source of emissions has been almost eliminated in the United States. Blood lead levels in children have paralleled this reduced exposure. However, other sources of lead in the environment retain the potential to affect children. The legacy of lead from gasoline remains in elevated lead content in urban dusts and soils. Even though lead-based paint has been banned in the United States since 1978, children can be exposed if they reside in older housing stock or in older homes that have been renovated. In addition, lead is still used extensively as an additive to gasoline in other parts of the world, resulting in widespread exposure of humans to lead.

Lead is not the only additive to gasoline that has become widely distributed in the environment. The additive methyl tertiary butyl ether (MTBE) has recently been recognized as a significant contaminant of groundwater supplies in the United States, raising concerns about the quality of drinking water (*EPA 2000b*). MTBE is potentially carcinogenic to humans and gives drinking water an unpleasant taste and odor. Ironically, the purpose of adding MTBE to gasoline was to reduce atmospheric pollution in order to comply with the 1990 amendments to the U.S. Clean Air Act.

Mercury is an example of an air toxic whose emissions stem from burning fossil fuels, especially in coal-fired power plants. Other sources include waste incinerators, landfills, copper and lead smelting operations, and cement manufacturing plants. Anthropogenic emissions have increased the global atmospheric burden of mercury by two- to fivefold (*CEC 2000*). A major proportion of the mercury present in the atmosphere is elemental mercury, which is extremely volatile and, in its gaseous form, has an estimated atmospheric residence time ranging from 3 months to 2 years. Organic methyl mercury can bioaccumulate through the food chain in aquatic systems to reach toxic levels in predatory fish, as was the case in Minimata Bay, Japan, where severe birth defects (e.g., cerebral palsy) resulted from fetotoxicity in pregnant women who consumed contaminated fish (Koos and Longo 1976). While the mercury in Minamata was not airborne in origin, it alerted the world to the dangers of mercury in the environment. Currently, 5 Canadian provinces and over 35 U.S. states have issued health advisories to reduce the consumption of certain freshwater fish that are known to contain excessive levels of mercury.

Table 13.2 Recent Trends in U.S. Greenhouse Gas Emissions and Sinks (MMTCE), 1990–1996

Gas/Source	1990	1991	1992	1993	1994	1995	1996
CO_2 (Totals)	1,348.3	1,333.2	1,353.4	1,385.6	1,408.5	1,419.2	1,471.1
Fossil fuel combustion	1,331.4	1,316.4	1,336.6	1,367.5	1,389.6	1,398.7	1,450.3
Natural gas flaring	2.0	2.2	2.2	3.0	3.0	3.7	3.5
Cement manufacture	8.9	8.7	8.8	9.3	9.6	9.9	10.1
Lime manufacture	3.3	3.2	3.3	3.4	3.5	3.7	3.8
Limestone and dolomite use	1.4	1.3	1.2	1.1	1.5	1.8	1.8
Soda ash manufacture and consumption	1.1	1.1	1.1	1.1	1.1	1.2	1.2
Carbon dioxide manufacture	0.2	0.2	0.2	0.2	0.2	0.3	0.3
Land use change and forestry (sink)[a]	(311.5)	(311.5)	(311.5)	(208.6)	(208.6)	(208.6)	(208.6)
CH_4 Totals	169.9	171.1	172.5	171.9	175.9	179.2	178.6
Stationary sources	2.3	2.3	2.4	2.3	2.3	2.4	2.5
Mobile sources	1.5	1.4	1.4	1.4	1.4	1.4	1.4
Coal mining	24.0	22.8	22.0	19.2	19.4	20.3	18.9
Natural gas systems	32.9	33.3	33.9	34.1	33.9	33.8	34.1
Petroleum systems	1.6	1.6	1.6	1.6	1.6	1.6	1.5
Petrochemical production	0.3	0.3	0.3	0.3	0.4	0.4	0.4
Silicon carbide production	+	+	+	+	+	+	+
Enteric fermentation	32.7	32.8	33.2	33.6	34.5	34.9	34.5
Manure management	14.9	15.4	16.0	16.1	16.7	16.9	16.6
Rice cultivation	2.5	2.5	2.8	2.5	3.0	2.8	2.5
Agricultural residue burning	0.2	0.2	0.2	0.2	0.2	0.2	0.2
Landfills	56.2	57.6	57.8	59.7	61.6	63.6	65.1
Wastewater treatment	0.9	0.9	0.9	0.9	0.9	0.9	0.9

Many "cobenefits" can be achieved by reducing the emissions of the fossil fuels that are a common source of criteria air pollutants, air toxics, and greenhouse gases. For example, if nations abided by the Kyoto Protocol to reduce their emissions of greenhouse gases (see Chapter 6), what might be the benefits in terms of the effects on human mortality? An interdisciplinary working group examined this question by combining models of energy consumption, carbon emissions, and associated emissions of PM. PM has been linked to increased mortality (Dockery and Pope 1994; Pope et al. 1995; Samet et al. 2000). Even at levels below the NAAQSs, PM increases daily rates of cardiorespiratory mortality and total mortality in

Gas/Source	1990	1991	1992	1993	1994	1995	1996
N$_2$O (Totals)	92.3	94.4	96.8	97.1	104.9	101.9	103.7
Stationary sources	3.7	3.7	3.7	3.8	3.8	3.8	4.0
Mobile sources	13.2	13.9	14.8	15.6	16.3	16.6	16.5
Adipic acid	4.7	4.9	4.6	4.9	5.2	5.2	5.4
Nitric acid	3.4	3.3	3.4	3.5	3.7	3.7	3.8
Manure management	2.6	2.8	2.8	2.9	2.9	2.9	3.0
Agricultural soil management	62.4	63.4	65.2	64.1	70.4	67.2	68.6
Agricultural residue burning	0.1	0.1	0.1	0.1	0.1	0.1	0.1
Human sewage	2.1	2.1	2.2	2.2	2.3	2.2	2.3
Waste combustion	0.1	0.1	0.1	0.1	0.1	0.1	0.1
HFCs, PFCs, and SF$_6$ (Totals)	22.2	21.6	23.0	23.4	25.9	30.8	34.7
Substitution of ozone-depleting substances	0.3	0.2	0.4	1.4	4.0	9.5	11.9
Aluminum production	4.9	4.7	4.1	3.5	2.8	2.7	2.9
HCFC-22 production	9.5	8.4	9.5	8.7	8.6	7.4	8.5
Semiconductor manufacture	0.2	0.4	0.6	0.8	1.0	1.2	1.4
Electrical transmission and distribution	5.6	5.9	6.2	6.4	6.7	7.0	7.0
Magnesium production and processing	1.7	2.0	2.2	2.5	2.5	3.0	3.0
Total Emissions	1,632.7	1,620.2	1,678.0	1,678.0	1,715.3	1,731.1	1,788.0
Net Emissions (Sources and Sinks)	1,321.2	1,308.7	1,334.2	1,469.4	1,506.7	1,522.5	1,579.5

Source: Environmental Protection Agency 2000a, ES-3.

Note: Totals may not sum due to independent rounding. +, does not exceed 0.05 MMTCE.

[a]Sinks are included only in net emissions total. Estimates of net carbon sequestration due to land use change and forestry activities exclude nonforest soils and are based partially upon projections of forest carbon stocks.

the United States and Europe (Katsouyanni et al. 1993). In children, PM levels also correlate with increased hospital admissions, school absences, and medication use, as well as a reduction in lung function (measured as peak respiratory flow rates) (Bascom et al. 1996). Under a climate policy scenario approximating that of the Kyoto Protocol (e.g., the United States emitting 7 percent below 1990 levels of greenhouse gas emissions), approximately 700,000 premature deaths due to exposure to PM might be averted annually by the year 2020 as compared with the business-as-usual forecast (Working Group on Public Health and Fossil-Fuel Combustion 1997). This analysis showed the potential near-term cobenefits of long-term policies to mitigate global climate change.

Climate Change and Levels of Air Pollutants

Secondary Air Pollutants

Fluctuations in weather have the most influence on pollutants that arise from chemical reactions in the atmosphere. These "secondary pollutants," such as ozone and acid rain, are derived from a mixture of pollutants directly emitted, which are termed *primary pollutants*. Even with the improving air quality in the United States and Canada, urban ozone pollution and regional acid rain persist as problems and have the potential to be exacerbated if air temperatures warm. Ozone, as described in Chapter 7, forms secondarily in the lower atmosphere by a temperature- and sunlight-dependent reaction between NO_x and VOCs, precursors of the formation of ozone. The major source of NO_x in North America is fossil fuel combustion used in transportation and electric utilities. Since 1970, the number of vehicle miles traveled in the United States has increased at a faster rate than that of the overall population (see Fig. 13.1). From 1970 to 1994, car ownership in Canada rose from 310 to 484 vehicles per 1,000 people (Last et al. 1998). Transportation is also a major source of anthropogenic VOCs. Other sources of VOCs include incinerators, gasoline vapors, paints and solvents, and some trees or other vegetative sources.

The Influence of Weather on Ozone Formation

Tropospheric ozone, as a photochemical oxidant and the main ingredient of urban smog, is pervasive and difficult to control. The reaction between NO_x and VOCs to form ozone is catalyzed by ultraviolet radiation and requires relatively high ambient air temperatures (see Chapter 7). Meteorological factors that could theoretically influence the levels of tropospheric ozone include ultraviolet radiation, air temperature, wind speed, and atmospheric mixing and transport.

In general, there is a direct correlation between temperature and levels of ozone (Kamens et al. 1982; Grey et al. 1987; Samson 1988). Increases in atmospheric temperature accelerate photochemical reaction rates in the atmosphere and tend to increase the rates at which tropospheric ozone and other oxidants (e.g., hydroxyl radicals) are produced (Hatakeyama et al. 1991; Morris et al. 1995). Studies of ambient temperature have reported that successive episodes of high temperatures characterize years with seasonally high levels of ozone. The relationship is nonlinear, and above a temperature of 32°C (90°F), there is a strong correlation between temperature and levels of ozone (Fig. 13.2). However, levels of ozone may not always increase with an increase in temperature. For example, the production of ozone is reduced when the ratio of VOCs to NO_x is low (see Fig. 7.8).

In the eastern United States and in Europe, the majority of days when the levels of ozone exceed air quality standards occur in conjunction with

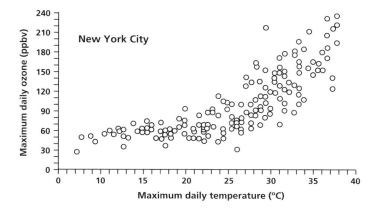

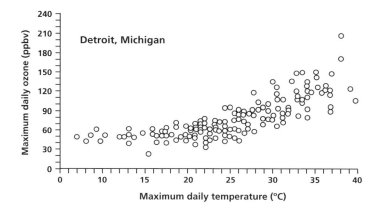

Figure 13.2 Relationship between maximum daily ozone concentration and maximum daily temperature in New York City and Detroit, Michigan. *Source:* Redrawn from Environmental Protection Agency 1996b, Figure 3.9.

slow-moving high-pressure systems around the time of the summer solstice (National Research Council [NRC] 1991). This is the period of greatest sunlight, when solar radiation is most intense and air temperatures are high. The relatively high levels of ozone in the United States during 1988 and 1995 were probably due in part to hot, dry, and stagnant conditions (Patz et al. 2000). During 1995, one of the hottest years on record, 32 percent of Americans, or 71 million people, resided in counties in which ozone levels exceeded the NAAQSs (EPA 1996a). Episodes of high levels of ozone last three to four days on average and extend over a large area (i.e., greater than 600,000 square kilometers).

Several factors associated with high-pressure systems are conducive to the photochemical production of ozone. Large high-pressure systems often create an inversion of the normal temperature profile, trapping pollutants in the shallow boundary layer at the earth's surface. Winds associated with major high-pressure systems are generally light, allowing greater accumulation of pollutants. Clear, warm conditions generally associated with large high-pressure systems afford ample sunlight to catalyze the photochemical production of ozone (NRC 1991).

Conditions in Los Angeles illustrate how the combined characteristics of meteorology and topography can adversely affect the air. This area is dominated by a persistent high-pressure system off the Pacific Ocean, which forms an inversion layer trapping pollutants over the city. Topography (river valleys or regions surrounded by mountains) also provide pockets to trap accumulations of primary pollutants and ozone, often capped by an inversion layer. Because of the region's low latitude and relatively clear skies, plenty of ultraviolet radiation is available for photochemical production of tropospheric ozone. A similar situation occurs in Mexico City (Box 13.2).

Trends in summer climatic conditions in the United States may have become more suitable for the formation of ground-level ozone. From 1949 to 1995, the number of summer heat waves (defined as more than three consecutive days with average apparent temperatures exceeding the 85th percentile) has increased by 20 percent, according to one study (Gaffen and Ross 1998). Climatic conditions could become more suitable for ozone formation under projections of global warming, but the extent of ozone formation will also depend on air mass and wind conditions.

The Influence of Weather on Ozone Transport

Urban smog has long been viewed as a local issue. However, both ozone and its precursors can travel relatively long distances in the atmosphere and be transported from region to region and across international boundaries. Such processes of transboundary pollution transport have implications for distant downwind populations. Analyses by such groups as the Ozone Transport Assessment Group (OTAG) have confirmed that, in the eastern United States, some areas experience a considerable influx of ozone across air boundaries and cannot meet air pollution standards by local measures alone (*Guinnup and Collom 1997*).

The distance that ozone can be transported ranges from 150 to 500 miles (roughly 240 to 800 kilometers). For example, ozone precursors from New York City are transported by prevailing winds on a roughly 200-mile northeast trajectory through Connecticut and as far as northeastern Massachusetts (Cleveland et al. 1976). Consequently, southwestern Connecticut has the highest concentrations of ozone in this region. The mid-Atlantic and northeastern United States are affected by even broader regional ozone transport. While the highest emissions of the ozone precursors,

BOX 13.2

Climate, Topography, and Fossil Fuel Emissions: The Special Case of Mexico City

Mexico City illustrates how geographic features can exacerbate local and regional air pollution. Mexico City is located in a basin that lies at a high altitude (2,240 meters above sea level) and is surrounded on the east, south, and west by mountain ranges that impede the movement of air over the basin. Frequent thermal inversions in the winter increase the accumulation of pollutants. Prevailing winds from the northeast tend to blow pollutants emitted from industrial sites and automobiles toward residential areas in the southwest.

One local health study examined biopsies of nasal mucosa from healthy residents of southwest Mexico City and compared the results to biopsies from residents of Veracruz, a nonpolluted Mexican port city. Of those subjects who had lived in southwest Mexico City for more than 60 days, 78 percent had evidence of precancerous squamous cell dysplasia, compared to 0 percent of those from Veracruz and 11 percent of those who had resided in southwest Mexico City for less than 30 days .

While more than 30,000 industries are located within the basin, emissions from automobiles are the greatest source of air contamination in Mexico City. The city's rapidly growing fleet numbered over four million automobiles in 1994, with growth rates in the late 1980s and early 1990s averaging about 5 percent annually. In the central city, where traffic congestion is greatest, the emissions of volatile organic compounds and nitrogen oxides, which are the precursors to the formation of ozone, are extremely high. The city's high altitude also leads to a relatively high influx of ultraviolet radiation, which promotes the formation of secondary air pollutants such as ozone. In 1992, the city's southwest area experienced more than 1,000 hours when the ambient concentrations of ozone exceeded the Mexican norm of 110 parts per billion for a maximum period of one hour.

Regionally, elevated levels of ozone, in combination with acid rain, have been blamed for extensive damage to the ecosystem in the surrounding mountain forests. Phytopathologists have documented reduced chlorophyll content and stunted growth in the pine forests, which may be attributable to ozone damage. Since the integrity of these mountainside forests is essential for the collection and filtering of water that eventually ends up as the city's drinking water, such disruption to the ecosystem has effects on economics and human health beyond the direct effects of the pollutants in the city.

VOCs and NO_x, come from large urban metropolitan areas, the largest elevated point source of NO_x emissions comes from the industrial Ohio River valley (*Guinnup and Collom 1997*). Based on spatial patterns and transport considerations, general control measures should clearly target urban nonattainment areas (areas that do not meet NAAQSs), as they contribute significantly to their own ozone problems. However, with the Ohio River valley's high density of NO_x point sources, NO_x controls targeted for this region may be an important measure for effectively reducing ozone pollution for the entire eastern United States.

The Health Effects of Ozone

Ozone causes symptoms of respiratory and mucous membrane (eyes, nose, and throat) irritation. As a strong oxidant, ozone can cause both cellular and structural damage in the lung. Three types of respiratory responses to ozone exposure are (1) irritative cough; (2) reduced lung capacity, as well as decreased expiratory flow rates; and (3) inflammation of the airway lining or submucosa (Bascom et al. 1996). An inflammatory response usually develops within an hour after exposure and persists for up to 24 hours. Exercise, of course, can increase the dose inhaled because of higher ventilation rates, and the most vulnerable populations include asthmatic children in the northeastern United States, where daily ozone levels may remain elevated for six to eight hours (Bascom et al. 1996).

Epidemiological studies of emergency room and hospital admissions indicate ozone's significant health effect. During the summertime (when many viral respiratory infections have subsided), exposure to ozone contributes an estimated 10–20 percent of all admissions to hospital due to respiratory conditions. On the days with the highest levels of ozone, exposure to ozone can account for nearly half of all respiratory admissions (EPA 1996a). Even levels below the NAAQSs can cause reversible decrements in respiratory function, including airway constriction, reduced lung volume, irritating cough, and chest pain (Bascom et al. 1996).

The effects of chronic exposure to ozone are less well understood. However, an analysis of pulmonary function data from the second National Health and Nutrition Examination Survey (NHANES II) revealed some association between ambient concentrations of ozone and the loss of lung function over time (Schwartz 1989). Some evidence also suggests that increased incidence and severity of asthma are related to exposure to ozone, although it is difficult to separate the effect of exposure to ozone from concomitant exposure to PM (EPA 1996a). In Mexico City, ozone has been linked to increased admissions to hospital for lower respiratory infections and asthma in children (Romieu et al. 1996). Ozone has been implicated in the exacerbation of asthmatic reactions by reducing the amount of allergen required to induce symptoms (Koren and Bromberg 1995; Koren and Utell 1997).

Acid Rain

SO_2 and NO_x oxidize via reactions with hydroxyl radicals and hydrogen peroxide in the atmosphere to form sulfuric acid and nitric acid, respectively (see Chapter 7). Acid deposition can be either wet or dry. Wet deposition occurs when emissions of sulfur oxides, primarily SO_2, and nitrogen oxides are transformed into acids in the atmosphere and then fall to earth as fog, rain, hail, or snow. Dry deposition occurs when these acid aerosols are brought to earth by gravity or other nonprecipitative means.

The National Surface Water Survey, which was conducted by the EPA, examined acid-sensitive areas of the United States comprising 1,180 lakes and 4,670 streams. Atmospheric deposition was found to be the dominant source of acid anions in 75 percent of the acidic lakes and in 47 percent of the acidic streams (Baker et al. 1991). According to the National Acid Precipitation Assessment Program (NAPAP), sulfur compounds from fossil fuel emissions are major precursors in the acidification of these surface water bodies (NAPAP 1991).

In the United States and Canada, coal-fired power plants, oil and gas processing, and the smelting of sulfur-rich ores account for about two-thirds of SO_2 emissions (*CEC 2000*). While global sulfur emissions from fossil fuel combustion are comparable to those from natural sources, over 90 percent of atmospheric sulfur in northern Europe and eastern North America is anthropogenic in origin. These findings have displaced earlier claims that volcanoes, trees, salt marshes, or other natural sources were primarily responsible for acidification in these regions (Schindler 1988).

Although the causes of many regional patterns of acid deposition remain uncertain, several of the factors that affect ozone formation also influence acid deposition (Penner et al. 1989). Higher temperatures accelerate the oxidation rates of SO_2 and NO_x to sulfuric and nitric acids, further potentiating acid formation. For example, high temperatures accelerate the production and concentration of hydrogen peroxide. This, in turn, increases the oxidation rate of SO_2 to sulfuric acid and ultimately the production of acid rain.

The Health Effects of Sulfur Dioxide

SO_2 is an irritant that can inflame the lining of the respiratory tract and may have been a key component in the smog disaster that killed thousands in London in 1952. In Ontario, Canada, sulfate levels have been found to correlate significantly with relative humidity in the summer (Bates and Sizto 1987), and the concentration of SO_2 was a predictor of hospital admissions for respiratory causes in the summer. On days with peak levels of pollution, summertime haze (comprising mostly ozone and acid aerosols) was associated with roughly half of all respiratory admissions (Thurston et al. 1994). These atmospheric chemical reactions are, therefore, essential in understanding the potential risk to health of changing climate conditions, including temperature and humidity.

Potential Effects of Climate Change on Air Pollution in the United States

As population growth and increases in energy consumption lead to increased fossil fuel emissions, the accumulation of CO_2, once thought to be a "clean" and harmless by-product of combustion, now poses indirect public health and ecological ramifications via global climate change (see "Possible Pathways of the Effect of Global Climate Change on Public Health," below). The change that is likely to have the most effect on conventional air pollutants is an increase in temperature.

Warmer temperatures due to climate change, all other things held constant, may increase concentrations of ozone and acid aerosols, as well as emissions of particulates and allergens (see under "Climate Variability, Biomass, Air Quality, and Health: Forest Fires and Allergens," below). A rise in temperature could affect concentrations of ozone by modifying the factors that affect the production of ozone (NRC 1991):

1. Elevated temperature—Ozone formation in the atmosphere is highly dependent on temperature. Urban areas tend to absorb and hold more heat, a process termed the *urban heat island effect*. The urban heat island effect can drive up air temperatures by 4–5°C or more (Landsburg 1981). The urban heat island may significantly contribute to elevation of ozone levels by adding thermal energy that can enhance the chemical formation of ground-level ozone (Quattrochi et al. 2000). Also, NO_x and VOC precursors of ozone are concentrated in congested cities. Urban air quality could therefore be particularly altered by higher global temperatures.
2. Increased frequency and intensity of stagnation periods—Climate change could result in more frequent or intense high-pressure systems, which provide favorable conditions for the reactions that produce ozone in the atmosphere.
3. Increased water vapor concentration—In addition to enhancing the formation of acid aerosols, increases in water vapor can potentiate the formation of ozone (Penner et al. 1989).
4. Increased emissions of precursor pollutants—Forests, shrubs, grasslands, and other sources of natural hydrocarbons (VOCs) emit greater quantities of these compounds at higher temperatures. Soil microbial activity may also increase with warmer temperatures, leading to an increase in NO_x emissions.

On the other hand, factors that could lead to reduced ozone concentrations include the following:

1. Increased cloud cover—A more vigorous hydrological cycle due to global warming could lead to an increase in cloudy days. More cloud cover, especially in the morning hours, could diminish reaction rates, thus lowering ozone formation.
2. Increased thickness of the boundary layer of air—Higher temperatures might be associated with greater convection, resulting in reduced atmospheric stability and higher wind speeds. If precursor pollutants

(VOCs or NO_x) are mixed in a greater volume of air, they will be less concentrated and less ozone will be formed (Smith and Tirpak 1989). The net effect of these factors is unclear; however, modeling studies combining these variables are being performed to assess the effects of climate change on air quality.

Modeling Studies of Air Quality and Climate Change

The most direct effect of climate change on air quality is the influence of increased temperature on the formation of ozone. Although studies have found that concentrations of ozone increase as temperature rises, the estimated magnitude of the effect varies considerably. This variation stems in part from limitations in the ability of atmospheric models to simulate complex photochemical reactions in the atmosphere. The presence of feedback effects between a rise in temperature and the formation of ozone is also a complicating factor. Therefore, the predictions of the modeling studies must be carefully interpreted in light of these limitations. Some of the modeling studies and their results are described below.

Grey et al. (1987) examined the effects of increased temperature and decreased stratospheric ozone (and thus more ultraviolet radiation [Chapter 7]) on the formation of ground-level ozone in eight U.S. cities: Los Angeles, New York, Philadelphia, Washington, D.C., Phoenix, Tulsa, Nashville, and Seattle. According to the model, the concentration of ground-level ozone would rise by about 2–4 percent for a 2°C increase in temperature and by about 5–10 percent for a 5°C increase in temperature; a loss of 15–30 percent of stratospheric ozone would have a greater effect. The most polluted cities showed the strongest response, implying that the effects of warming would be worse in those places where the concentration of ground-level ozone is already relatively high. A follow-up study of a possible 4°C rise in temperature indicated that the size of the area out of compliance with national standards for ozone would double in the San Francisco Bay region and nearly triple in the midwestern and southeastern United States (Morris et al. 1989).

A broader study of ground-level ozone in Memphis, Dallas, Philadelphia, Baton Rouge, and Atlanta included the influence of temperature on biogenic emissions of hydrocarbon (Morris et al. 1992). The study estimated that the control of VOC emissions required to attain the ozone NAAQS would increase approximately in proportion to the local increases in temperature. If perturbations to the stratospheric ozone layer were combined with increases in temperature, the stringency of the required controls would increase even further. In a subsequent study, model simulations showed that the concentrations of ozone in the northeastern United States would increase if temperatures rose by 4°C without the added complication of stratospheric ozone depletion (Morris et al. 1995).

The analysis considered the amount of upward penetration of emission plumes, the mass of hydrocarbons emitted from biogenic sources, and the response of evaporative emissions of hydrocarbons in gasoline from motor vehicles.

These studies of the effects of climate change on air quality must be considered indicative but by no means definitive. Many aspects of weather affect air quality, and ultimately climate is largely a general aggregate of weather patterns. Important local weather factors may not be adequately represented in these models. These models do, however, include such important factors as temperature stratification (or thermal inversions) and clouds, which play an important role in mixing and redistributing air pollutants (Oke 1987). The models used to simulate changes in air quality as a result of increases in temperature and ultraviolet radiation do not address issues such as the frequency of episodes of stagnant weather, which is associated with the highest concentrations of ozone observed over broad areas. Regional-scale changes in simulations of climate are inconsistent from model to model, and thus predicted changes in weather patterns carry much uncertainty. However, the trend in all models of global climate change is to predict higher maximum and minimum temperatures (Houghton et al. 1996), and associated summer episodes of weather stagnation have been observed more frequently in the United States (Gaffen and Ross 1998).

Climate Variability, Biomass, Air Quality, and Health: Forest Fires and Allergens

Extremes in climate, such as heat waves or severe droughts, are difficult to predict but are anticipated to become more prevalent under climate change scenarios. Severe droughts with accompanying forest fires have the potential to affect air quality. For example, in 1997–98, El Niño–driven droughts (see Chapter 8) contributed to the development of expansive forest fires in Indonesia, Mexico, Central and South America, Florida, Canada, and several other parts of the world. The extended duration of biomass burning affected air quality over regions far greater than the areas where burning actually occurred. The exposure of humans to fire smoke has been associated with irritation of the throat, lungs, and eyes. In addition, fire smoke carries a large amount of fine particles, which exacerbate respiratory problems, such as asthma and chronic obstructive pulmonary disease (Duclos et al. 1990).

In June and July of 1998, 2,277 fires burned approximately 500,000 acres in both rural and urban areas in Florida (Karels 1998). An analysis of records from several hospitals in Volusia and Flagler Counties showed that visits to the emergency department increased for asthma (by 91%), bronchitis (132%), and chest pain (37%) (Centers for Disease Control and Prevention 1999).

During the spring of 1998, low rainfall in Guatemala, Nicaragua, Honduras, El Salvador, Costa Rica, and several parts of Mexico caused tropical forested areas, usually too humid to burn on their own, to become vulnerable to fire (Agency for International Development 1998). By May 1998, thousands of fires had burned more than one million acres in Central America and emitted large quantities of smoke into the atmosphere. The smoke plume traveled northward along the Gulf coast and affected air quality extending well up into the midwestern United States. During the smoke event, according to the Texas Natural Resources Conservation Committee (TNRCC), the maximum PM measured in Brownsville, Texas, was 580 micrograms per cubic meter, almost four times the health-based NAAQS of 155 (*TNRCC 2000*). A survey conducted by the Texas Department of Health showed increased numbers of visits to doctor's offices and hospital emergency rooms for respiratory illnesses during the smoke event, mostly by patients with chronic, preexisting conditions (*TNRCC 1998*).

Another interaction between climate variability and biomass may be found in the release and dispersal of spores and pollen, which can act as allergens. Allergens are not criteria air pollutants but are included because of their temperature-dependent effect on air quality for allergic people. Pollen counts from birch trees (the main cause of seasonal allergies in northern Europe) seem to increase with increasing temperature (Ahlholm et al. 1998). Pollen counts for Japanese cedar significantly increase in years when summertime temperatures are unusually high (Takahashi et al. 1996; Tamura et al. 1997).

The Synergy between Heat Stress and Pollutants Affecting Health

The incidences of a range of illnesses are potentially associated with increases in temperature. Cardiovascular deaths are associated with temperature. The Chicago heat wave of July 1995 led to more than 700 excess deaths in the metropolitan area. In Athens, Greece, during a heat wave in 1987, the daily number of deaths increased by more than 40 when the mean 24-hour air temperature exceeded 30°C (Katsouyanni et al. 1993).

Mortality curves assume a classic J or V-shape, with highest mortality occurring at both high and low extremes of temperature. Generally, populations in warmer regions tend to be more vulnerable to cold (Eurowinter Group 1997), and those residing in cold climates are more sensitive to heat (Kalkstein and Greene 1997). In temperate regions, mortality rates are highest during the winter.

Physiological responses to both extreme heat and cold are not straightforward. For example, blood viscosity and cholesterol have been found to increase with high temperatures (Keating et al. 1986). On the other hand, blood pressure and fibrinogen levels increase during winter, although outdoor temperature does not seem to determine the seasonal variation of fibrinogen (van der Bom et al. 1997).

Climatologists project a doubling in the frequency of heat waves associated with a rise of 2–3°C in average summer temperature. A study of 44 U.S. cities found that, after adjusting for some expected acclimatization, heat-related mortality might increase by 70–150 percent (Kalkstein and Greene 1997). A meta-analysis of studies on 20 international cities, however, found a reduction in overall mortality due to fewer deaths during winter (Martens 1998).

Both heat waves and air pollution kill people. In addition to their individual effects, however, is there evidence for interaction between the effects of heat and air pollution? Recent studies have examined the health effects of exposure to extreme heat and air pollution to determine whether there are potential synergistic effects related to simultaneous exposure.

Katsouyanni et al. (1993) investigated the potential synergistic effects of air pollution and air temperature on excess mortality in Athens, Greece. The increased mortality during the major heat wave of July 1987 was compared to the number of deaths in July for the six previous years. There was a greater increase in the number of deaths in Athens (97%) during the July 1987 heat wave than in all other less polluted urban areas (33%) or in all nonurban areas (27%). The authors found interactions between high levels of air pollution and high temperature (30°C) that were statistically significant for SO_2 and suggestive for ozone and particulates.

Sartor et al. (1995) found that mortality during the Belgium heat wave of 1994 was higher than expected; an increase of 9.4 percent was observed among those younger than 65 years (236 excess deaths) and an increase of 13.2 percent was observed among those older than 65 years (1,168 excess deaths). Daily death figures were mostly correlated with mean daily temperature and 24-hour ozone concentration from the previous day. A synergistic interaction between the effects of temperature and ozone on mortality was determined across age groups and explained 39.5 percent of the variance for daily deaths for those over age 65. The authors concluded that elevated ambient temperatures combined with high concentrations of ozone were likely to have been responsible for the unexpected excess mortality.

These studies are far from conclusive, but other studies at least show seasonality in health effects. Sunyer et al. (1996) found that SO_2 in Barcelona, Spain, was associated with respiratory deaths in the summer; levels of ozone and NO_2 during the summer also positively correlated with cardiovascular mortality and mortality in the elderly population. On the other hand, Samet et al. (1998) found little evidence that weather conditions modified the effect of pollution. These interactions are methodologically difficult to study, but it will become increasingly important to determine whether there are any synergistic interactions between exposure to high temperatures and exposures to high levels of pollution.

Possible Pathways of the Effect of Global Climate Change on Public Health

Although the focus of this chapter has been on global climate change and air pollution, it is important to recognize that the potential health effects of global climate change go far beyond adverse effects of air pollution. Global climate change may affect human health via multiple pathways (Fig. 13.3). Further, the dynamics of global climate change occur in a broader context of the interplay between natural and anthropogenic forces in our global ecosystem that are changing how land, water, air, and energy are used. As such, an exploration of the possible pathways by which global climate change affects public health appears in multiple contexts throughout this book. Illustrations of these pathways selected from this book are discussed below. More in-depth reviews on this subject can be found in the literature (Strzepek and Smith 1995; McMichael et al. 1996; Patz et al. 1996; Patz 1998; World Health Organization 1996).

The Rise in Temperature

One of the main expectations of global climate change is a rise in global mean temperature (Chapter 7). Therefore, one of the priorities for understanding the health consequences of global climate change is an assessment of health problems that will grow worse as temperature increases. Under warmer conditions, the rate of heat-related fatalities is likely to grow, especially among elderly populations (see "The Synergy between Heat Stress and Pollutants Affecting Health," above). Increasing heat may exacerbate the effects of conventional air pollutants, most notably ground-level ozone, as described under "The Influence of Weather on Ozone Formation" (above). Increased heat may also lead to greater problems with allergies to spores and pollen (see "Climate Variability, Biomass, Air Quality, and Health: Forest Fires and Allergens," above).

The quality of food may also deteriorate. For example, cholera, which is an infectious disease that may be transmitted through food, seems to be more frequently transmitted at warmer temperatures (see Chapters 10 and 11). In addition, temperature is a critical factor in the cycle of transmission of many parasitic diseases. For example, the parasite that causes malaria is transmitted from person to person by the bite of a mosquito; both the survival of the mosquito vector and the development of the parasite within the mosquito vector depend on temperature (see Chapters 10 and 12). In

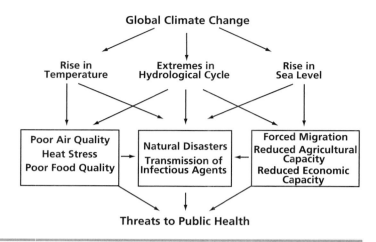

Figure 13.3 Linkages between global climate change and threats to public health.

many cooler settings at higher latitudes and higher altitudes, an increase in temperature could affect the chance of transmission. Similar arguments apply to insect pests and plant pathogens that affect agricultural crops (see Chapter 10).

Extremes in the Hydrological Cycle

Another major concern about the effects of global climate change is an increase in the frequency of extremes in the hydrological cycle, that is, more floods and more droughts (see Chapters 7 and 9). Extreme flooding has direct effects on people as a natural disaster, causing accidental deaths (e.g., drowning, burial in mudslides) and forcing people from their homes (see Chapter 14). Flooding also has less direct effects through an increase in the incidence of some infectious diseases, such as leptospirosis (see Chapter 14).

Increased rainfall can affect the incidence of infectious diseases through changes in wildlife populations that are the reservoirs of infection. For example, hantavirus pulmonary syndrome in the southwestern United States is transmitted by a virus circulating in rodent populations whose numbers increase during unusually wet periods (Glass et al. 2000) because of abundant growth of the vegetation that provides a source of food (see Chapters 2 and 10).

Droughts, at the other extreme, can lead to the loss of agricultural productivity and contribute to the forced migration of people from rural areas (see Chapter 6). Forced migration for whatever reason can result in refugee camps with extreme health problems, including waterborne diseases (see Chapters 6 and 14).

Shortages of water because of diversion also provide evidence of the possible effects of droughts on health. The system for distributing drinking water in Nukus, Uzbekistan, in the Aral Sea basin was made more vulnerable to biological contamination because of low pressure in the pipes, which in turn was due to the diversion of water to irrigate cotton fields (see Chapter 14). Diversion of water for cotton fields has also been responsible for shrinkage of the Aral Sea, which has increased the number of dust storms in the region (see Chapter 14). On a smaller scale, the diversion of water from lakes in California to support the metropolitan growth of Los Angeles has increased the dustiness of the air in violation of U.S. standards for air quality (see Chapter 14).

The Rise in Sea Level

The third major physical effect of global climate change is a global rise in sea level (Chapter 7), a process that has been evident over the past century (Chapter 9). A rise in sea level could inundate land and contaminate freshwater aquifers with seawater, leading to the forced migration of populations as well as reductions in agricultural and economic activity (see Chapter 9). As noted under "Extremes in the Hydrological Cycle" (above), forced migration itself can lead to a variety of serious health problems (see Chapters 6 and 14). A rise in sea level can also increase the risk of flooding and hence all of the health consequences associated with floods.

The Relationship with Stratospheric Ozone Depletion

Global warming at the earth's surface is associated with a cooling of the stratosphere, which tends at high latitudes to accelerate the destruction of stratospheric ozone that protects the earth from excessive ultraviolet-B (UV-B) radiation (Chapter 7) (Shindell et al. 1998). The relationship is due to intrinsic physical characteristics of the flow of energy and does not depend on whether the cause of global warming is natural or anthropogenic.

Studies of the health effects of global climate change have focused on the effects of a rise in temperature, extremes in the hydrological cycle, and a rise in sea level (see Fig. 13.3). Studies of the health effects of stratospheric ozone depletion have focused on the effect of UV-B radiation on various organs, especially skin, eyes, and the immune system (see Chapter 7). Consideration of illnesses induced by UV-B can deepen an appreciation for the relationship between global warming and stratospheric ozone depletion, but including these illnesses in a total accounting of global climate change can also foster confusion. Stratospheric ozone depletion occurs because many industrial chemicals act directly to break down stratospheric ozone and is not simply a consequence of global warming.

Conclusion

This chapter has contrasted the ramifications of increases in greenhouse gas emissions against the backdrop of improving trends in conventional air pollution in North America. Greenhouse gas emissions causing climate change have emerged as a new air pollution problem with direct and indirect health implications. Classical urban air pollution problems still plague much of the developing world. In developed countries with cleaner air, climate change has the potential, because of the effects of temperature and weather conditions on the formation and transport of air pollution, to reverse some of the gains achieved in air quality.

We discussed four key themes in this chapter:

1. Historically, air pollution problems and their remediation have generally diffused local hazards into more regional and now global distributions.
2. Fossil fuel use is a major source of both conventional air pollutants (criteria pollutants in EPA terminology) and greenhouse gases, such as CO_2.
3. As a "secondary" air pollutant, regional ozone formation and transport is influenced by climatic factors, such as temperature, wind, and UV radiation. Ozone, or "photochemical smog," remains a pervasive problem in North America, which may be exacerbated by future global warming or stratospheric ozone depletion.
4. Global climate change may affect public health through a diversity of pathways, including heat waves, air pollution (especially ozone), vector- and waterborne diseases, and sea level rise or weather extremes that could alter agriculture and water supplies or displace human populations.

SUGGESTED STUDY PROJECTS

Suggested study projects provide a set of options for individual or team projects that will enhance interactivity and communication among course participants (see Appendix A). The Resource Center (see Appendix B) and references in all of the chapters provide starting points for inquiries. The process of finding and evaluating sources of information should be based on the principles of information literacy applied to the Internet environment (see Appendix A).

PROJECT 1: Global Climate Change and Air Pollution

The objective of this project is to demonstrate an understanding of aspects of the relationships between global climate change and air pollution.

Task 1. Describe the key sources of greenhouse gases, criteria air pollutants, and air toxics in the United States.

Task 2. Describe the evidence linking increased temperatures to higher concentrations of ground-level ozone.

Task 3. Describe projections of the effects of global climate change on ground-level ozone.

PROJECT 2: Unintended Consequences of Programs to Control Harmful Emissions

The objective of this project is to demonstrate an understanding of how programs to control harmful emissions may cause unintended environmental health problems.

Task 1. Select a historical, current, or projected program for controlling harmful emissions (e.g., hydrocarbons, acid rain).

Task 2. Describe the benefits and unintended negative consequences (realized or potential) for human health and the environment.

PROJECT 3: Possible Pathways of the Effect of Global Climate Change on Public Health

The objective of this project is to demonstrate an understanding of the possible pathways of the effect of global climate change on public health.

Task 1. Select two or more linkages in the possible pathways of the effect of global climate change on public health (see Fig. 13.3) and find examples of each. At least one example of each linkage must be found in the literature outside of this book.

Task 2. Select two or more geographic areas. For each area, identify the possible pathways of the effect of global climate change that are most relevant to the population in that area.

Task 3. For each geographic area selected in task 2, describe other environmental changes that could interact with the possible pathways of the effect of global climate change in that area.

Task 4. Summarize your results.

Acknowledgments

Special thanks go to Anne Grambsch, U.S. EPA, for her information on climate change and air pollution modeling studies. We are also grateful to Dr. David Engelberg, University of British Columbia, and Dr. Jonathan Samet, Johns Hopkins School of Public Health, for reviews of the manuscript.

References

Agency for International Development. 1998. *Mexico and Central America—Fires, Situation Report #9.* Bureau for Humanitarian Response. Office of U.S. Foreign Disaster Assistance, Washington.

Ahlholm JU, Helander ML, Savolainen J. 1998. Genetic and environmental factors affecting the allergenicity of birch (*Betula pubescens* ssp. czerepanovii [Orl.] Hamet-Ahti) pollen. *Clin Exp Allergy* 28:1384–88.

Baker LA, Herlihy AT, Kaufmann PR, Eilers JM. 1991. Acidic lakes and streams in the United States: The role of acidic deposition. *Science* 252:1151–54.

Bascom R, Bromberg PA, Costa DA, Devlin R, Dockery DW, Frampton MW, Lambert W, Samet JM, Speizer FE, Utell M. 1996. State of the art: Health effects of outdoor air pollution. *Am J Respir Crit Care Med* 153:3–50.

Bates DV, Sizto R. 1987. Air pollution and hospital admissions in southern Ontario: The acid summer haze effect. *Environ Res* 43 (2): 317–31.

Burnett RT, Cakmak S, Brook JR. 1998. The effect of the urban ambient air pollution mix on daily mortality rates in 11 Canadian cities. *Can J Public Health* 89 (3): 152–56.

Calderon-Garciduenas L, Osorno-Velazques A, Bravo-Alvarez H, Delgado-Chavez R, Barrios-Marquez R. 1992. Histopathologic changes of the nasal mucosa in southwest metropolitan Mexico City inhabitants. *Am J Pathol* 140 (1): 225–32.

Centers for Disease Control and Prevention. 1999. Surveillance of morbidity during wildfires—central Florida, 1998. *MMWR* 47 (4).

Cleveland W, Kleiner B, McRae J, Warner J. 1976. Photochemical air pollution: Transport from the New York City area into Connecticut and Massachusetts. *Science* 191:179–81.

Dockery D, Pope C. 1994. Acute respiratory effects of particulate air pollution. *Annu Rev Public Health* 15:107–32.

Duclos P, Sanderson LM, Lipsett M. 1990. The 1987 forest fire disaster in California: Assessment of emergency room visits. *Arch Environ Health* 45 (1): 53–58.

Environmental Protection Agency. 1996a. *National Air Quality and Emissions Trends Report, 1995*. Environmental Protection Agency, Office of Air Quality Planning and Standards, Washington.

———. 1996b. *Air Quality Criteria for Ozone and Related Photochemical Oxidants*. Vol. 1 of 3, EPA/600/P-93/004af. Environmental Protection Agency, Office of Research and Development, Washington.

———. 1997. National Air Pollutant Emissions Trends Report, 1900–1996. EPA-454/R-97–011, U.S. Environmental Protection Agency, Research Triangle Park, N.C. (December).

Eurowinter Group. 1997. Cold exposure and winter mortality from ischaemic heart disease, cerebrovascular disease, respiratory disease, and all causes in warm and cold regions of Europe. *Lancet* 349:1341–46.

Gaffen DJ, Ross RJ. 1998. Increased summertime heat stress in the U.S. *Nature* 396 (10): 529–30.

Glass GE, Cheek JE, Patz JA, Shields TM, Doyle TS, Thoroughman DA, Hunt DK, Enscore RE, Gage KL, Ireland C, Peters CJ, Bryan R. 2000. Using remotely sensed data to identify areas of risk for hantavirus pulmonary syndrome. *Emerg Infect Dis* 63 (3): 238–47.

Grey M, Edmond R, Whitten G. 1987. *Tropospheric Ultraviolet Radiation: Assessment of Existing Data and Effect on Ozone Formation*. Environmental Protection Agency, Research Triangle Park, N.C.

Hatakeyama S, Izumi K, Fukuyama T, Akimoto H, Washida N. 1991. Reactions of OH with alpha-pinene and beta-pinene in air: Estimate of global CO production from the atmospheric oxidation of terpenes. *J Geophys Res* 96:947–58.

Houghton JT, Meira Filho LG, Callander BA, Harris N, Kattenberg A, Maskell K, eds. 1996. *Climate Change, 1995—The Science of Climate Change: Contribution of Working Group I to the Second Assessment Report of the Intergovernmental Panel on Climate Change*. Cambridge University Press, Cambridge.

Jensen TK, Toppari J, Keiding N, Skakkebaek NE. 1995. Do environmental estrogens contribute to the decline in male reproductive health? *Clin Chem* 41:1896–1901.

Kalkstein LS, Greene JS. 1997. An evaluation of climate/mortality relationships in large U.S. cities and possible impacts of a climate change. *Environ Health Perspect* 105 (1): 2–11.

Kamens R, Jeffries H, Sexton K, Gerhardt A. 1982. *Smog Chamber Experiments to Test Ox-idant-Related Control Strategy Issues.* Environmental Protection Agency, Research Triangle Park, N.C.

Karels J. 1998. Wildland fire season in review. *Fla Fire Service Today* 6:8–19.

Katsouyanni K, Pantazopoulou A, Touloumi G, Tselepidaki I, Moustris K, Asimakopoulos D, Poulopoulou G, Trichopoulos D. 1993. Evidence for interaction between air pol-lution and high temperature in the causation of excess mortality. *Arch Environ Health* 48 (4): 235–42.

Keating WR, Coleshaw SR, Easton JC, Cotter F, Mattock MB, Chelliah R. 1986. Increased platelet and red cell counts, blood viscosity, and plasma cholesterol levels during heat stress, and mortality from coronary and cerebral thrombosis. *Am J Med* 81 (5): 795–800.

Koos BJ, Longo LD. 1976. Mercury toxicity in the pregnant woman, fetus, and newborn infant. *Am J Obstet Gynecol* 126:390–409.

Koren HS, Bromberg PA. 1995. Respiratory responses of asthmatics to ozone. *Int Arch Al-lergy Immunol* 107 (1–3): 236–38.

Koren HS, Utell MJ. 1997. Asthma and the environment. *Environ Health Perspect* 105 (5): 534–37.

Landsberg HE. 1981. *The Urban Climate.* Academic Press, New York.

Last J, Trouton K, Pengelly D. 1998. *Taking Our Breath Away.* David Suzuki Foundation, Vancouver.

Logan WPD. 1953. Mortality in the London fog incident, 1952. *Lancet* 1:336–38.

Martens W. 1998. Climate change, thermal stress and mortality changes. *Soc Sci Med* 46 (3): 331–44.

McMichael AJ, Ando M, Carcavallo R, Epstein P, Haines A, Jendritzky G, Kalkstein L, Odongo R, Patz J, Pever W. 1996. Human population health. In *Climate Change 1995—Impacts, Adaptations, and Mitigation of Climate Change: Scientific-Technical Analysis. Contribution of Working Group II to the Second Assessment Report of the In-tergovernmental Panel on Climate Change* (Watson RT, Zinyowera MC, Moss RH, eds.). Cambridge University Press, Cambridge, pp. 561–84.

McMichael AJ, Baghurst PA, Wigg NR, Vimpani GV, Robertson EF, Roberts RJ. 1988. Port Pirie cohort study: Environmental exposure to lead and children's abilities at the age of four years. *N Engl J Med* 319 (8): 468–75.

Monastersky R. 1998. Asian pollution drifts over North America. *Sci News* 154 (Decem-ber 12): 374.

Morris RE, Gery MW, Liu MK, Moore GE, Daly C, Greenfield SM. 1989. *Sensitivity of a Regional Oxidant Model to Variations in Climate Parameters.* Environmental Protec-tion Agency, Washington.

Morris RE, Guthrie PD, Knopes CA. 1995. Photochemical modeling analysis under global warming conditions. In *Annual Meeting of the Air and Waste Management Associa-tion, June 18–23, 1995, San Antonio, Texas,* Paper 95-WP-4B02, Vol. 3A, pp. 1–20.

Morris RE, Whitten GZ, Greenfield SM. 1992. Preliminary assessment of the effects of global climate change on tropospheric ozone concentration. In *Conference on Tropo-spheric Ozone and the Environment II, Atlanta, November 4–7, 1992.* Air and Waste Management Association [Code TR-20], pp. 422–38.

National Acid Precipitation Assessment Program. 1991. *Acidic Deposition: State of Science and Technology.* National Acid Precipitation Assessment Program, Washington.

National Research Council. 1991. *Rethinking the Ozone Problem in Urban and Regional Air Pollution.* National Academy Press, Washington.

Needleman HL, Gunnoe C, Leviton A, Reed R, Peresie H, Maher C, Barett P. 1979.

Deficits in psychologic and classroom performance of children with elevated dentine lead levels [published erratum appears in *N Engl J Med* 331 (9): 616, 1994]. *N Engl J Med* 300 (13): 689–95.

Oke TR. 1987. Air pollution in the boundary layer. In *Boundary Layer Climates* (Oke TR, ed.). Cambridge, Cambridge University Press, pp. 304–423.

Patz JA. 1998. Climate change and health: New research challenges. *Health Environ Digest* 12 (7): 49–53.

Patz JA, Epstein PR, Burke TA, Balbus JM. 1996. Global climate change and emerging infectious disease. *JAMA* 275 (3): 217–23.

Patz JA, McGeehin MA, Bernard SM, Ebi KL, Epstein PR, Grambsch A, Gubler DJ, Reiter P, Romieu I, Rose JB, Samet JM, Trtanj J. 2000. The potential health impacts of climate variability and change for the United States: Executive summary of the report of the health sector of the U.S. National Assessment. *Environ Health Perspect* 108:367–76.

Penner JE, Connell PS, Wuebbles DJ, Covey CC. 1989. *Climate Change and Its Interactions with Air Chemistry: Perspectives and Research Needs.* Environmental Protection Agency, Washington.

Pope C, Bates D, Razienne M. 1995. Health effects of particulate air pollution: Time for reassessment? *Environ Health Perspect* 103:472–80.

Quattrochi DA, Luvall JC, Rickman DL, Estes MG Jr, Laymon CA, Howell BF. 2000. A decision support information system for urban landscape management using thermal infrared data. *Photogrammetric Engineering and Remote Sensing* 66:1195–1207.

Romieu I. 1991. Urban air pollution in Latin America and the Caribbean. *J Air Waste Manage Assoc* 41:1166–70.

Romieu I, Meneses F, Ruiz S, Sienra JJ, Huerta J, White MC, Etzel RA. 1996. Effects of air pollution on the respiratory health of asthmatic children living in Mexico City. *Am J Respir Crit Care Med* 154 (2 Pt. 1): 300–307.

Samet JM, Dominici F, Curriero FC, Coursac I, Zeger SL. 2000. Fine particulate air pollution and mortality in 20 U.S. cities, 1987–1994. *N Engl J Med* 343:1742–49.

Samet J, Zeger S, Kelsall J, Xu J, Kalkstein L. 1998. Does weather confound or modify the association of particulate air pollution with mortality? An analysis of the Philadelphia data, 1973–1980. *Environ Res* 77 (1): 9–19.

Samson P. 1988. *Linkages between Global Climate Warming and Ambient Air Quality.* Global Climate Linkages Conference, Washington, D.C.

Sartor F, Snacken R, Demuth C, Walckiers D. 1995. Temperature, ambient ozone levels, and mortality during summer 1994, in Belgium. *Environ Res* 70 (2): 105–13.

Schindler DW. 1988. Effects of acid rain on freshwater ecosystems. *Science* 239:149–57.

Schwartz J. 1989. Lung function and chronic exposure to air pollution: A cross-sectional analysis of NHANES II. *Environ Res* 50:309–21.

Shindell DT, Rind D, Lonergan P. 1998. Increased polar stratospheric ozone losses and delayed eventual recovery owing to increasing greenhouse-gas concentrations. *Nature* 392:589–92.

Smith JB, Tirpak DA, eds. 1989. Appendix F: Air quality. In *The Potential Effects of Global Climate Change on the United States,* Report 230–05–89–056. U.S. Environmental Protection Agency, Office of Policy, Planning and Evaluation, Washington.

Strzepek KM, Smith JB, eds. 1995. *As Climate Changes: International Impacts and Implications.* Cambridge University Press, Cambridge.

Sunyer J, Castellsague J, Saez M, Tobias A, Anto JM. 1996. Air pollution and mortality in Barcelona. *J Epidemiol Community Health* 50 (Suppl. 1): s76–80.

Takahashi Y, Kawashima S, Aikawa S. 1996. Effects of global climate change on Japanese cedar pollen concentration in air—estimated results obtained from Yamagata City and its surrounding area. *Arerugi* 45 (12): 1270–76.

Tamura Y, Kobayashi Y, Watanabe S, Endou K. 1997. Relationship of pollen counts of Japanese cedar to weather factors in Isehara City, Kanagawa. *Nippon Jibiinkoka Gakkai Kaiho* 100 (3):326–31.

Thurston GD, Ito K, Hayes CG, Bates DV, Lippman M. 1994. Respiratory hospital admissions and summertime haze air pollution in Toronto, Ontario: Consideration of the role of acid aerosols. *Environ Res* 65 (2): 271–90.

Tong S, Baghurst P, McMichael A, Sawyer M, Mudge J. 1996. Lifetime exposure to environmental lead and children's intelligence at 11–13 years: The Port Pirie cohort study [published erratum appears in *BMJ* 313 (7051): 198, 1996]. *BMJ* 312 (7046): 1569–75.

United Nations Environment Programme. 1991. *Urban Air Pollution.* United Nations Environment Programme, Nairobi.

van der Bom JG, de Maat MP, Bots ML, Hofman A, Kluft C, Grobbee DE. 1997. Seasonal variation in fibrinogen in the Rotterdam study. *Thromb Haemost* 78 (3): 1059–62.

Working Group on Public Health and Fossil-Fuel Combustion. 1997. Short-term improvements in public health from global-climate policies on fossil-fuel combustion: An interim report. *Lancet* 350 (9088): 1341–49.

World Bank. 1997. *Clear Water, Blue Skies.* World Bank, Washington.

World Health Organization. 1996. *Climate Change and Human Health.* World Health Organization, Geneva.

Electronic References

Commission for Environmental Cooperation. 2000. Continental Pollutant Pathways: An Agenda for Cooperation to Address Long-Range Transport of Air Pollution in North America. 1997. http://www.cec.org (Date Last Revised 3/20/2000).

Environment Canada. 2000. Environmental Priority. Clean Air. http://www.ec.gc.ca/ envpriorities/cleanair_e.htm (Date Last Revised 2/18/2000).

Environmental Protection Agency. 1999a. National Air Quality Trends Report, 1996. http://www.epa.gov/oar/aqtrnd96 (Date Last Revised 1/11/1999).

———. 1999b. National Air Quality Trends Report, 1997. http://www.epa.gov/oar/ aqtrnd97 (Date Last Revised 6/10/1999).

———. 2000a. U.S. Emissions, 1998. Inventory of U.S. Greenhouse Gas Emissions and Sinks, 1990–1996 (March 1998). EPA 236-R-98–006. Publications—GHG Emissions. http://www.epa.gov/globalwarming/publications/emissions/us1998/ index.html (Date Last Revised 1/14/2000).

———. 2000b. MTBE (methyl tertiary butyl ether). Office of Underground Storage Tanks. http://www.epa.gov/swerust1/mtbe/ (Date Last Revised 2/8/2000).

Guinnup D, Collom B. 1997. Telling the OTAG Ozone Story with Data. Final Report, Vol. 1: Executive Summary. OTAG Air Quality Analysis Workgroup. June 2, 1997. http://capita.wustl.edu/otag/reports/aqafinvol_I/animations/ v1_exsumanimb.html (Date Last Revised 6/2/1997).

Marland G, Boden TA, Andres RJ, Brenkert AL, Johnston C. 1999. Global, Regional, and National CO2 Emissions. In Trends: A Compendium of Data on Global Change. Carbon Dioxide Information Analysis Center, Oak Ridge National Laboratory, Oak Ridge, Tenn. http://cdiac.esd.ornl.gov/trends/emis/tre_usa.htm (Date Last Revised 5/26/1999).

Texas Natural Resource Conservation Commission. 1998. Texas Natural Resource Conservation Commission. Executive Office. Agency Communication, Natural Outlook.
    TNRCC safeguards public health with statewide response to smoke event.
    http://www.tnrcc.state.tx.us/publications/pd/020/98–02/smokin.html (Date Last
    Revised 7/30/1998).
———. 2000. Texas Natural Resource Conservation Commission. Monthly Summary
    Report by Site. http://www.tnrcc.state.tx.us/cgi-bin/monops/select_month (Date
    Last Revised 3/27/2000).

CHAPTER 14

Too Little, Too Much

How the Quantity of Water Affects Human Health

Les Roberts, Ph.D.,

Ulisses E. C. Confalonieri, M.D., D.V.M., D.Sc.,

and Joan L. Aron, Ph.D.

The purpose of this chapter is to present a broad look at the importance of water resources to public health and the way that ecosystem changes are threatening those resources and public health on a global scale. In spite of the image of our blue planet, about one-third of the earth's land surface receives less than 250 millimeters of rainfall per year and is therefore classified as arid or semiarid (see World Health Organization [WHO] 1992, 109 and Fig. 8.2b). Only a very small fraction of the earth's water supply is fresh and unfrozen (see Table 9.1 and Fig. 9.2). An increase in population combined with an increase in per capita demand for water have resulted in a sixfold to sevenfold increase in the withdrawal of water during the twentieth century (see Fig. 9.4). Given the (essentially) fixed volume of fresh water available, water resources are being stressed, often with accompanying degradation of quality, diminished acceptability for human uses, and adverse ecological effects. Excess runoff and flooding due to deforestation or the construction of impermeable surfaces are growing challenges to the management of water resources (see Chapter 9). Projections of climate change indicate additional stress from global warming, accelerated rise in sea level, and more hydrological extremes of intense precipitation and drought (see Chapter 9). Variability in the flow of water may also contribute to poor water quality, since shortages and excesses may lead to biological and chemical contamination.

Increased use of water has been strongly associated with improvements in human health. Domestic (municipal) water use in the United States soared in the late nineteenth and early twentieth centuries, increasing to over 100 gallons (or 378.5 liters) per person per day in major cities by 1920. Concomitantly, water treatment increased and human morbidity and mortality decreased. More recently, among populations as diverse as those from nations of moderate income and those from impoverished refugee camps, increased use of water has led to diminished rates of diarrhea, improved growth of children, and lower rates of mortality. Some of the mechanisms responsible for the linkage between increased use of water and improvements in human health are increased washing of hands, increased washing of food, increased washing of clothes, increased production of food, and more effective removal of waste via drainage and sewers. Improvements in sanitation often accompany improvements in water supply. A combination of improvements in water supply and sanitation worldwide could reduce mortality among infants and children by more than 50 percent and almost eliminate morbidity due to some infectious diseases, such as cholera, typhoid, and leptospirosis (WHO 1992, 122).

In this chapter, four geographic examples illustrate how the quantity of water affects human health in diverse ways. The examples include aspects of water quality insofar as poor quality is related to having too little or too much water; problems in water quality may arise for several reasons (see Chapter 9). The first example describes a situation of extreme water scarcity in a refugee camp created by the forced displacement of many thousands of people during a war in Africa. The second example demonstrates the effects of excess water with a focus on flooding in Brazil. The third example broadens the scope of water-related health problems to include several adverse effects of the diversion of water for the irrigation of cotton fields in the Aral Sea basin. The problems range from water scarcity and degradation of water quality to greater frequency of dust storms, loss of fish from the diet, and loss of an economic base for the region. The fourth example examines how the diversion of water for the metropolitan growth of Los Angeles has increased the amount of particulate matter in the air of the Great Basin of California. The problem in California demonstrates that even a wealthy country with extensive natural resources must pay attention to the adverse consequences of the diversion of water. It is also noteworthy that the issue of air quality rather than water quality triggered the regulatory apparatus of the state and federal governments.

Extreme Water Scarcity and Mortality in a Refugee Population in Goma, Zaire

Rwanda, formerly Africa's most densely populated country, underwent an extraordinarily violent civil war over a three-month period beginning in April 1994, after a mysterious plane crash killed the presidents of Rwanda and Burundi. An estimated 800,000 people were killed and an additional two million fled their homeland. This level of violence in a population of only seven million essentially destroyed the social fabric of the tiny nation. While press reports at the time characterized the conflict as ethnically motivated, issues of environmental degradation, overpopulation, and international economics all conspired to set the stage for an eruption of civil strife in Rwanda. Likewise, the cholera epidemic that gripped the refugees fleeing at the close of the war developed because of an unfortunate confluence of hydraulic, geological, immunological, and behavioral factors.

The root of the ethnic conflict began some 600 years earlier when the ancestors of the present Tutsi minority invaded from the north. Compared to the ethnic Hutu population of the region, the Tutsis were taller and thinner, and Tutsis tended to maintain a separate culture. The Belgian colonial government only exacerbated these cultural differences by employing primarily Tutsis in its local civil service. At the time of independence in 1962, the Tutsi minority possessed a marked advantage in terms of education and governmental experience. Since independence, several cycles of violence have forced Tutsis to flee to neighboring countries for refuge. Particularly in the late 1980s, ethnic inequity in university admissions, military promotions, and government employment led the newly elected government to promote policies strongly favoring the Hutu majority. Ethnic tensions flared, a rebel Tutsi group gained control of a northern province, and observers from the United Nations were brought in to help keep the peace.

When the president of Rwanda's plane crashed in 1994, extremist elements within the government found it very easy to convince the Hutu majority that their elected leader had been killed by Tutsi rebels. Local officials had apparently prepared for a campaign of ethnic cleansing and, within days, the genocide of Tutsis and the killing of presumed Tutsi sympathizers were under way. Approximately 800,000 people were slaughtered. Tutsi rebels went on the offensive and gained control of much of eastern Rwanda by the end of May. From that point on, the front line of fighting moved steadily to the west, culminating in the exodus of 500,000–800,000 people into Goma, Zaire, between July 14 and July 17, 1994. The country of Zaire has since been renamed the Democratic Republic of the Congo, but the name Zaire is retained in this chapter to describe events that occurred before 1997.

Goma is a town in the African Great Lakes region at the northern end of Lake Kivu, where the eastern border of the Democratic Republic of the Congo meets the northwestern border of Rwanda (*Norwegian Council for Africa 2000*). A series of volcanoes runs from northwestern Rwanda into the Democratic Republic of the Congo (*Volcano World 2000*). These volcanoes are famed for their mountain gorillas and lava outflows, some of which have occurred as recently as the 1990s. The lava outflows have formed a sheet of volcanic rock extending over 50 kilometers from the northern end of Lake Kivu in a northwesterly direction. Refugees who fled Rwanda by the roads passing north of Lake Kivu were forced to settle onto this sheet of volcanic rock, through which rainfall cannot permeate into the ground and wells cannot be dug. Thus, when the refugees arrived, fresh water existed in very few places other than Lake Kivu, and those without access to Lake Kivu would require water from the lake via tanker trucks.

Lake Kivu itself is unique. Volcanic gases emitted into the lake from subterranean vents cause the water to have a high pH of about 8.1 and the water column to be very clear by tropical standards. Although the water column is clear, the hydraulic retention time of the lake (the average time it takes a water molecule to pass through the lake) is 265 years (Marshall 1991), so the lake has very little flow relative to its size. In some African lakes, heavier cold water sinks and warm water flowing into the lake can just skim along the surface and generate a considerable horizontal surface flow in spite of a long average retention time. By chance, during the month of July, there is no thermocline (vertical temperature gradient) in Lake Kivu, making it stagnant while appearing bright blue and pristine. When the refugees arrived in Goma, Zaire, after months on the road, Lake Kivu was an irresistible temptation for bathing, swimming, and obtaining drinking water.

Outbreaks of cholera had occurred in many parts of neighboring Burundi in 1992 and 1993. Several dozen imported cases (cases with an origin outside of the country) had been reported in Kigali, Rwanda's capital. In recent decades, cholera had been endemic only in those areas of Rwanda surrounding Lake Kivu and the Ruzizi River, which flows out of Lake Kivu to the south. Most refugees coming from central Rwanda were immunologically naive (no prior exposure) and susceptible to the illness. *Vibrio cholerae,* the bacterium that causes cholera (see Chapter 11), does not survive well in acidic waters but can survive for days or weeks in water of high pH, like that found in Lake Kivu.

Thus, the arrival of over half a million exhausted and susceptible refugees into the small town of Goma without a safe water supply in an area where cholera was endemic made conditions ideal for an outbreak of cholera. The relief group Médecins Sans Frontières (MSF) reported that many people were suffering from what they suspected to be cholera by July 18, 1994. Laboratory confirmation of the outbreak as the El Tor biotype (see Chapter 11) occurred about three days later. The plight of these Rwandan refugees was seen in news reports throughout the world. The volcanic terrain prevented refugees from burying their dead, resulting in a ghastly accumulation of bodies along the edges of roads. The number of bodies collected in the Goma area from July 18 to September 19 started at a high point of almost 35 bodies per 10,000 population per day and declined through August and September (Roberts and Toole 1995). Body collection may not have been an ideal source of mortality data because some bodies may have been hidden or buried and because there was typically a delay of one or two days between death and body collection. Nevertheless, estimates of mortality based on the body count were similar to estimates based

Table 14.1   Mortality Estimates for Refugees in the Area of Goma, Zaire, in 1994

Source of Information on Deaths	End of Survey Period[a]	Estimated Population (thousands)	Crude Mortality Rate[b]	% Population Dying during Survey Period[c]
Katale survey	Aug 4	80	41.3	8.3 (7.1–9.5)
Kibumba survey	Aug 9	180	28.1	7.3 (6.2–8.4)
Mugunga survey	Aug 13	150	29.4	9.1 (7.9–10.3)
Count of dead bodies[d]	Aug 14	500–800	31.2–19.5	9.7–6.0

Source: Reproduced with permission from Goma Epidemiology Group 1995, Table 1. © The Lancet Ltd.
[a]The survey period began on July 14, 1994.
[b]The crude mortality rate is a daily rate per 10,000 population (number of deaths/10,000/day).
[c]The numbers in parentheses represent the 95% confidence interval for the estimate.
[d]The count of dead bodies applies to all refugee populations in the Goma area.

on surveys conducted in three refugee camps (Table 14.1). By any measure, the level of mortality was extraordinarily high, with 6–10 percent of the population dying during the first month of the crisis.

By July 19, 1994, the United Nations High Commissioner for Refugees (UNHCR) had secured locations for the refugees to settle away from the town of Goma. As refugees and relief groups moved to these camps, rehydration centers were established in the camps and in Goma. Medical services were completely overwhelmed. Records of attendance at approximately 22 clinics and 6 rehydration centers were often based on estimates rather than counts and sometimes aggregated patients with watery diarrhea (cholera), bloody diarrhea (dysentery), and dehydration into a general category of diarrheal disease. According to the best estimates of the numbers of patients with diarrheal disease seen in rehydration centers and clinics each day in the Goma area during the outbreak, the peak in the number of cases of diarrheal disease occurred on July 26, 1994 (Goma Epidemiology Group 1995). The peak in the number of estimated cases of diarrheal disease preceded the peak in the number of bodies collected by two days. Relief workers estimated that on July 24, 1994, approximately one-third of all deaths in the camp clinics were primarily from exhaustion without evidence of vomiting or diarrhea (Roberts and Toole 1995). Moreover, much of the diarrheal illness was from dysentery rather than cholera.

Surveys in the camps found that about half (47%) of the 23,800 individuals who died of cholera in Zaire during the crisis had never sought medical help (Goma Epidemiology Group 1995). However, medical assistance probably had minimal effect in reducing mortality, since approximately 12,600 (35%) of an estimated 35,500 patients with cholera seen at health facilities died. Assuming that cholera was fatal in similar proportions for those who did and did not seek medical treatment, the 35,500 patients with cholera represent about 67,000 cases in total. With estimates of the refugee population at the time ranging from 500,000 to 800,000, the cholera attack rate was probably between 8.4 percent (67,000 cases in 800,000 refugees) and 13.4 percent (67,000 cases in 500,000 refugees). Given a typical ratio of 1:9 or 1:10 for the number of cases of cholera to the number of infections that remain asymptomatic, it is likely that the epidemic of cholera subsided simply because the entire population had been infected. The fact that the availability of water was still less than two liters per person per day when the epidemic was almost over strongly implies that the reason for the termination of the epidemic was not related to improvements in the water supply. In fact, the outbreak of cholera was quickly followed by a lethal outbreak of dysentery due to shigella that had developed resistance to some types of antibiotics (Goma Epidemiology Group 1995). The UNHCR recommends a daily minimum of 15–20 liters of water per person (Goma Epidemiology Group 1995).

The outbreak of cholera in Goma is notable not only for its magnitude, but especially for its rapidity. About 10 days spanned the average arrival date of the refugees and the peak of the outbreak. Relief officials repeatedly refer to this crisis as a worst case scenario or an outrageous combination of interweaving factors. This may be true. The million or more refugees who fled to Tanzania and Burundi were not similarly affected. Had the water in Lake Kivu had a pH of 6 or looked filthy and unpalatable, disease transmission in the town of Goma may have been much less and the outbreak may have been spread over a longer period. If the geology of the area had been more typical of Africa and water had been available in many places, such as wells, the outbreak might have been less extensive and fewer people might have died from simple dehydration. Had the population come from an area where cholera was endemic or gone to an area without cholera present, the outbreak may not have occurred at all. But it did.

At any one moment, there are typically more than 40 armed conflicts under way in the world. In the last decades of the twentieth century, hundreds of thousands of people have fled from Afghanistan, Angola, Bosnia, Bhutan, Burma, Burundi, Ethiopia, Eritrea, Haiti, Iraq, Liberia, Mozambique, Somalia, and Sudan, to name just a few. Only in Goma did such an explosive outbreak of cholera take place. In Somalia and Ethiopia, however, equally horrifying famines occurred. In Bosnia, outrageous human rights violations (thousands of rapes, Serb snipers selectively shooting children) were committed. Less dramatic but equally deadly outbreaks have occurred in smaller refugee populations: *Escherichia coli* O157 in Swaziland, hepatitis E in Kenya, meningitis in Burundi and Nepal. In this context, Rwanda should be seen as a dramatic event amid an ongoing series of unfortunate events.

While death during the Goma crisis was mediated primarily by either water pollution or water shortage, the underlying forces for conflict were more likely to have been the fundamental causes. During the month after the war in Rwanda, the United Nations Children's Fund (UNICEF) held a series of sessions with traumatized children in Kigali. The UNICEF project asked children who were 10–12 years of age to draw on a piece of paper the relative importance of three factors in starting the recent war: Hutu-Tutsi problems, not enough land, and bad government. The preponderance of children drew the ethnic strife as a small part of the cause of the war, while the other two factors were considered to be largely responsible for the conflict. As the world becomes a more crowded place, crises surrounding water are likely to arise in many new and terrifying ways.

Flooding and Health in Brazil

Floods, simply defined as the overflow of areas that are not normally submerged, can affect human health in many ways. Flooding is the most common of all environmental hazards and adversely affects about 75 million people worldwide, regularly claiming over 20,000 lives per year (Smith 1996). The most important source of flood hazard is heavy rains, although floods can also result from breaching of sizable dams or levees. The extent of the flood hazard is defined not only by the magnitude of the flooding but also by its frequency and duration. Another important factor determining risks to humans is the onset of the flooding phenomenon. Flash flood, which is the sudden overflow of the water in a river channel or other drainage way, can be devastating, since it provides little warning and often hits with great velocity (Tobin and Montz 1997). On the other hand, some drainage basins may take several weeks to flood, providing ample time for warning.

Floods, however, are not always a hazard, since they may be an essential component of social and ecological systems, providing the basis for the regeneration of plant and aquatic life and of livelihoods derived from them (Blaikie et al. 1994). This is the case in the vast floodplains in the Amazon basin, subject to seasonal flooding that provides a huge sediment load and a rich nutrient base for an average four to seven months each year, nurturing the large plant and animal diversity and productivity of this ecosystem (Goulding et al. 1996). At the other end of the spectrum are urban areas that, because of their design and composition, are particularly prone to floods—especially flash floods—during intense downpours. Concrete and asphalt make the surfaces of a city virtually impervious to water and, therefore, elaborate storm water sewer systems are required to transport runoff to nearby natural drainages (Morgan and Moran 1997). Small floods may be increased up to 10 times by urbanization (Hollis 1975). Another flood-intensifying condition arising from changes in land use is deforestation. It seems to be a likely cause of flood runoff plus an associated decrease in channel capacity due to sediment deposition (Smith 1996).

Floods are a complex mix of natural hazards and human action, and thus vulnerability to floods is partly a product of environments created by people, though the risks are experienced in varying degrees by different groups. There is evidence that the number of people vulnerable to hydrological hazards is increasing as populations rise and the lack of alternative settlement sites, as well as economic and social pressures, force marginalized people into flood-prone urban locations (Blaikie et al. 1994). The main mechanisms by which heavy rains and floods affect human health are

1. accidents, such as drowning, mudslides, and electrocution;
2. exposure to contaminated flood water or drinking water, resulting in conditions such as gastroenteritis and leptospirosis;
3. vector-borne diseases due to an increase in breeding sites for disease vectors (see Chapter 10) and the migration of refugees from flooded areas where the diseases are endemic;
4. mental health effects, usually transient psychological reactions to stress and strain, especially among the elderly and the very young (Noji 1997);
5. disruption and destruction of livelihoods, such as the loss of assets or ability to work, which can result in famine; and
6. infections (respiratory, skin) from closely packed, unsanitary living in refugee settlements.

Many factors can contribute to injuries, illnesses, and deaths associated with floods, such as the social vulnerability and other circumstances that surround the event. Differential causes and levels of mortality have been observed in association with individual flood events in various areas of the world (Noji 1997). In the United States and its possessions, most deaths linked to floods are caused by drowning, usually inside vehicles (Centers for Disease Control and Prevention 1993, 1994; French et al. 1983; Staes et al. 1994), while most of the mortality associated with a flood episode in Bangladesh is due to diarrheal diseases (Siddique et al. 1991). The effects of a flood may be numerous, as demonstrated by Mozambique, in which a major flood episode in 2000 caused many drowning deaths; an increase in the incidence of malaria, cholera, and dysentery; and the destruction of a national economy (Jeter 2000). As the waters started to recede, a South African entomologist noted that conditions in Mozambique were right for a major epidemic of malaria, which was already beginning to become evident (Tren 2000). As of that reporting, it seemed likely that malaria would kill more people during the following few months than had been lost in the floods.

In the city of Rio de Janeiro, Brazil, most deaths due to flood episodes result either from mudslides associated with the storms that caused the flooding or from leptospirosis, an infectious disease transmitted from the urban sewage rat. Incidence of leptospirosis is associated with heavy rainfall that occurs in a short period of time; a large outbreak in Rio's Jacarepaguá district occurred when heavy rainfall fell on three consecutive days in February 1996 (Table 14.2). In this city, although records of flood episodes resulting in property damage and loss of lives date back to the mid–eighteenth century, there is evidence that the frequency and intensity of storms have increased in the last 50 years and that flooding has increased markedly since the early sixties (Brandão 1997). There were 14 events with 30 percent above average precipitation for the rainy season of the years 1941 to 1990, as opposed to 8 events from 1851 to 1900. If only the rainiest month is considered (February), the positive deviation (30%) episodes were 19 for the period 1941–90, compared with 7 episodes for 1851–1900 (Brandão 1997).

As regards flood episodes in Rio, there are records of severe storms resulting in flooding (some of them catastrophic) in at least 50 percent of the years of the twentieth century. It is not well established whether this change is associated with global climate changes. This rise in the frequency of torrential rains, associated with a greater encroachment on hillsides by low-income settlements, not only increases the rates of runoff but also exposes the inhabitants to landslides and has resulted in an increasing number of victims from storms and floods in Rio de Janeiro. Currently, about 28 percent of the population of the city (about 1.5 million people) live in areas that are either flood-prone or vulnerable to mudslides (Brandão 1997).

Flooding can be associated with major climatic phenomena such as the El Niño/Southern Oscillation (ENSO), an interannual variability in weather patterns originating in the Pacific Ocean, which ultimately affects most continents (see Chapter 8). In 1982–83, the El Niño phase of ENSO caused floods in several parts of the world, including southern South America, where it claimed 170 lives. Rio Grande do Sul, which is Brazil's southernmost state, received 1,180 millimeters of rainfall from June to October in 1982. (Floods related to El Niño events do not affect Rio de Janeiro.) In the 1997–98 El Niño event, Rio Grande do Sul received 1,533 millimeters of rainfall from June to October 16 in 1997, and rainfall for southern Brazil was 300 percent above average in October 1997. The Uruguay River on the border of Argentina and Brazil rose to 10 meters above its usual level, while the Ibicui River in Rio Grande do Sul was elevated by 12 meters. This resulted in disaster for 70 municipalities in this Brazilian region, where 3,000 houses were destroyed and 9,000 people were made homeless; 8 deaths were reported during the event. The northeastern region of neighboring Argentina experienced 18 deaths and 60,000 homeless, and 60 millimeters of rainfall in Buenos Aires resulted in the

Table 14.2 Cases of Leptospirosis and Heavy Rainfall in Rio de Janeiro, Brazil, 1990–1998

Year	No. Cases				Rainfall >100 mm		Rainfall >50 mm and ≤100 mm	
	Annual	Jan	Feb	Mar	No. Days	Date [Amount (mm)]	No. Days	Date [Amount (mm)]
1990	18	n.a.	n.a.	n.a.	0	—	0	—
1991	19	1	2	2	0	—	3	Jan 12 [53.1] Feb 19 [83.3] Mar 27 [77.1]
1992	9	5	1	0	0[a]	—	0[a]	—
1993	12	1	2	1	0[b]	—	1[b]	Mar 6 [75.2]
1994	9	0	0	4	1	Mar 27 [104.4]	0	—
1995	3	0	1	1	0	—	0	—
1996	811	4	397	382	2	Feb 13 [110.6] Feb 14 [135.3]	3	Jan 7 [74.8] Feb 4 [60.4] Feb 15 [55.5]
1997	22	12	4	2	0	—	0	—
1998	26	23	0	3	1	Mar 12 [327.2]	1	Jan 8 [93.2]

Source: Data from Secretaria Municipal da Saúde do Rio de Janeiro and Instituto Nacional de Meteorologia (INMET), sixth district.

Note: Only January, February, and March, which are the rainiest months of the year, are included in rainfall data. Data are for only the Jacarepaguá district of the city.

[a]Excluding January.

[b]Incomplete records for February.

deaths of 2 people. Recent studies suggest that, in a scenario of global warming with a doubling of the atmospheric concentration of carbon dioxide (see Chapter 7), the average frequency of El Niño events may increase from every five years to every three years. This, in turn, would increase the likelihood of flood episodes in several parts of the world.

Consequences of the Diversion of Water

Ecological Deterioration on a Massive Scale in the Aral Sea Basin

The Great Salt Lake, the Dead Sea, and the Aral Sea are among the best known of the world's major lakes that do not have an outflow of surface water. In these lakes, a dynamic equilibrium is struck between the inflow of surface water, the evaporation of water, and the infiltration of water to the groundwater table. The processes underlying the dynamic equilibrium of a lake determine both its size and its saltiness. When the inflow is particularly great, the lakes swell. Conversely, when the inflow is reduced, the lakes shrink. While water infiltrating into the ground can carry dissolved minerals (solutes), water that evaporates leaves the solutes behind and makes the remaining water saltier. Earth's oceans, which have a salinity of about 30,000–35,000 milligrams per liter, are subject to the same process. All three of the outletless lakes mentioned above are saltier than the oceans and lose the vast majority of their inflow to evaporation.

The Aral Sea is the catchment reservoir for two major river basins: the Amu Darya and the Syr Darya. These river basins extend eastward, encompassing almost all of the country of Uzbekistan and much of Kazakhstan and Turkmenistan (see *MSF 2000*). The Amu Darya flows though the Kara Kum and Kyzyl Kum Deserts, the latter of which the Syr Darya skirts as well. In spite of the arid climate, these rivers are used extensively for cultivation of crops, especially cotton. In 1992, Uzbekistan, with its population of only 23 million and an area less than 5 percent of the size of the United States, was the third largest exporter of cotton in the world. The combination of an arid climate and extensive diversion of water has contributed to a problem of increasing salinization in the Amu Darya and Syr Darya Rivers (Smith 1991). Water that has greater than 1 percent salinity (equivalent to 1,000 milligrams per liter) is generally considered nonpotable (Viessman and Hammer 1985). In an attempt to compensate for the Amu Darya's salinization problem, Nukus, the major city in the region, now extracts its water more than 200 kilometers upriver and pipes it down to the city.

The salinization is brought about primarily through the use of water for irrigation. When a field is flooded, a small volume of water can bathe a large surface area and thus can dissolve a large amount of material from rocks and minerals. If the soils are well drained, the dissolved salts are carried down through the soil column and water that seeps into the ground may or may not become salinized. But when the drainage is poor, much of the water leaves the soil by evaporation, which carries salts from within the soil column to the soil surface. In extreme cases, a white crust of salt can form on the soil surface, as exhibited by much of the agricultural land along the Amu Darya. Salinization of soils can diminish or destroy their agricultural productivity and leads to severe salinization of any water that pours onto the field and flows off without infiltrating into the ground.

The loss of water to evaporation is so severe in the case of the Amu Darya basin that the river now dries up some 50 kilometers before it reaches the former Aral Sea shoreline. The flow of the Syr Darya also reduces dramatically before it reaches its end. Because the balance between the inflow and the outflow of the Aral Sea depends on the inputs from the two rivers and evaporative withdrawals, the diminishing inputs and relatively constant evaporation have reduced its surface area by over 40 percent since 1960 (*MSF 2000; Johnson Space Center 1999*). This reduction in volume has not only made the sea saltier but also left a thick crust of salt behind on those areas of dried seabed. The strong winter winds pick up these salt particles and blow them eastward over the Amu Darya and Syr Darya river basins. Thus, a vicious cycle of salt buildup is in place, choking off plant and animal life in the river basins and the Aral Sea. Government officials in Uzbekistan report diminished rainfall in the Nukus region in recent years, perhaps indicating a microclimatic change in which the smaller Aral Sea reduces the amount of rainfall available to replenish itself.

In the Aral Sea, a fishing fleet that caught 45,000 metric tons in 1960 now lies beached in Muynak, Uzbekistan, about 60 kilometers from the water where virtually no fish are found now (Giller Institute 1995). Agricultural productivity in the river basins has diminished, but the exact extent to which salinity contributes to these reductions is not well documented. Agricultural productivity has also been influenced by the economic downturn that accompanied the independence of the central Asian states of the former Soviet Union in 1991. Diminished availability of capital has reduced the application of agrochemicals and the availability of fuel in the region. Even before independence, the use of technology for harvesting and irrigation had stagnated (Tsukatani 1998). Harvesting is performed manually by a disorganized workforce, including people from the city. During the harvest season in Uzbekistan, for example, universities and government offices are closed while students and workers are required to pick cotton for only a token salary.

The combination of limited resources and the increase in salinization has led to reductions in the quantity and quality of drinking water. Facilities for the treatment and distribution of water have been deteriorating. A study of diarrheal disease in Nukus has clearly demonstrated that its mu-

Table 14.3   Comparison of Diarrheal Burden among Three Study Groups in Nukus, Uzbekistan, in 1996

Study Group	No. Households	Relative Risk	95% Confidence Interval
No home chlorination	58	1.00 (reference)	—
Home chlorination	62	0.15	0.07–0.31
Piped water	120	0.45	0.34–0.59

Source: Reproduced with permission from Semenza et al. 1998, Table 2.

Note: The comparison uses a household rate of diarrhea, which is defined as the ratio of the number of episodes of diarrhea in a household to the number of people in the household. The study period of 9.5 weeks began in late June of 1996. The first two study groups listed did not have piped water in the household.

nicipal water system was a major source of transmission for waterborne diarrheal diseases in 1996 (Semenza et al. 1998). The implication of the water system in transmission seems counterintuitive at first glance because the water was chlorinated where it was drawn from the Amu Darya over 200 kilometers upstream and then chlorinated again at a booster station as it arrived in Nukus. But households connected to the municipal water supply (piped water) reported an average of one hour per day without running water, and about one-third of those households had no residual chlorine in samples of their tap water taken in July and August of 1996. People who did not have municipal water but chlorinated their water at home had only one-third as much diarrhea as those drinking municipal water without home treatment (Table 14.3). Among people in houses with piped water, those without residual chlorine experienced 60 percent more cases of diarrhea than did those with residual chlorine, although the difference was not statistically significant.

Because the municipal drinking water was chlorinated twice before distribution, the likely explanation for these observations is that leaky sewer lines or unsanitary ground water contaminated the pipes carrying water for drinking (cross-contamination). A shortage of water in a distribution system creates negative pressure and makes cross-contamination more likely. Nukus experienced a shortage of water for several reasons: leaks in the pipes, excessive use of water for people's gardens, a lack of alternative potable water sources near Nukus, and the fixed volume that could pass through the lengthy pipe needed to secure the water supply. Thus, an indirect effect of the salinization of water in the Aral Sea basin was a considerable amount of infectious waterborne disease!

The water shortage in Nukus has also contributed to waterborne disease via another route. Because the water supply is insufficient for the entire distribution system at present, expansion of the system is unlikely in the near future. Twenty percent of the Nukus population, some 40,000 people, have no piped water and slim prospects of acquiring piped water in the coming years. In the absence of home chlorination, households without piped water experience more episodes of diarrhea than do households with piped water (Table 14.3). Although the immediate cause of disease may be related to the lack of water for sanitary activities, the lack of potable water regionally is probably the underlying cause.

Water shortages and biological contamination of drinking water are only part of a much larger picture of threats to public health in the Aral Sea basin. The air factor is also important for public health in the region (Elpiner 1999). Associated with the drying of the Aral Sea has been an increase in the number of days with dust storms and an increase in the removal of salt from the dried sea bottom. A joint concern for the quality of both air and water has been the historically heavy use of pesticides and agricultural chemicals in the region, especially in association with the production of cotton. Smith (1991) documented that 44 percent of all agricultural chemicals used in the Soviet Union in 1980 were used in small Uzbekistan. Some pesticides and their metabolites can persist for years in the environment and accumulate in the fatty tissue of living organisms, including humans. In Kazakhstan, levels of polychlorinated biphenyls (PCBs) and organochlorine pesticides in samples of breast milk taken in 1994 revealed broad similarity with levels in industrialized Europe for most contaminants, but the levels of β-hexachlorocyclohexane (β-HCH) were among the highest reported in the published literature (Hooper et al. 1997). In addition, the samples of breast milk taken in 1994 and follow-up samples in 1996 and 1997 demonstrated evidence of prior exposure to dioxin in agricultural villages in southern Kazakhstan near the border with Uzbekistan; the highest levels were found in state farms adjacent to a reservoir that receives agricultural runoff from cotton fields where aircraft sprayed pesticides and defoliants between 1965 and 1985 (Hooper et al. 1998, 1999).

Elpiner's (1999) framework for analyzing public health in the region includes a nutritional factor in addition to the air and water factors. The amount of land devoted to the production of industrial crops far exceeds the amount devoted to the production of food crops, which include grain, potatoes, vegetables, melons, and animal fodder. The shrinking of the Aral Sea has also led to precipitous declines in the local fisheries and consequently in the amount of fish in the local diet.

Not surprisingly, several indicators point to a general decline in public health in the region (Elpiner 1999). Two of the most important are the infant mortality rate (IMR) and the prevalence of anemia. (See also Chapter 2, under "Descriptive Measures of Health Status.") The IMR increased by about 20 percent between 1980 and 1989, with an increasing share of deaths occurring in the first weeks of life. In 1988, anemia was found in more than 80 percent of women of childbearing age in Uzbekistan's autonomous republic of Karakalpakstan, whose capital is Nukus. However, against a backdrop of massive ecological deterioration and a host of socioeconomic problems, it is difficult to fill in the details of the links between environmental changes and public health as suggested by Elpiner's framework of multiple factors. Epidemiological research is hampered by inadequate systems for surveillance of disease. Cases of a disease are often recognized clinically without the aid of laboratory diagnostics. Even sophisticated laboratory diagnostics are not always consistent in the detection of minute residues of chemicals such as pesticides. Limited numbers of sampling sites make it difficult to draw inferences about the region as a whole and about geographic variation within the region. (See also Chapter 2 under "Potential Sources of Error in Epidemiological Studies.")

One optimistic sign in the region is the establishment of a health project by the international relief group MSF (*MSF 2000*). With staff in Tashkent, Nukus, and Muynak, the first priority of the MSF is to improve the surveillance and control of tuberculosis, acute respiratory infections, diarrheal diseases, and anemia. Perhaps this effort can develop over time in parallel with the International Program on the Health Effects of the Chernobyl Accident (see Chapter 2). Both the nuclear power accident at Chernobyl and the shrinking of the Aral Sea are now recognized as catastrophic examples of damage to the environment caused by human actions. Indeed, people in the former Soviet Union often use the term *Quiet Chernobyl* to refer to the Aral Sea basin (Glantz 1998). This central Asian example of the interplay between human health and water is much less direct than that observed by John Snow for cholera a century and a half ago (see Chapters 2 and 11) but is no less tangible. The delicate equilibrium of the Aral Sea ecosystem has been disrupted. In the end, the costs of reestablishing that equilibrium may exceed the profits reaped over several decades of growing cotton in the desert. If so, poor ecological planning ironically will have inflicted the greatest cost on those who partook in the world's greatest experiment with a planned economy.

Exposure to Dust in the Great Basin Unified Air Pollution Control District of California

Although the problems of the Aral Sea are half a world away from the United States, those problems are now invoked in the development of policies regarding the diversion of water to support the city of Los Angeles, California. Two foci of concern are Mono Lake and Owens Lake, which lie in Mono County and Inyo County, respectively, in the Great Basin area of California (*California Air Resources Board [CARB] 2000;* Patton and Ono 1995; Ono and Schade 1997). The similarities between the Aral Sea and these lakes are striking. Mono Lake, like the Aral Sea, does not have an outflow of surface water. After the inflow to the lake was tapped to provide wa-

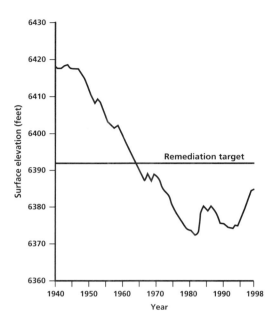

Figure 14.1 History of the level of Mono Lake, California, 1940–98. The level of the lake is from October 1 of the year indicated. The target level for remediation efforts is set at 6,392 feet (25 feet below the level in 1941). *Source:* Data from *Mono Lake Committee 1999.*

ter for Los Angeles in 1941, the level of the lake began to drop (Fig. 14.1). The history of Owens Lake is similar, but the diversion of water began earlier and the lake bed is almost entirely dry.

The public health problems caused by the diversion of water from Mono Lake and Owens Lake are much less severe than the public health problems in the Aral Sea basin for two reasons. First, the two lakes in California are much smaller than the Aral Sea. Second, the two lakes in California are not critical to the provision of food and drinking water for the human population living near the lakes. However, the dust generated as the lakes have been drying up violates the standards of the U.S. Clean Air Act as amended in 1990 (see Chapter 4). The primary pollutant of concern is PM_{10}, particulate matter less than 10 microns in diameter (the standards for even smaller particles came into force after California had developed plans to control the dust near the lakes). The standards in California for 24-hour exposures and annual average exposures are even more stringent than the corresponding federal standards (*CARB 2000*).

The control of particulates should prevent excess deaths from short-term exposures, the exacerbation of symptoms in sensitive patients with respiratory disease, and seasonal declines in pulmonary function. Anecdotal evidence from physicians in the area points to a variety of pulmonary problems as well as eye irritation caused by dust blown off the lake beds. Although the standards for exposure to particles do not consider their chemical composition, the planning report for Mono Basin also expresses concern about "desert lung syndrome," which is characterized by deposits of sandy dust in the lungs (Patton and Ono 1995, 22). The planning report for Owens Valley concludes that the levels of cadmium, arsenic, and other toxic metals exceed standards under California's air toxics program and pose an unacceptably high increase in the risk of cancer (Ono and Schade 1997, 3–13). (Chapter 4 explains the terminology of "air toxics" used in the U.S. Clean Air Act.)

Under state and federal laws, the Great Basin Unified Air Pollution Control District was required to prepare plans to reduce the levels of dust around Mono Lake and Owens Lake. For Mono Lake, the plan is embodied in Mono Lake Basin Water Right Decision 1631. Decision 1631 specifies that water be allowed to return to the lake to achieve a partial restoration, which should be sufficient to cover up the sources of dust (Patton and Ono 1995; *Mono Lake Committee 1999*). The level of the lake has already started to rise since Decision 1631 became effective in 1994 (see Fig. 14.1). Decision 1631 also involves plans for the conservation and reclamation of water in southern California (*Mono Lake Committee 1999*). A remarkable feature of the decision is the number of major participants in the proceedings: Los Angeles Department of Water and Power, Mono Lake Committee, National Audubon Society, California Trout, California Department of Fish and Game, U.S. Forest Service (Inyo National Forest), California Department of Parks and Recreation, State Lands Commission, Great Basin Unified Air Pollution Control District, U.S. Fish and Wildlife Service, Sierra Club Legal Defense Fund, and Upper Owens River Landowners. Public health was one of many factors in the decision. The plan to control dust around Owens Lake does not involve restoration of the lake, but rather a combination of shallow flooding, managed vegetation, and the application of gravel (Ono and Schade 1997). Any possible restoration of Owens Lake is hampered not only by the intrinsic difficulty of restoring a lake that became virtually dry by 1930, but also by a complicated history of ownership and use of the land. The most salient issue is a long-running legal dispute between the City of Los Angeles and the Natural Soda Products Company, which leased thousands of acres of dry lake bed for mining. The City of Los Angeles is prohibited from flooding the lake bed and interfering with mining interests.

This example of the diversion of water in California demonstrates that even a wealthy country suffers adverse consequences of massive schemes to use water for new purposes, such as the growth of a major metropolitan area. Solutions to these problems lie outside the traditional sphere of public health and must involve a diverse set of actors from the public sector and the private sector.

Conclusion

Water resources are important for public health not only for the direct uses of the water but also for broader ecological benefits. Population growth and movement, economic development, and climate change have all been associated with threats to the water resources necessary to sustain public health. Dealing with those threats requires collaboration by traditional specialists in public health and others who have a broad interest in the global ecosystem. The challenge for the twenty-first century is to develop a better understanding of the linkages between water resources and health so that those resources can be managed wisely.

SUGGESTED STUDY PROJECTS

Suggested study projects provide a set of options for individual or team projects that will enhance interactivity and communication among course participants (see Appendix A). The Resource Center (see Appendix B) and references in all of the chapters provide starting points for inquiries. The process of finding and evaluating sources of information should be based on the principles of information literacy applied to the Internet environment (see Appendix A).

The objective of all three projects is to arrive at a clearer understanding and presentation of the diversity of water-related health problems.

PROJECT 1: Analysis of Ecosystem Changes, Water Resources, and Public Health

The objective of this project is to deepen the understanding of the interrelated forces affecting emerging water-related health problems.

Task 1. Analyze the information of the chapter's four examples in a report, with special attention given to the diversity of water resources operating in various health problems.

Task 2. Find an example of a water-related health issue not in the chapter and discuss how it is affected by global ecosystem changes. Compare your example with the four examples in the chapter.

PROJECT 2: Reference Chart of Emerging Water-Related Health Problems

The objective of this project is to build a reference chart (written and graphic) providing an overview of the scope of emerging water-related health problems, which can be expanded as additional information becomes available.

Task 1. Summarize the information transmitted in the examples of this chapter and your own example from project 1, using key words and phrases.

Task 2. Develop a clear format in which to present an overview of this information (e.g., water scarcity, excess water, water/air relation).

Task 3. Find additional information (examples) to expand or modify your reference chart.

PROJECT 3: Design of Project to Study Global Ecosystem Change and Water-Related Health Issues

The objective of this project is the application of a wider perspective regarding the scope of emerging water-related health problems. This project can be pursued independently as an in-depth research paper, but it also can be assigned as a continuation and refinement of the information chart.

Acknowledgments

We thank Luciane Medeiros of the National School of Public Health in Rio de Janeiro for reviewing reports of the effects of the 1997–98 El Niño event in the state of Rio Grande do Sul in Brazil. Carol Rubin and Alden Henderson of the Centers for Disease Control and Prevention gave valuable comments on the source material and the manuscript, Clive Shiff of the Johns Hopkins School of Public Health sent a news report on flooding in Mozambique, Mickey Glantz of the National Center for Atmospheric Research sent prepublication material from a book on the Aral Sea, Kim Hooper of California's Department of Toxic Substances Control discussed and sent publications on the Aral Sea region, and Ellen Hardebeck of the Great Basin Unified Air Pollution Control District Office in Bishop, California, provided copies of the state implementation plans for Mono Basin and Owens Valley. We also appreciate the assistance of Tony Van Curen of the California Air Resources Board in Sacramento and the Mono Lake Committee Information Center in Lee Vining, California, for identifying useful sources of information.

References

Blaikie P, Cannon T, Davis I, Wisner B. 1994. *At Risk: Natural Hazards, People's Vulnerability, and Disasters*. Routledge, London.

Brandão AMPM. 1997. As chuvas e a ação humana: Uma infeliz coincidência [In Portuguese: Rainfall and human action: An unhappy coincidence]. In *Tormentas Cariocas* (Rosa LP, Lacerda WA, eds.). Coordenação de Programas de Pós-Graduação em Engenharia, Universidade Federal do Rio de Janeiro, Rio de Janeiro, pp. 21–38.

Centers for Disease Control and Prevention. 1993. Flood-related mortality—Missouri, 1993. *MMWR* 42 (48): 941–43.

———. 1994. Flood-related mortality—Georgia, July 4–14, 1994. *MMWR* 43 (29): 526–29.

Elpiner LI. 1999. Public health in the Aral Sea coastal region and the dynamics of changes in the ecological situation. In *Creeping Environmental Problems and Sustainable Development in the Aral Sea Basin* (Glantz MH, ed.). Cambridge University Press, Cambridge, pp. 128–56.

French J, Ing R, Von Allmen S, Wood R. 1983. Mortality from flash floods: A review of National Weather Service reports, 1969–1981. *Public Health Rep* 98 (6): 584–88.

Giller Institute. 1995. *Social Problems of the Kazakhstan Priaral and Its Population Appraisal*. Giller Institute, Almaty, Kazakhstan.

Glantz MH. 1998. Creeping environmental problems in the Aral Sea basin. In *Central Eurasian Water Crisis: Caspian, Aral, and Dead Seas* (Kobori I, Glantz MH, eds.). United Nations University Press, Tokyo, pp. 25–52.

Goma Epidemiology Group. 1995. Public health impact of Rwandan refugee crisis: What happened in Goma, Zaire, in July 1994? *Lancet* 345 (8946): 339–44.

Goulding M, Smith NJH, Mahar DJ. 1996. *Floods of Fortune: Ecology and Economy along the Amazon.* Columbia University Press, New York.

Hollis GE. 1975. The effect of urbanization on floods of different recurrence intervals. *Water Resources Res* 11:431–34.

Hooper K, Chuvakova T, Kazbekova G, Hayward D, Tulenova A, Petreas MX, Wade TJ, Benedict K, Cheng Y-Y, Grassman J. 1999. Analysis of breast milk to assess exposure to chlorinated contaminants in Kazakhstan: Sources of 2,3,7,8-tetrachlorodibenzo-p-dioxin (TCDD) exposures in an agricultural region of southern Kazakhstan. *Environ Health Perspect* 107 (6): 447–57.

Hooper K, Petreas MX, Chuvakova T, Kazbekova G, Druz N, Seminova G, Sharmanov T, Hayward D, She J, Visita P, Winkler J, McKinney M, Wade TJ, Grassman J, Stephens RD. 1998. Analysis of breast milk to assess exposure to chlorinated contaminants in Kazakstan: High levels of 2,3,7,8-tetrachlorodibenzo-p-dioxin (TCDD) in agricultural villages of southern Kazakstan. *Environ Health Perspect* 106 (12): 797–806.

Hooper K, Petreas MX, She J, Winkler J, McKinney M, Mok M, Sy F, Garcha J, Gill M, Stephens RD, Semenova G, Sharmanov T, Chuvakova T. 1997. Analysis of breast milk to assess exposure to chlorinated contaminants in Kazakstan: PCBs and organochlorine pesticides in southern Kazakstan. *Environ Health Perspect* 105 (11): 1250–54.

Jeter J. 2000. U.S. joins in Mozambique rescue. *Washington Post,* 6 March 2000.

Marshall BE. 1991. Abundance of *Limnothrissa mioden* in Lake Kivu. *J Fish Biol* 39:641.

Morgan MD, Moran JM. 1997. *Weather and People.* Prentice Hall, Upper Saddle River, N.J.

Noji EK, ed. 1997. *The Public Health Consequences of Disasters.* Oxford University Press, New York.

Ono D, Schade T. 1997. *Owens Valley PM10 Planning Area Demonstration of Attainment State Implementation Plan.* Great Basin Unified Air Pollution Control District, Bishop, Calif. (July).

Patton C, Ono D. 1995. *Mono Basin Planning Area PM-10 State Implementation Plan.* Great Basin Unified Air Pollution Control District, Bishop, Calif. (May).

Roberts L, Toole MJ. 1995. Cholera deaths in Goma. *Lancet* 346 (8987): 1431.

Semenza JC, Roberts L, Henderson A, Bogan J, Rubin CH. 1998. Water distribution system and diarrheal disease transmission: A case study in Uzbekistan. *Am J Trop Med Hyg* 59 (6): 941–46.

Siddique AK, Baqui AH, Eusof A, Zaman K. 1991. 1988 floods in Bangladesh: Pattern of illness and causes of death. *J Diarrheal Dis Res* 9 (4): 310–14.

Smith DR. 1991. Growing pollution and health concerns in the Lower Amu Dar'ya Basin, Uzbekistan. *Soviet Geog* 31 (8): 553–65.

Smith K. 1996. *Environmental Hazards: Assessing Risk and Reducing Disaster,* 2d ed. Routledge, London.

Staes C, Orengo JC, Malilay J, Rullan J, Noji E. 1994. Deaths due to flash floods in Puerto Rico, January 1992: Implications for prevention. *Int J Epidemiol* 23 (5): 968–75.

Tobin GA, Montz BE. 1997. *Natural Hazards: Explanation and Integration.* Guilford Press, New York.

Tren R. 2000. Malaria will now take its toll in Mozambique. *Wall Street Journal Europe,* 6 March 2000.

Tsukatani T. 1998. The Aral Sea and socioeconomic development. In *Central Eurasian Water Crisis: Caspian, Aral, and Dead Seas* (Kobori I, Glantz MH, eds.). United Nations University Press, Tokyo, pp. 53–74.

Viessman W, Hammer M. 1985. *Water Supply and Pollution Control,* 4th ed. Harper & Row, New York.

World Health Organization. 1992. *Our Planet, Our Health.* Report of the WHO Commission on Health and Environment. World Health Organization, Geneva.

Electronic References

California Air Resources Board. 2000. California Air Resources Board. http://www.arb.ca.gov/ (Date Last Revised 3/17/2000).

Johnson Space Center. 1999. Water Features and Water Issues: Aral Sea. http://eol.jsc.nasa.gov/newsletter/html_Mir/aral.html (Date Last Revised 9/13/1999).

Médecins Sans Frontières. 2000. The MSF Aral Sea Project Web Site. http://www.msf.org/aralsea/ (Date Last Revised 3/2000).

Mono Lake Committee. 1999. Yearly Lake Levels. History of Mono Lake's Fluctuating Past. http://www.monolake.org/library/lakelevel/yearly.htm (Date Last Revised 12/12/1999).

Norwegian Council for Africa. 2000. Congo-Zaire: Security and Conflict. http://www.africaindex.africainfo.no/pages/Country_pages/Congo-Zaire/Security_and_Conflict/ (Date Last Revised 3/21/2000).

Volcano World. 2000. Volcanoes in the African Region. http://volcano.und.edu/vwdocs/volc_images/africa/ (Date Last Revised 3/17/2000).

Stimulating Inquiry

**Textbooks and Information Literacy
in the Internet Environment**

Erika G. Feulner, M.A.

The Traditional Role of the Textbook

Textbooks have played an important role in our educational systems since about 1830. From first grade primers to sophisticated texts for graduate programs, they have assisted teachers and instructors on every level in content provision and assimilation. Textbooks have been the core and basis for assignment schedules and test materials. Frequently they have supplemented lecture and classroom material, and occasionally the study of their content has fulfilled total curriculum requirements. Textbooks today still fulfill these functions to a large extent, but their role is adapting to a rapidly changing learning environment.

Changes in the Learning Environment

Our concept of learning in the United States has been dominated for nearly a half-century by the behavioral methods and theories of B. F. Skinner. In our traditional educational settings, we have focused on the acquisition and feedback of information largely disseminated by instructors. Within this framework, the teacher, instructor, or lecturer functions as the core information and referral center. From specific assignments to independent study projects, the source material used will in some way link back with the information provision or referral of the classroom instructor. This has put the responsibility of content integrity and evaluation criteria of information entirely on individual course designers. Students are responsible for the correct feedback demonstrating their understanding of the information received. The learning process is regarded as a stimulus-response cycle (Skinner 1953), with quizzes and tests acting as checkpoints for the projected success of the cycle. This leaves little time for interpretation or application of information (i.e., individual knowledge formation). The focus on information processing—from teacher to student and back to the teacher—has created barriers to independent thinking and individual expression, thereby causing an alarming lack of motivation to learn.

The emerging new learning environment is shaped by today's information explosion, the advent of the Internet, which provides almost unlimited access to information, and by rapid advances in brain research, which have led to the discovery of a new, uncharted potential of human learning capabilities. On the basis of these events, our perception of the learning process is changing with equal rapidity. We have become sensitive to different learning styles (Porter 1992; Gardner 1993) and the adaptation of content to diverse learning modes (Campbell et al. 1992). Our perception of a student is transforming according to new theories and knowledge regarding human intelligence and how it works (Feuerstein et al. 1990).

New learning methods have surfaced steadily in experimental educational settings over the last 30 years but now are welcomed quite often as appropriate tools for meeting the demand of change. The electronic magazine on teaching and learning in higher education "DeLiberations" (*London Guildhall University 2000*) offers articles, comments, and discussions on several new learning methods: effective learning, flexible learning, resource-based learning, and collaborative learning. These new learning methods are designed to meet the demands of the work force operating in a growing global economic environment. Calls for lifelong learners, motivated learners, independent thinkers, and persons with interdisciplinary skills and decision-making capabilities appear steadily in newspaper articles, corporate lectures, and training magazines and have a direct effect on our educational framework. In the words of Peter F. Drucker, "The disciplines and the methods that produced knowledge for two hundred years are no longer fully productive. . . . The rapid growth of cross-disciplinary and interdisciplinary work should indeed argue that new knowledge is no longer obtained from within the disciplines around which teaching, learning, and research have been organized in the nineteenth and twentieth centuries" (Drucker 1989, 252).

The New Role of the Textbook

The changing learning environment has a special effect on institutions of higher learning, since their graduates are expected to meet not some but all of the demands stated above. Textbooks in this emerging new learning environment are changing their role to handbooks for students and professors alike, aiding independent learning processes and evaluative knowledge formation. This textbook has been designed to be a handbook for today's researchers pursuing multiple inquiries into global environmental health issues. It provides numerous starting points for short- and long-term research projects (see "Introduction: How to Use This Book").

Mental Skills Development

Research strategies and procedures have developed over time into strictly logical linear thought processes. The fact that linear thinking is only part of a larger whole-brain thinking process has been forgotten. Therefore, our customary research tools often do not meet the increasing demands of interdisciplinary inquiries.

Recent developments in brain research have not only provided new insights into and understandings of the vast abilities of our learning potential and how our brain works but also have opened new ways and methods to activate our learning potential. Heightened mental skills can have a decisive effect on the design of research approaches in interdisciplinary research projects. One technique of learning how to maximize our brain's untapped potential and use "radiant thinking" (Buzan 1996) is mind mapping, a unique way of note taking and organizing thoughts and ideas. Instead of writing ideas, notes, and memos in columns, like a shopping list, the notation starts with a central idea put into a circle in the center of a sheet of paper and then develops major and secondary branches reaching outward, connecting all pertinent information to the central topic. This notation system reflects exactly how our "whole brain" works. It promotes better memory and quicker realizations of connections in establishing cause and effect, and it develops a wider scope of perception. Tony Buzan, the originator of mind mapping, has written several books on the subject and has developed mind-mapping software, which is being increasingly used in planning and development functions (*MindJet LLC 2000*).

Information Literacy on the Internet

The American Library Association and the library science departments of several universities have assumed leadership in adapting traditional information literacy to the new Internet environment (*American Library Association 1999*). Their aim is continually to develop and refine appropriate standards and criteria for searching, evaluating, and citing information gleaned from the vast availability of traditional and electronic sources (see Appendix B) in order to guarantee sound research procedures. The tasks of evaluating, selecting, organizing, and indexing materials from credible sources have traditionally been the venue of the librarian, in close conjunction with course designers and researchers. Today, every student, researcher, lecturer, and writer who is quoting sources must learn to perform these tasks, referring to the librarian only as guide but not as source. Information literacy skills are becoming essential for the performance of serious research work. The researcher must be able to separate the wheat from the chaff among the multitude of available sources, especially when using computer-based or Web-based resources. According to Lida L. Larsen,

> the term *information literacy* means the ability of people to
>
> — know when they need information
> — find information
> — evaluate information
> — process information
> — use information to make appropriate decisions
>
> The Internet has added a new dimension to traditional information literacy issues—especially in the exploding growth of the World Wide Web. Nearly a mix between all other media, the Web democratizes information ownership, provision, and retrieval. The Web allows us to speak directly to the purveyors of information in every imaginable field. Few reference librarians, teachers, publishers, or other mediating forces stand between us and information on the Internet, and specifically, the Web. While this does have great advantages in expanding our information base and providing more accurate and timely information at the "click of a mouse," it also means, perhaps, more intellectual effort on the part of the information consumer to develop valuable critical thinking skills and to evaluate the sources, quality, and quantity of that information. It also means serious attention should be paid to intellectual property and appropriate use issues. (Reproduced with permission from *Larsen 1999*)

Information literacy—the development of standards and criteria for selecting credible sources and validating information in an environment where information proliferates explosively and unchecked—is an ongoing effort by library scientists in official positions of recognized educational institutions. Part of the process is raising awareness of the need for information literacy and communicating the necessary skills to students on every level.

Evaluating Information

Proving the reliability and authenticity of information in traditional settings has been done by obtaining the author's name and background or institutional affiliation, the date of the publication, and peer review if available—all traceable through the publishing agency. Today, in a world of website creations and self-publishing, new questions are being raised for information validation, and a number of elements will help the researcher in identifying resources of value using the World Wide Web: content scope, authority and bias, accuracy, timeliness, permanence, value-added features, presentation (*Larsen 1999*). Abbreviated comments on each criterion, based on L. Larsen's online information, are found in Box A.1.

Criteria for Evaluating Information on the Web

Content Scope: Determine how much material is covered.

Authority and Bias: Look for credentials of the information provider, institutional affiliation, and points of view being "sold" to the viewer.

Accuracy: Match appropriateness of topic to site. Look for credentials of author/creator/publisher, peer review, and other resources on the topic.

Timeliness: Check posting and revision dates and policy statements for information maintenance.

Permanence: Look for explicit statements of temporary or changing location of servers or files and author's relationship to the server infrastructure.

Value-Added Features: Look for Web sites moderated by trained professionals who receive and respond to feedback. Determine whether other sites evaluate or rate informational content. Check whether computer tools are enhancing the transmission of information.

Presentation and Organization: Look for clarity in the page or site layout, the site's organizational design, and the help/example sections.

Note: These criteria have been adapted with permission from *Larsen 1999.* Further guidelines for evaluating Web sources can be accessed at the University of Maryland Libraries at http://www.lib.umd.edu/UMCP/UES/evaluate.html.

Citing Electronic Information

Citing electronic information is an integral part of establishing standards for the use of electronic information. Several guidelines have been developed by different institutions. Citations in this appendix are following the standards put forth in *Land (1998).*

Browsing the Web

Using the Web as a research tool can be extremely rewarding, but it also can cause time delays due to sidetracking and losing focus. Research on the Web requires a detective's mind—open to all possibilities for solving the case—and strict search discipline. Considering time constraints and academic requirements, the following guidelines will maximize the results of Web searches:

1. Have a good plan for beginning your interdisciplinary inquiries.
2. Set a time for browsing (2 hours).
3. Keep a log and summarize findings.
4. While searching, write down new facts you learn (including negative results) and new questions that emerge.
5. Link information received to other sites and traditional sources.
6. Evaluate information as you search (see criteria in Box A.1).
7. Be sure to keep a written account of your search.

Interactivity and Communication

The concept of research being a solitary pursuit still lingers. Within the context of our information age, a new paradigm emerges: shared research will produce deeper knowledge. The classroom of today has to share ideas and findings of each individual and arrive at knowledge drawn from multiple perspectives.

Study projects can be carried out in a variety of ways—from individual assignments to group projects in short and long time frames. Many combinations are possible. An individual paper could serve as the basis for a group discussion. Documentation of Web searches could serve as a basis for group discussion. Conversely, a summary of group discussion could serve as the basis for an individual paper. Presentations of group efforts could generate shared knowledge and create the basis for future research. The key to successful results and the continuation of motivated research is the establishment of a communication system between class members and facilitators of information.

References

Buzan T. 1996. *The Mind Map Book.* Plume, BBC Books, Penguin Books, London.
Campbell L, Campbell B, Dickinson D. 1992. *Teaching and Learning through Multiple Intelligences.* Campbell & Associates, Washington, D.C.
Drucker P. 1989. *The New Realities.* Harper & Row, New York.
Feuerstein R, Presseisen B, Sternberg R, Fischer K, Knight C. 1990. *Learning and Thinking Styles: Classroom Interaction.* National Education Association of the United States, Washington.
Gardner H. 1993. *Frames of Mind.* Basic Books, Harper Collins Publisher, New York.
Porter B. 1992. *Quantum Learning.* Dell Publishing, New York.
Skinner BF. 1953. *Science and Human Behavior.* Macmillan, New York.

Electronic References

American Library Association. 1999. Information Power-Building Partnerships for Learning. [WWW abstract] http://www.ala.org/aasl/ip_toc.html (Date Last Revised 3/16/1999).
Land T [a.k.a. Beads]. 1998. Web Extension to American Psychological Association Style Guide (WEAPAS) (Rev.1.5.2) [WWW document] http://www.beadsland.com/weapas (Date Last Revised 10/15/1998).
Larsen L. 1999. Information Literacy: The Web Is Not an Encyclopedia. Online Information Resources Office of Information Technology. University of Maryland, College Park, Md. http://www.umd.edu/literacy (Date Last Revised 8/1999).
London Guildhall University. 1995–2000. Deliberations on Teaching and Learning in Higher Education. Educational Development Unit, London Guildhall University, London. http://www.lgu.ac.uk/deliberations (Date Last Revised 7/31/2000).
MindJet LLC. 2000. Mind Manager: The Ultimate Organization Tool. http://www.mindman.com/index.html (Date Last Revised 2000).

The Resource Center

Compiled by Erika G. Feulner, M.A.

The Resource Center offers a selection of information sources as first contact points for launching interdisciplinary research projects. In conjunction with Chapter 1, the individual chapter references, and suggested study projects, the Resource Center provides a unique linkage to traditional and electronic resources within the contents of the book. It also assists the user in accessing outside resources, thereby expanding on the themes of this book.

References appear at the end of each chapter. Electronic references are listed as chapter references but might reappear in the Resource Center.

The most useful function of the Resource Center is its guidance for searching and utilizing website information. Searching electronic resources has become an integral part of conducting modern-day research.

Websites are listed in three categories:

1. Annotated websites: comprehensive sites dealing with the themes of the book on an international/global basis and providing extensive links.
2. Topically arranged websites: sites that are grouped under specific topics (such as climate, land use, and population growth) with a focus on a particular theme or regional and local information. They are detail oriented and often relate to electronic chapter references.
3. Useful directories and online libraries: sites that provide either one or a combination of the following services: free access to large library collections online; information network systems providing cross-referenced resources; e-mail addresses, fax numbers, and phone numbers for contacting experts.

All listed entries were checked for currency and accessibility during January and February 2000.

The Resource Center does not claim to be all-comprehensive but presents a carefully selected choice of resources to encourage innovative and motivated research. Its main objective is to make links of information more transparent. The quality standards of the American Library Association served as selection criteria. The criteria for evaluating electronically transmitted information sources are outlined in Appendix A.

Annotated Websites

The following websites were selected for their cross-reference linkages supporting the themes of the book, their capacity to provide access to multiple information systems, their timely update practice, and their international or global dimension. Many of the sites feature links to each other. This can prove helpful when a server does not bring the desired site on screen via its uniform resource locator (URL)—access can be obtained quite frequently through another site. Some sites refer primarily to environmental issues; some refer primarily to health issues. A few sites feature cross-cutting themes and allow access to environmental and health issues (i.e., CIESIN, United Nations, World Bank, World Resources Institute).

The first eight URL listings are U.S. government agencies with primary concerns for the environment and human health. They are followed by two United States–based sites with strong international links.

The next 10 listings were selected for their specific global and international essence and interdisciplinary research goals. The selection starts with the United Nations site, which has worldwide locations for various member groups or programs, of which five were selected for annotation. The remaining four listings are programs sponsored by the International Council of Scientific Unions (ICSU); all are located in Europe, but they are globally linked. They constitute a close international research network. The site functioning as an umbrella for the ICSU-sponsored programs is the International Geosphere-Biosphere Programme (IGBP), accessible via two addresses.

The next small cluster of four listings with home pages in the United States and Europe were chosen for their emphasis on linkages.

The compilation ends with four selections offering services or perspectives that reinforce the global and interdisciplinary aspects of environmental research.

U.S. Government Agencies

URL: http://www.epa.gov/epahome/research.htm
Environmental Protection Agency (EPA)
EPA's mission is to protect human health and the environment. The information on this site is extensive and provides access to publications and

technical documents, ongoing research programs, environmental data and tools for scientific inquiry, laboratories and research centers, and a host of links to other resources. As the gateway to environmental information in the United States, it is linked to other federal and state government sites, thereby providing the opportunity to cross-reference a vast collection of available information on a theme. The environmental page of the Department of Energy's Energy Information Administration (EIA, http://www.eia.doe.gov/energy/environ.html), particularly, features EPA/EIA cross-cutting references. The EPA site is updated regularly and current within the span of a month.

URL: http://www.hhs.gov/
U.S. Department of Health and Human Services (HHS)
HHS is the federal government's principal agency for protecting the health of all Americans and providing essential human services. The department includes more than 300 programs, carried out by 13 agencies. Of the 8 public health operating divisions, 2 provide particularly appropriate resources for the themes of this book: Centers for Disease Control and Prevention (CDC) and National Institutes of Health (NIH). Both are annotated separately.

URL: http://www.cdc.gov/
Centers for Disease Control and Prevention (CDC)
CDC is located in Atlanta, Georgia, and performs many of the administrative functions for the Agency for Toxic Substances and Disease Registry (ATSDR). CDC's mission is to promote health and quality of life by preventing and controlling disease, injury, and disability. The CDC includes 11 centers, institutes, and offices and produces a wide variety of health reports. The National Center for Environmental Health (NCEH), the National Center for Health Statistics (NCHS), and the National Center for Infectious Diseases (NCID) are especially valuable for the researcher of environmental health issues in providing resource material. They are easily accessible through the CDC site and feature updated information.

URL: http://www.nih.gov/
National Institutes of Health (NIH)
The National Institutes of Health are the federal government's medical research centers. NIH provides a large pool of resources such as consumer health publications, clinical trials, health hotlines, MEDLINE, and the NIH Information Index (a subject-word guide to diseases and conditions under investigations at NIH). The National Library of Medicine (NLM), a part of NIH, offers any researcher in a health-related field the most extensive resource collection on health issues. MEDLINE is NLM's premier database, covering the fields of medicine, nursing, dentistry, veterinary

medicine, the health care system, and the preclinical sciences and may be accessed free of charge on the World Wide Web. MEDLINE contains bibliographic citations and author abstracts from more than 3,900 biomedical journals published in the United States and 70 foreign countries and more than nine million records dating back to 1966. NLM can be accessed over the NIH site or separately at http://www.nlm.nih.gov/.

URL: http://www.nasa.gov/
National Aeronautics and Space Administration (NASA)
NASA has an extensive site that provides the searcher with special search help, allowing access to HTML, PDF, and Word files. NASA is deeply committed to spreading the unique knowledge that flows from its aeronautics and space research and offers extensive information in many formats. The environmental researcher will find the results and publications from NASA's Earth Science projects most helpful, especially the "Global Change Master Directory" (http://gcmd.gsfc.nasa.gov), which is linked to neonet's ceos-idn (see under "Directories," below).

URL: http://www.usgs.gov/
U.S. Geological Survey (USGS)
USGS provides reliable scientific information on biological resources, geology, national mapping, and water resources. Data can be accessed via the National Geospatial Data Clearinghouse, which is a component of the National Spatial Data Infrastructure (NSDI). USGS is committed to the management of water, biological, energy, and mineral resources. A special program is the USGS Global Change Research Program, annotated below.

URL: http://geochange.er.usgs.gov/
USGS Global Change Research Program
The USGS Global Change Research Program is a component of the U.S. Global Change Research Program (USGCRP), complementing research and observations on oceanic, atmospheric, and biological processes in other federal agencies. USGS global change research examines terrestrial and marine processes and the natural history of global change. USGS documents the character of environments in the past and present and the interactions of processes involved in environmental change.

URL: http://www.nsf.gov/
National Science Foundation (NSF)
NSF is an independent U.S. government agency responsible for promoting science and engineering through programs that invest $3.3 billion per year in almost 20,000 research projects in science and engineering. NSF partnerships with other government agencies and in the private sector are extensive through the sponsorship of these research projects. NSF is strongly

involved in environmental research, especially via its cross-cutting programs. This website is extensive and provides many valuable links.

United States–based Organizations

URL: http://www.ciesin.org/
Center for International Earth Science Information Network (CIESIN)
CIESIN was established in 1989 as a not-for-profit, nongovernmental organization to provide information for a better understanding of our changing world and is United States–based, located at Columbia University in New York. CIESIN conducts projects with a broad array of national and international sponsors. It maintains two major programs: the Global Change Research Information Office (GCRIO) and the Socioeconomic Data and Applications Center (SEDAC). In terms of supporting the themes of this book, CIESIN offers a particularly helpful resource in thematic guides to key environmental issues, annotated separately below.

URL: http://www.ciesin.org/TG/thematic-home.html
CIESIN Thematic Guides
Thematic guides offer overviews of some of the key topics and issues that pertain to the human dimensions of global change, including agriculture, human health, land use, ozone depletion, political institutions, remote sensing, environmental treaties, resource indicators, and integrated assessment modeling of climate change. Through the use of links within editorial essays that provide structure and direction, thematic guides offer online access to the full texts of journal articles, book sections and chapters, papers from proceedings, governmental reports, maps and images, and other relevant materials. The human health guide includes climate change-related subsections: changes in the incidence of vector-borne diseases attributable to climate change, potential increases in mortality due to global warming, and health effects from increased exposure to ultraviolet-B radiation.

URL: http://www.wri.org/
World Resources Institute (WRI)
Founded in 1982 and based in Washington, D.C., WRI is an independent center for policy research and technical assistance on global environmental and development issues. The institute's particular concern is to build bridges between ideas and action, meshing the insights of scientific research, economic and institutional analyses, and practical experience with the need for open and participatory decision making. WRI publishes "World Resources: A Biennial Report on the Global Environment," a comprehensive data source combining a broad array of environmental, economic, and social data for 148 countries with supporting essays on key trends and timely issues. WRI policy initiatives include health and envi-

ronment indicators, climate protection, environmental health education, reducing pollution in developing countries, and environmental performance indicators. The site is updated daily and provides a comprehensive set of links to other organizations, electronic media, and data sources.

International Organizations

URL: http://www.un.org/
United Nations
The home page of this world organization lists the United Nations's five major concerns: peace and security, economic and social development, international law, human rights, and humanitarian affairs. Fourteen buttons serve as pathfinders of which the "UN around the World" is most helpful in finding specific U.N. programs or affiliations. The click on "Web Sites in the UN System" produces a world map and a listing of the acronyms of all U.N. programs and affiliations, whose geographic locations are indicated by connective lines to the map. Five of these programs were selected to be annotated for this collection (WHO, The World Bank Group [IBRD, MIGA], UNEP, FAO, WMO).

URL: http://www.who.int/
World Health Organization (WHO)
WHO was founded in 1948 at the peak of efforts, which had lasted more than 100 years, to combat worldwide health hazards like cholera, plague, smallpox, and yellow fever. A specialized agency of the United Nations with 191 member states, the World Health Organization leads the world alliance for Health for All. WHO has four main functions: to give worldwide guidance in the fields of health, to set global standards for health, to cooperate with governments in strengthening national health programs, and to develop and transfer appropriate health technology, information, and standards. WHO's mission is the attainment of the highest possible level of health by all peoples, which translates into more than a dozen different objectives and specialized functions. Health promotion and the environment is one objective, which has put WHO in research partnerships for gathering current data on conditions and needs, particularly in developing countries. Information is widely available and accessible through the pages on "Information Sources" and "Health-Related Sites Hosted by the World Health Organization." One of WHO's regional offices is singled out for separate annotation (see below).

URL: http://www.paho.org/
Pan American Health Organization (PAHO)
In the historical sequence of efforts culminating in the founding of the World Health Organization, the Pan American Sanitary Bureau—fore-

runner of today's PAHO—was established in 1902; it later became WHO's regional office for the Americas. In 1950, PAHO was recognized as a fully autonomous and specialized inter-American organization, thereby becoming a component of the United Nations and the inter-American systems. It is based in Washington, D.C. PAHO's mission statements coincide closely with those of the WHO but concentrate on the Americas. PAHO has documented the changes and advances in health achieved by the countries in the region for close to a century, thus providing invaluable information resources. PAHO's country profile database functions as the official regional information source on mortality in the Americas. PAHO issues a quadrennial publication, "Health in the Americas," as well as the annual series, "Health Statistics from the Americas."

URL: http://worldbank.org/
The World Bank Group
The World Bank's formal relationship with the United Nations is defined by a 1947 agreement that recognizes the bank as an independent specialized agency of the United Nations—as well as a member and observer in many U.N. bodies. United Nations/World Bank cooperation dates back to the founding days of the two organizations (1944 and 1945, respectively). The Global Environment Facility (GEF), launched in 1990, is jointly administered by the bank, the U.N. Development Programme (UNDP), and the U.N. Environment Programme (UNEP). GEF projects are administered in four program areas:

1. conserving biodiversity and improving forest, farmland, coastal, mountain, marine, and wildlife management
2. saving energy and promoting open markets for renewable energy
3. addressing the degradation of oceans, coastlines, lakes, wetlands, and rivers caused by the loss of habitats and pollution and stopping overfishing
4. assisting nations in eastern Europe and the Russian Federation in phasing out the use of chemicals that deplete the ozone layer

The Environmentally and Socially Sustainable Development (ESSD) Network enhances and maintains strategic alliances with key partners (governments, foundations, regional development banks, bilateral and multilateral agencies, U.N. agencies, nongovernmental organizations, and specific constituents) to advance environmentally and socially sustainable development in client countries. ESSD oversees the integration of economic, environmental, rural, and social criteria in World Bank–financed projects.

The bank's site map lists an index of all development topics and subtopics (e.g., environment or health, nutrition and population) that provide access to specialized information.

URL: http://www.unep.org/
United Nations Environment Programme (UNEP)
UNEP is a comprehensive, always currently updated site addressing far-reaching environmental issues, including human health and well-being. Global, national, and regional data on environmental issues are collected and documented. The Global Environmental Outlook (GEO) facilitates a comprehensive report on the state of the environment based on the input of 20 regional centers. The list of environmental issues is organized under 7 headings: access to environmental information, chemicals, environmental law and enforcement, forest fires, industry and environment, trade and environment, and youth and environment; together, these include 30 subheadings. Part of the UNEP is the United Nations System-wide Earthwatch (http://www.unep.ch/earthw.html).

URL: http://www.fao.org/
Food and Agriculture Organization of the United Nations (FAO)
FAO was founded in 1945 with a mandate to raise levels of nutrition and standards of living, improve agricultural productivity, and better the condition of rural populations. It is the largest autonomous agency within the U.N. system. The organization offers direct development assistance and acts as an international forum for debate on food and agriculture issues. FAO is active in land and water development, plant and animal production, forestry, fisheries, economic and social policy, investment, nutrition, food standards and commodities, and trade. A specific priority of FAO is sustainable development, a long-term strategy for the conservation and management of natural resources.

Within FAO's extensive website, three locations are of special interest in the context of this book's resource collection: Global Information and Early Warning System on Food and Agriculture (GIEWS), Emergency Prevention System for Transboundary Animal and Plant Pests and Diseases (EMPRES), and Sustainable Development (SD) Dimensions Specials. All three locations can be accessed from FAO's home page.

URL: http://www.wmo.ch/
World Meteorological Organization (WMO)
WMO is a U.N. specialized agency, founded in 1950. The organization features nine major programs: Applications of Meteorology Programme (AMP), Atmospheric Research and Environment Programme (AREP), Education and Training Programme (ETR), Global Climate Observing System (GCOS), Hydrology and Water Resources Programme (HWR), Technical Cooperation Programme (TCO), World Climate Programme (WCP), World Climate Research Programme (WCRP), and World Weather Watch (WWW). All nine programs provide vital information relating to

the topics of this book. WMO's programs are frequently cosponsored with other organizations, especially in regard to climate research, thereby providing natural links to other websites of this collection.

URL: http://neonet.nlr.nl/ceos-idn/campaigns/IGBP.html
International Geosphere-Biosphere Programme (IGBP)
IGBP is an interdisciplinary scientific activity established and sponsored by the International Council of Scientific Unions (ICSU) since 1986. The IGBP secretariat was established at the Royal Swedish Academy of Sciences in 1987. The program is focused on acquiring basic scientific knowledge and on the interactive processes of the biology and chemistry of the earth as they relate to global change. IGBP's 11 program elements are organized in 8 broadly discipline-oriented core projects, covering such topics as atmospheric science, terrestrial ecology, oceanography, hydrology, and links between the natural and social sciences. The site is an umbrella for detailed sites on the 11 program elements, each of which features selective links to specifically pertinent sites within the program. The following sites have been selected from the 11 program elements.

URL: http://neonet.nlr.nl/ceos-idn/datacenters/LUCC.html
Land Use and Land Cover Change (LUCC), IGBP
LUCC is a central program element and core project of IGBP, cosponsored by the International Council of Scientific Unions (ICSU), the International Human Dimensions Programme on Global Environmental Change (IHDP), and the International Social Science Council (ISSC). LUCC is an interdisciplinary program aimed at improving understanding of land use and land cover change dynamics and their relationships with global environmental change. One LUCC project is in southeast Asia, with teams in Malaysia, the Philippines, Indonesia, and Thailand. Research assistance is provided by IGBP and START (see below). The main objective of this project is to gain better understanding of the driving forces underlying forest conversions in the region, which will link to an overall scheme on climatic change. Enlargement of the project's scope in the future will include investigation of how land cover changes affect human activities. The site is often accessed more easily via http://www.start.or.th/LUCC/.

URL: http://neonet.nlr.nl/ceos-idn/campaigns/START.html
Global Change System for Analysis, Research and Training,
IGBP (START)
START is a joint project of the International Geosphere-Biosphere Program (IGBP), the International Human Dimensions Program (IHDP), and the World Climate Research Program (WCRP), a U.N. affiliate, developing the concept of a global system of regional networks of institutions. START's mission is

— to develop a system of regional networks of collaborating scientists and institutions,

— to conduct research on regional aspects of global change,

— to assess the causes and effects of global regional change, and

— to provide relevant information to policymakers and governments.

START has established Regional Research Networks (RRN) with affiliated Regional Research Sites (RRS) and at least one Regional Research Center (RRC).

URL: http://neonet.nlr.nl/ceos-idn/campaigns/IHDP.html
International Human Dimensions Programme on Global Environmental Change (IHDP)
IHDP is an international, interdisciplinary, nongovernmental, social science program dedicated to promoting and coordinating research aimed at describing, analyzing, and understanding the human dimensions of global change. IHDP links researchers, policymakers, and stakeholders by identifying new research priorities and facilitating the dissemination of research results. IHDP emphasizes four primary projects: land use and land cover change, global environment and human security, institutional dimensions of global change, and industrial transformation. It is involved in other IGBP core projects. The site also provides access to online working papers, program updates, upcoming conferences, and other links.

Focus on Linkages

URL: http://www.igbp.kva.se/intergov.html
Intergovernmental Organizations
This site is part of the IGBP address. It lists well-developed links between IGBP and U.N. bodies on global change issues. The U.S. Global Change Research Information Office is listed with the notation of "offering many links to related web pages."

URL: http://www.gcrio.org/
U.S. Global Change Research Information Office (GCRIO)
GCRIO provides access to data and information on global change research, adaptation/mitigation strategies and technologies, and global change–related educational resources on behalf of the U.S. Global Change Research Program (USGCRP). GCRIO's global change resources include the "Compendium of Global Change Information Resources."

URL: http://www.usgcrp.gov/usgcrp/GCRPINFO.html
U.S. Global Change Research Program (USGCRP)
The USGCRP was formalized in 1990 by the Global Change Research Act of 1990. Its mission and goals focus on the scientific study and compre-

hensive investigation of Earth system processes and their interactions, with the participation of an extensive community of international scientists from a wide range of scientific disciplines. USGCRP research is organized around a framework of observing, documenting, understanding, and predicting global change, and its activities are coordinated with other related national and international research programs. Included in the program is the development of tools and capabilities to conduct integrated assessments to synthesize and communicate the findings of this research.

URL: http://www.gcdis.usgcrp.gov/
Gateway to Global Change Data
This site is one of the tools developed by the U.S. Global Change Research Program (USGCRP) and Global Change Data and Information System (GCDIS) to communicate global change data. The site is linked to pertinent programs and research conducted in other U.S. agencies and disseminates the latest data and abstracts on new publications.

Reinforcement of Interdisciplinary Perspectives

URL: http://sdgateway.net/noframe/events/en_21980.htm
SD Gateway
SD Gateway integrates the online information developed by members of the Sustainable Development Communications Network (see IISD under "Directories," below). Information is indexed alphabetically and by concept. Its services include the "SD Primer," with more than 1,600 links, listings of mailing lists, sustainable development news sites, library collections, and the calendar of events. The latter features past and future events from May 1997 through the year 2000, thus providing an excellent overview of current scientific information addressing sustainable development. Future services will include collaborative work on sustainable livelihoods, environmental law, public participation, sustainable cities, and water.

URL: http://www.fni.no/
The Fridtjof Nansen Institute (FNI)
FNI is an independent foundation engaged in applied social science research on international issues concerning energy, resource management, and the environment. The academic approach is multidisciplinary and collaborative with other research institutions in Norway and abroad. FNI carries out its research within several programs and activities, on which the "Yearbook of International Co-operation on Environment and Development" provides extensive information. It combines independent, high-quality analysis and updated reference material. The "Country Profiles" section of the yearbook, especially, features comprehensive data on environmental research.

URL: http://www.iiasa.ac.at/
The International Institute for Applied Systems Analysis (IIASA)
IIASA is a nongovernmental research organization located in Austria and is sponsored by scientific "national member organizations" in nations of Europe, North America, and Asia. The institute conducts interdisciplinary scientific studies on environmental, economic, technological, and social issues in the context of human dimensions of global change. It has been the site of successful international scientific collaboration addressing such areas of concern as energy, water, environment, risk management, and human settlement. In addition to the research plans, the institute offers the Young Scientists Summer Program and the Young Postdoctoral Fellows Program.

URL: http://www.ncar.ucar.edu/
The National Center for Atmospheric Research (NCAR)
NCAR carries out important research activities in several scientific divisions and programs—working with member universities on a broad range of investigations to better understand our earth's climate systems. Among the nine programs, the Environmental and Societal Impacts Group (ESIG) is mostly concerned with environmental change and the effects associated with the complex relationship of the atmosphere, environment, and society. ESIG is singled out for separate annotation below.

URL: http://www.esig.ucar.edu/
The Environmental and Societal Impacts Group (ESIG)
ESIG provides extensive information and links to related sites in its quest to document various effects of environmental change. One of its outstanding services is the listing of publications. ESIG offers a comprehensive bibliography of its staff publications by year, selected publication abstracts, and online newsletters, journals, and web-only publications.

Topically Arranged Websites

The following addresses are samples of smaller sites with a stronger focus on a particular theme or regional and local concerns. The updating of these sites might not always be dependable, but they will serve as excellent information sources for research inquiries, providing specifics for cross-cutting analyses. A list of the chapters in this book that include URL references is added for easier access in topically designed research.

Atmospheric Changes—Climate

URL: http://www.pik-potsdam.de/
Potsdam Institute for Climate Impact Research (PIK)

URL: http://www.igc.apc.org/climate/Eco.html
The Climate Action Network Newsletter

URL: http://www.me3.org/issues/climate/reports.html
Sustainable Minnesota's Climate Change

Land Use—Agriculture

URL: http://www.mluri.sari.ac.uk/
Macauly Land Use Research Institute

URL: http://www.crle.uoguelph.ca/iale/
The International Association for Landscape Ecology

URL: http://www.ecostudies.org/
The Institute of Ecosystem Studies

Water

URL: http://www.den.doi.gov/wwprac/
Western Water Policy Review Advisory Commission

URL: http://www.hydroweb.com/iaeh.html
International Association for Environmental Hydrology

URL: http://www.irn.org/
International Rivers Network

Population—Urbanization-Migration

URL: http://www.cnie.org/pop/urban.htm
The National Library for the Environment

URL: http://www.state.gov/www/global/prm/
U.S. State Department-Bureau of Population, Refugees, and Migration

URL: http://opr.princeton.edu/
Office of Population Research-Princeton University

Environmental Policies

URL: http://www.ifc.org/enviro/
IFC Environment Division

URL: http://www.econ.ag.gov/briefing/con_env/
Domestic Conservation and Environmental Policies Briefing Room

URL: http://www.heartland.org/
The Heartland Institute

Chapters Listing URLs

— Chapter 1: Information on Issues of Global Change
— Chapter 2: Epidemiological Study Designs
— Chapter 3: Geographic Information Systems
— Chapter 4: The Science/Policy Interface
— Chapter 5: Integrated Assessment
— Chapter 6: Human Populations in the Shared Environment
— Chapter 7: The Changing Chemistry of Earth's Atmosphere
— Chapter 8: An Earth Science Perspective on Global Change
— Chapter 9: Water Resources Management
— Chapter 12: Malaria and Global Ecosystem Change
— Chapter 13: Global Climate Change and Air Pollution: Interactions and Their Effects on Human Health
— Chapter 14: Too Little, Too Much: How the Quantity of Water Affects Human Health

Online Directories and Library Services

The main criteria for listing an information source as a directory are the volume and comprehensiveness of the information and the presentational format, that is, compilations of lists and information in digest form leading to full texts and cross-references. Annotations will give both directories and library services more specific qualifications for the user.

Directories

URL: http://gcmd.gsfc.nasa.gov
Global Change Master Directory
A comprehensive, always updated directory of information about earth science data, including broad coverage of the oceans, atmosphere, hydrosphere, solid earth, biosphere, and human dimensions of global change.

URL: http://neonet.nlr.nl/directory/index.html
NEONET Directory

— Atmosphere: Atmospheric institutes and meteorological information. Special topic: Atmospheric chemistry.
— Land: Agriculture, land use, forestry, solid Earth (geodesy).
— People: Who is in remote sensing?
— Water: Water management, coastal zones, research on oceans and climate change.
— Education and Training: Universities and courses on remote sensing.
— Library: Online periodicals and publications, access to remote sensing libraries.

— Organizations: Related remote sensing organizations, European projects (Committee on Earth Observation Satellites—CEOS) and space agencies.
— Providers: Links to NEONET providers of information on remote sensing.
— Technology: About sensors and platforms.

URL: http://neonet.nlr.nl/ceos-idn/
Committee on Earth Observation Satellites—International Directory Network (CEOS-IDN)
This is the Dutch operating node of the CEOS-IDN. For further information refer to the European Coordinating Node at the European Space Agency's Earth Observation Informatics Services (ESA/ESRIN) or to NASA's Global Change Master Directory. CEOS-IDN contains metadata references to datasets, datacenters, sensors, projects, and satellites. The IDN can be searched by performing a keyword search, by browsing through the page lists, or by browsing through the index tree.
Page lists

— Campaigns (121 entries, 105 new or modified in last update)
— Datacenters (150 entries, 148 new or modified in last update)
— Datasets (7,803 entries, 7,803 new or modified in last update)
— Sensors (114 entries, 30 new or modified in last update)
— Sources (183 entries, 13 new or modified in last update)

URL: http://www.globalchange.org/
Global Change—Electronic Edition
"Global Change" is published by the Pacific Institute for Studies in Development, Environment, and Security. The electronic edition provides extensive current listings of global change information, links to other online resources, and access to Global Change Archives and the Global Change Digest Archives. Although the primary focus is on climate and ozone depletion, this source presents a wide spectrum of information on forces contributing to global change.

URL: http://www.ncdc.noaa.gov/
National Climatic Data Center (NCDC)
NCDC is the world's largest active archive of weather data with multiple feed-in sources. It is part of the U.S. Department of Commerce National Oceanic and Atmospheric Administration (NOAA)'s National Environmental Satellite, Data, and Information Service (NESDIS). NCDC's mission is global in nature and promotes global environmental stewardship by describing, monitoring, and assessing the climate and supporting efforts to predict changes in the earth's environment.

URL: http://iisd.ca/
The International Institute for Sustainable Development (IISD)
IISD is located in Canada. IISDnet, the home page of IISD, is a state-of-the-art Internet information server. Users will find the most up-to-date and extensive resource information on sustainable development, research, new trends, global activities, and contacts. It features several information sites (directories) within the institute and is linked to sustainable development research centers around the world. Two IISD directories are annotated below.

URL: http://www.iisd.ca/linkages/
Linkages
Linkages is a multimedia resource for environment and development policymakers. It includes up-to-date information on past, present, and upcoming U.N. negotiations pertaining to sustainable development, hypertext index to all issues of the "Earth Negotiations Bulletin," links to the official U.N. documents supporting each of the negotiations, and full text versions of selected interventions and background documents.

URL: http://iisd1.iisd.ca/ic/
A list of key international sustainable development organizations and institutes
This list and other information sources from the IISD Sourcebook and Publication Catalogue can also be accessed through the IISD home page by clicking "Information Centre."

URL: http://gssd.mit.edu/Gssd/gssd.nsf?Open
The Global System for Sustainable Development at MIT—Knowledge Meta-Networking for Decision & Strategy (GSSD)
GSSD is a project of the Global Accords Consortium for Sustainable Development housed at the Massachusetts Institute of Technology (MIT). It consists of an evolving, quality-controlled, and highly cross-referenced index to some of the best resources and materials on sustainability to be found on the Internet. A set of knowledge management, search, and navigation tools allows users to customize their own site inputs into the system or tailor specific retrieval queries over the GSSD knowledge base, which consists of more than 2,500 abstracted, indexed, and cross-referenced sites. These sites are drawn from more than 250 institutions worldwide. Users accessing the GSSD knowledge base are assisted by a guide to research outlines of 14 core concepts and a sophisticated threefold search mechanism.

URL: http://www.cnie.org/
The National Council for Science and the Environment (CNIE)
The CNIE introduces its services as "The National Library for the Environment." However, CNIE's information dissemination includes topics be-

yond library services and takes on the broader scope of a major directory. A major asset is the inclusion of all Congressional Research Service (CRS) reports on environmental issues, as well as reference resources generated by CRS.

Library Services

URL: http://www.nap.edu/
National Academy Press
This online reading room enables users to read more than 1,500 books online. Titles can be accessed via a category browser.

URL: http://www.nlm.nih.gov/
National Library of Medicine (NLM)
NLM has the world's largest selection of medical literature and documents. Access through the main address offers overviews to health information and library services, including specialized services (SIS) and research programs, as well as a link to MEDLINE.

URL: http://www.ncbi.nlm.nih.gov/PubMed/
PubMed
PubMed is NLM's search service to access the nine million citations in MEDLINE and Pre-MEDLINE (with links to participating journals) and in other related databases.

URL: http://www.idealibrary.com
International Digital Electronic Access Library (IDEAL)
IDEAL is an online library, specializing in professional journals. It currently offers full-text electronic access to nearly 250 Academic Press, W. B. Saunders, and Churchill Livingstone journals.

Abbreviations

AAOE	Antarctic Atmospheric Ozone Experiment
ADEOS	Advanced Earth Observing System (Japan)
Ae. aegypti	*Aedes aegypti*
AIDS	acquired immune deficiency syndrome
An. albimanus	*Anopheles albimanus*
An. arabiensis	*Anopheles arabiensis*
An. funestus	*Anopheles funestus*
An. gambiae	*Anopheles gambiae*
An. melas	*Anopheles melas*
An. quadrimaculatus	*Anopheles quadrimaculatus*
atm	atmosphere (unit of pressure)
ATSDR	Agency for Toxic Substances and Disease Registry
AVHRR	advanced very high resolution radiometer
AWMA	Air and Waste Management Association
BAD	biologically accumulated dosage
BCC	basal cell carcinoma
CARB	California Air Resources Board
CDC	Centers for Disease Control and Prevention (see HHS)
CDR	crude death rate
CEC	Commission for Environmental Cooperation
CEOS	Committee on Earth Observation Satellites
CEOS-IDN	Committee on Earth Observation Satellites–International Directory Network
CFC	chlorofluorocarbon
CH_4	methane
CHAART	Center for Health Applications of Aerospace Related Technologies (see NASA)
CIESIN	Center for International Earth Science Information Network
CIIT	Chemical Industry Institute of Toxicology
CMR	crude mortality rate
CNIE	National Council for Science and the Environment
CO	carbon monoxide
CO_2	carbon dioxide
CRS	Congressional Research Service
Cx. pipiens	*Culex pipiens*

DEM	digital elevation model
DLG	digital line graph
DNA	deoxyribonucleic acid
DOQ	digital orthophoto quadrangle
dT	diurnal temperature difference
DU	Dobson unit
EIA	Energy Information Administration
EKMA	empirical kinetic modeling approach
ENSO	El Niño/Southern Oscillation
EOS	Earth Observing System (see NASA)
EPA	Environmental Protection Agency
ESA	European Space Agency
ESIG	Environmental and Social Impacts Group (see NCAR)
ESSD	Environmentally and Socially Sustainable Development
EUMETSAT	European Meteorological Satellite organizations
EUROGI	European Umbrella Organisation for Geographical Information
FAO	Food and Agriculture Organization (see U.N.)
FCCC	Framework Convention on Climate Change (see U.N.)
FGDC	Federal Geographic Data Committee
FNI	Fridtjof Nansen Institute
GCDIS	Global Change Data and Information System
GCM	global circulation model
GCRIO	Global Change Research Information Office
GDP	gross domestic product
GEF	Global Environment Facility
GIS	geographic information system
GOME	Global Ozone Monitoring Experiment (see ESA)
GPCP	Global Precipitation Climatology Project
GPS	global positioning system
GSFC	Goddard Space Flight Center (see NASA)
GSSD	Global System for Sustainable Development
GWP	global warming potential
ha	hectare
HCFC	hydrogenated chlorofluorocarbon
β-HCH	β-hexachlorocyclohexane
HFRS	hemorrhagic fever with renal syndrome
HGE	human granulocytic ehrlichiosis
HHS	Department of Health and Human Services

HIV	human immunodeficiency virus
H_2O	water
HPS	hantavirus pulmonary syndrome
hr	hour
IAI	Inter-American Institute for Global Change Research
IAM	integrated assessment model
ICDDRB	International Centre for Diarrhoeal Disease Research, Bangladesh
ICSU	International Council of Scientific Unions
IDEAL	International Digital Electronic Access Library
IGBP	International Geosphere-Biosphere Program
IHDP	International Human Dimensions Programme on Global Environmental Change
IIASA	International Institute for Applied Systems Analysis
IISD	International Institute for Sustainable Development
IMAGE	Integrated Model to Assess the Greenhouse Effect
IMR	infant mortality rate
IPCC	Intergovernmental Panel on Climate Change
IPHECA	International Program on the Health Effects of the Chernobyl Accident
IQ	intelligence quotient
IR	infrared
ISSC	International Social Science Council
IUCN	World Conservation Union
km	kilometer
LPS	lipopolysaccharide
LUCC	land use and land cover change
mg	milligram
MIR	mid infrared
mm	millimeter
MMTCE	million metric tons of carbon equivalents
MMWR	*Morbidity and Mortality Weekly Report*
MODIS	moderate resolution imaging spectroradiometer
mPa	millipascal
MSF	Médecins Sans Frontières [Doctors Without Borders]
MTBE	methyl tertiary butyl ether
NAAQS	National Ambient Air Quality Standards
NADP	National Atmospheric Deposition Program
NAP	National Academy Press

NAPAP	National Acid Precipitation Assessment Program
NASA	National Aeronautics and Space Administration
NCAR	National Center for Atmospheric Research
NCEH	National Center for Environmental Health (see CDC)
NCDC	National Climatic Data Center (see NOAA)
NCHS	National Center for Health Statistics (see CDC)
NCI	National Cancer Institute (see NIH)
NCID	National Center for Infectious Diseases (see CDC)
NDSC	Network for the Detection of Stratospheric Change
NE	nephropathia epidemica
NGO	nongovernmental organization
NHANES	National Health and Nutrition Examination Survey
NIH	National Institutes of Health
NIR	near infrared
NLM	National Library of Medicine
NMHC	nonmethane hydrocarbon
NO	nitric oxide
N_2O	nitrous oxide
NO_2	nitrogen dioxide
NO_x	nitrogen oxides
NOAA	National Oceanic and Atmospheric Administration
nm	nanometer
NPP	net primary productivity
NRC	National Research Council
NSDI	National Spatial Data Infrastructure
NSF	National Science Foundation
O_2	molecular oxygen
O_3	ozone
OECD	Organization for Economic Cooperation and Development
OGC	Open GIS Consortium
OR	odds ratio
ORS	oral rehydration salts
OTAG	Ozone Transport Assessment Group
P. falciparum	*Plasmodium falciparum*
P. malariae	*Plasmodium malariae*
PAH	polycyclic aromatic hydrocarbons
PAHO	Pan American Health Organization (see WHO)
PAN	peroxyacetyl nitrate
Pb	lead

PCB	polychlorinated biphenyl
PCR	polymerase chain reaction
PFC	perfluorinated compound
Pg	petagram
pH	potential of hydrogen
pK_a	acid dissociation constant
PM	particulate matter
PM_{10}	particulates smaller than 10 microns
$PM_{2.5}$	particulates smaller than 2.5 microns
POP	persistent organic pollutant
ppb	parts per billion
ppbv	parts per billion by volume
ppm	parts per million
ppmC	parts per million of carbon
ppmv	parts per million by volume
PSC	polar stratospheric cloud
R_0 (sometimes R)	basic reproduction ratio
RR	relative risk
SAMC	Southern Africa Malaria Control
SeaWiFS	Sea-viewing Wide Field-of-View Sensor
SEDAC	Socioeconomic Data and Applications Center
SEM	scanning electron microscopy
SO_2	sulfur dioxide
SOI	Southern Oscillation Index
SPOT	Système probatoire d'observation de la terre [System for Observation of the Earth]
spp.	species (plural)
SSH	sea surface height
SST	sea surface temperature
SSTA	sea surface temperature anomaly
START	Global Change System for Analysis, Research and Training (see IGBP)
STD	sexually transmitted disease
TIGER	Topologically Integrated Geographic Encoding and Referencing System
TIN	triangulated irregular network
TIR	thermal infrared
TNRCC	Texas Natural Resource Conservation Commission
TOMS	Total Ozone Mapping Spectrometer
TSP	total suspended particulates

U.N.	United Nations
UNCHS	United Nations Centre for Human Settlements (Habitat)
UNDP	United Nations Development Programme
UNEP	United Nations Environment Programme
UNESCO	United Nations Educational, Scientific and Cultural Organization
UNHCR	United Nations High Commissioner for Refugees
UNICEF	United Nations Children's Fund
URL	uniform resource locator
USGCRP	U.S. Global Change Research Program
USGS	U.S. Geological Survey
UTM	Universal Transverse Mercator
UV	ultraviolet
UV-A	ultraviolet-A
UV-B	ultraviolet-B
UV-C	ultraviolet-C
V. cholerae	*Vibrio cholerae*
VC	vectorial capacity
VMT	vehicle miles traveled
VOC	volatile organic compound
WCRP	World Climate Research Program
WHO	World Health Organization (see U.N.)
Wm^{-2}	Watts per square meter
WMO	World Meteorological Organization (see U.N.)
WRI	World Resources Institute

Glossary

Joan L. Aron, Ph.D., and Mark L. Wilson, Sc.D.

Accelerated global warming: *See* enhanced global warming.

Aedes aegypti: Species of mosquito that is a vector of dengue virus. *See* vector.

Air toxic: Any air pollutant, usually a carcinogen or an irritant, whose emissions are regulated according to guidelines in the U.S. Clean Air Act that are different from the guidelines affecting criteria air pollutants. Examples are benzene and mercury. Air toxics may also be called hazardous air pollutants. *See* Clean Air Act, criteria air pollutant.

Algae: Plant found mostly in aquatic environments, having no true root or stem. *See* phytoplankton.

Algal bloom: Rapid growth and increase of algae. *See* algae, phytoplankton bloom.

Anopheles: Genus of mosquitoes containing species that are vectors of malaria. *See* anopheline, vector.

Anopheline: A mosquito of the genus *Anopheles,* usually in reference to a species that can be a vector of malaria parasites. *See* vector.

Anthroponosis: Disease of humans caused by an infectious agent that normally circulates among humans and for which humans are the natural reservoir. *See* zoonosis, reservoir.

Bacteriophage: Virus whose host is a bacterial cell.

Basic reproduction ratio: A theoretical value representing the average number of new infections that arise during the period of infectiousness of a single infectious individual who has entered a population of totally susceptible people.

Biome: Regional ecosystem defined by biogeographic and climatic characters. Tropical rain forest and desert are examples of terrestrial biomes (Allaby 1998).

Biosphere: The part of the earth's environment where living organisms are found and interact. The biosphere may also be termed the ecosphere.

Burden of disease: Mortality, morbidity, and disability in a population caused by a disease. *See* mortality rate, morbidity, disability.

Carbon cycle: Movement of carbon on a global scale through oceans, atmosphere, and biological organisms.

Chromophore: Molecule that absorbs solar radiation.

Clean Air Act: Legislation passed by the U.S. Congress to control emissions of air pollutants and improve air quality in the United States. *See* criteria air pollutant, air toxic.

Cold episode: *See* La Niña.

Confidence interval: A numeric interval, or range, that has a probability (usually 95%) of containing the true value of the quantity being estimated.

Criteria air pollutant: Any of six common air pollutants—small particles, carbon monoxide, sulfur dioxide, nitrogen dioxide, ground-level ozone, and lead—whose emissions are regulated according to guidelines in the Clean Air Act that require the preparation of a "criteria document" by the U.S. Environmental Protection Agency. In the case of ground-level ozone, the regulation of emissions applies to volatile organic compounds, which are precursors to the formation of ozone. *See* Clean Air Act, air toxic.

Dimer: A molecule formed by the combination of two smaller, identical molecules.

Disability: A restriction of the ability to perform an activity considered normal. *See* morbidity, disease.

Disease: Recognized symptoms of ill health. An infectious disease is a result of an infectious microorganism present in or on the body of the host. *See* infection.

Dose: Quantity of material entering an exposed person. Dose is not the same as exposure. *See* exposure.

El Niño: A shift in climatic conditions worldwide associated with a warming of the central and eastern equatorial Pacific Ocean, which typically occurs every two to seven years. *See* El Niño/Southern Oscillation, La Niña.

El Niño/Southern Oscillation: Climatic variability that produces warming and cooling of the ocean and an oscillation in atmospheric pressure in the equatorial Pacific region. *See* El Niño, La Niña, Southern Oscillation Index.

Endemic transmission: Relatively stable transmission of an infectious agent in a human population. *See* epidemic transmission, enzootic transmission.

Enhanced global warming: Warming of the earth caused by anthropogenic emissions of greenhouse gases. *See* greenhouse gas.

Enzootic transmission: Relatively stable transmission of an infectious agent in an animal population. Enzootic transmission is analogous to endemic transmission in humans. *See* epizootic transmission, endemic transmission.

Epidemic transmission: Transmission of an infectious agent in a human population in excess of that normally observed in a region during a particular period. *See* endemic transmission, epizootic transmission.

Epizootic transmission: Transmission of an infectious agent in an animal population in excess of that normally observed in a region during a particular period. Epizootic transmission is analogous to epidemic transmission in humans. *See* enzootic transmission, epidemic transmission.

Evapotranspiration: Combination of evaporation and transpiration, two processes that release water into the atmosphere as water vapor.

Exposure: Contact with a material at a potential portal of entry into the body—the skin, the respiratory tract, or the gastrointestinal tract. Exposure is not the same as dose. *See* dose.

Extrinsic incubation period: The development time of an infectious microorganism in a vector, which is the time interval between the acquisition of the microorganism and the attainment of the capacity to transmit the microorganism. *See* vector.

Fossil fuel: Coal, oil, and natural gas, which are carbon-based sources of energy derived from the fossilized remains of dead plants and animals.

Framework convention: Treaty in international diplomacy under which participating nations agree to work with one another and take action as necessary in an open-ended arrangement. In contrast, most treaties commit participating nations to a specific set of rules.

Framework Convention on Climate Change: International agreement for nations to work together to avoid damage to the global environment caused by enhanced global warming. *See* framework convention, global climate change, enhanced global warming.

Genetic drift: Random fluctuation of gene frequencies in populations because of chance events, particularly evident in very small and isolated populations (Allaby 1998). *See* natural selection.

Global climate change: Natural or anthropogenic changes to the climate of the earth. Global climate change is often used loosely to mean the consequences of enhanced global warming. *See* enhanced global warming.

Global warming: Warming of the earth that may occur due to natural or anthropogenic forces. Global warming is often used loosely to mean enhanced global warming. *See* enhanced global warming.

Global warming potential: A measure of the warming potential of a greenhouse gas over a period of time relative to the same weight of carbon dioxide. A time period of 100 years is typically used in reports directed at policymakers. *See* greenhouse gas, million metric tons of carbon equivalents.

Greenhouse effect: Warming of the atmosphere induced by greenhouse gases. *See* greenhouse gas.

Greenhouse gas: Gas that absorbs infrared radiation emitted from the earth and warms the atmosphere. *See* enhanced global warming.

Ground-level ozone: Ozone that is in the lower portions of the atmosphere near the surface of the earth. *See* tropospheric ozone, criteria air pollutant.

Hazard: An environmental exposure associated with increased risk of an adverse health outcome. The term is used in the context of environmental impact assessment.

Herd immunity: The proportion of a population that is immune to an infectious agent, or the prevalence of immunity among members of a population. Sometimes, the term *herd immunity* applies specifically to a population whose proportion immune exceeds a particular herd immunity threshold. *See* herd immunity threshold.

Herd immunity threshold: The minimum level of herd immunity that prevents sustained transmission of an infectious agent in a population. The value of the threshold depends on characteristics of the infectious agent, characteristics of the host population, and environmental conditions. *See* herd immunity.

Host: Any animal or plant that, under natural conditions, provides sustenance or shelter to an infectious agent. For parasites, primary or definitive hosts are those where sexual reproduction normally occurs, and all other hosts are considered secondary or intermediate.

Hydrological cycle: Movement of water molecules on a global scale through oceans, atmosphere, surface water, and ground water.

Incidence rate: Ratio of the number of new events during a defined time period to the population at risk of experiencing the event (Last 1995). Incidence is often used loosely to mean incidence rate. *See* mortality rate, prevalence rate.

Infection: The colonization and replication of a parasitic microorganism that gains sustenance or shelter from the body of a host whether or not that host experiences ill health. Infection is not the same as disease. *See* disease.

Killed vaccine: A vaccine prepared from a killed microorganism that can no longer reproduce but still produces an immune response. *See* live attenuated vaccine.

La Niña: A shift in climatic conditions worldwide associated with a cooling of the central and eastern equatorial Pacific Ocean. *See* El Niño.

Leptospirosis: An infectious disease caused by bacteria in the genus *Leptospira*, usually found in nonhuman animals such as the urban rat.

Live attenuated vaccine: A vaccine made from living, reproducing microorganisms that have been selected to produce milder effects than the natural (wild) strain. *See* killed vaccine.

Metric ton: 10^6 grams, or approximately 2,205 pounds.

Million metric tons of carbon equivalents: Million metric tons of a greenhouse gas multiplied by its global warming potential and by 12/44, which is the fraction by weight of carbon in carbon dioxide. The global warming potential is calculated using a period of 100 years. *See* greenhouse gas, global warming potential.

Morbidity: Any departure, subjective or objective, from a state of physiological or psychological well-being (Last 1995). *See* disease, disability, mortality rate, burden of disease.

Mortality rate: Ratio of the number of deaths during a defined time period to the number of persons at risk of dying during the period (Last 1995). A mortality rate is in effect an incidence rate for which deaths are counted instead of new cases of a disease. *See* incidence rate.

Multiobjective: Handling multiple objectives, applied to a system for decision analysis.

Mycobacterium tuberculosis: Species of bacterium that causes human tuberculosis.

Natural selection: A complex mechanism by which the genetic structure of a population changes over time because of survival of particular members of that population in different environments (Allaby 1998). *See* genetic drift.

Net primary productivity: Amount of plant growth produced by photosynthesis per area per unit of time.

Oral vaccine: A vaccine administered by mouth. *See* parenteral vaccine.

Ozone: A reactive gas composed of three atoms of oxygen (O_3), which is both harmful in the troposphere (increasing respiratory illness) and beneficial in the stratosphere (filtering some ultraviolet radiation). *See* stratospheric ozone, tropospheric ozone.

Parenteral vaccine: A vaccine administered by injection. *See* oral vaccine.

Phytoplankton: Aquatic plant life that is free-floating and usually microscopic. *See* zooplankton.

Phytoplankton bloom: Rapid growth and increase of phytoplankton. *See* algal bloom.

Plankton: Aquatic biological organisms that are free-floating and usually microscopic. *See* phytoplankton, zooplankton.

Plasmodium: Genus of malaria parasites. Four species of malaria parasites infect humans: *Plasmodium falciparum, Plasmodium vivax, Plasmodium malariae, Plasmodium ovale.*

Plasmodium falciparum: One species of malaria parasites, which causes the most severe clinical illness in humans.

Population attributable risk: Proportion of outcomes (cases of disease) attributable to a specific exposure in a population of exposed and unexposed people.

Precautionary principle: A rationale for taking action to avoid catastrophic damage.

Prevalence rate: Ratio of the total number of individuals who have an attribute or disease at a particular time to the population at risk of having the attribute or disease at that point in time (Last 1995).

Radical: Chemical species that contains an odd number of electrons and is thus especially reactive.

Red Book: An influential report entitled "Risk Assessment in the Federal Government: Managing the Process," which was produced by the U.S. National Research Council in 1983.

Remote sensing: Use of electromagnetic radiation sensors to record attributes of the environment (Patterson 1998).

Reservoir: The host (vertebrate, invertebrate) or substance (water, soil) in which a parasitic organism normally resides and which is typically requisite for parasite replication or development. For most human infections, humans or other vertebrate species serve as the reservoir.

Risk: Probability that an event will occur in a specified period of time. The term *risk* is also used loosely as a synonym for risk factor and hazard. *See* uncertainty, surprise.

Risk assessment: The use of the factual base to define the health effects of exposure of individuals or populations to hazardous materials and situations. Risk assessment has qualitative and quantitative aspects.

Risk factor: An attribute or exposure associated with increased risk of an adverse health outcome.

Schistosoma: Genus of parasitic worm that causes schistosomiasis.

Sea surface temperature: Temperature of the sea surface. *See* sea surface temperature anomaly.

Sea surface temperature anomaly: Difference between sea surface temperature and average of sea surface temperature over many years, usually calculated in a particular region. *See* sea surface temperature.

Southern Oscillation Index: Difference between measurements of sea-level atmospheric pressure in Tahiti and Darwin, Australia (Tahiti minus Darwin). El Niño is related to the negative phase of the index. *See* El Niño, El Niño/Southern Oscillation.

Steady state: A configuration of a dynamic system in which all of the variables remain unchanged over time.

Stratosphere: Atmospheric shell around the earth between the troposphere and mesosphere, in which temperature tends to increase with altitude due to the absorption of solar ultraviolet radiation by ozone. The stratosphere typically extends from altitudes of about 7–15 kilometers at the tropopause up to about 55 kilometers at the stratopause (top of the stratosphere, at which temperatures reach a local maximum). *See* stratospheric ozone, troposphere, and tropopause.

Stratospheric ozone: Ozone in the stratosphere. Stratospheric ozone prevents harmful ultraviolet light from reaching the earth. *See* stratosphere and tropospheric ozone.

Surprise: Situation in which events or outcomes are very different from prior expectations. *See* risk and uncertainty.

Thermocline: Zone in a lake or an ocean where the temperature changes rapidly from warm to cool with increasing depth.

Transovarial transmission: A mode of transmission of an infectious agent from a female arthropod vector to her offspring via infection of her eggs.

Tropopause: Boundary between the troposphere and stratosphere at which temperatures in the lower atmosphere reach a minimum. The tropopause typically occurs at an altitude of about 7–15 kilometers, being higher at the equator and lower at the poles. *See* troposphere and stratosphere.

Troposphere: Lowest atmospheric shell around the earth, extending from the surface of the earth up to the tropopause at altitudes of about 7–15 kilometers. In the troposphere, temperature tends to decrease with altitude. *See* tropospheric ozone, tropopause, and stratosphere.

Tropospheric ozone: Ozone in the troposphere. Excessive tropospheric ozone is harmful to human health and the environment. *See* ground-level ozone, troposphere, criteria air pollutant, stratospheric ozone.

Uncertainty: Situation in which the risk of a specified event or outcome is unknown. *See* risk, surprise.

Vaccine: Biological preparation administered to a person to induce immunity to an infectious agent. *See* oral vaccine, parenteral vaccine, killed vaccine, live attenuated vaccine.

Vector: The organism, usually an invertebrate (insect, tick, or snail, for example), that carries and transmits a pathogen to a host, usually through a bite or environmental or food contamination.

Vector competence: The inherent, species-specific ability of a vector (most often applied to a blood-sucking arthropod) to support the development and transfer of an infectious microbe.

Vectorial capacity: Potential number of parasite inoculations into the human population by the vector population that are produced from an individual infective human host per day of infectivity.

Vibrio: Genus of bacterium containing species that cause cholera and some other infectious diseases.

Vibrio cholerae: Species of bacterium that causes cholera.

Warm episode: *See* El Niño.

Water cycle: *See* hydrological cycle.

Zoonosis: Disease of humans caused by an infectious agent that normally circulates among nonhuman animal reservoirs. *See* anthroponosis, reservoir.

Zooplankton: Aquatic animal life that is free-floating and usually microscopic. Copepods are a major component. *See* phytoplankton.

Zooplankton bloom: Rapid growth and increase of zooplankton. *See* phytoplankton bloom.

References

Allaby M. 1998. *A Dictionary of Ecology.* Oxford University Press, New York.

Last JM. 1995. *A Dictionary of Epidemiology,* 3d ed. Oxford University Press, New York.

Patterson M. 1998. Glossary. In *People and Pixels: Linking Remote Sensing and Social Science* (Liverman D, Moran EF, Rindfuss RR, Stern PC, eds.). National Academy Press, Washington, App. B.

Index